Advanced Timber Structures

Yves Weinand (Ed.)

Advanced Timber Structures

Architectural Designs and Digital Dimensioning

Birkhäuser
Basel

Table of contents

How can a schedule and a technologically innovative process shift the perspective of the construction industry toward sustainability?

Yves Weinand

Timber construction has a promising future—especially in relation to climate change and our need to find sustainable solutions for the construction industry and to implement the use of appropriate building materials. Timber's comparatively low energy consumption has been a known fact for years. Challenges around sustainability in the construction industry also touch on the question of architectural form. At the Department of Timber Structures, IBOIS, EPFL Lausanne, we posed the question of how a formal and technologically innovative process might be developed from a sustainable perspective. The renewal of construction technologies and technical procedures in timber, taking into account the innate qualities of the material, could lead to the increased use of timber in contemporary construction. This is not about the principle of longevity or permanence, on which modernity is based, "modern" implying that which lasts, or that which remains. Timber is generally regarded as being a traditional building material. This preconceived image is advantageous to socially legitimate research that is about finding complex shapes, or creating free-form structures, when it is done in timber.

Choosing to approach complex geometries from the perspective of the designer, rather than merely in terms of generating forms, can be seen as stepping away from fashionable trends in architecture, especially the free, amorphous forms of so-called "Blob" architecture. Many such free-form buildings completely ignore the problem of sustainability, partly due to their choice of materials and partly due to their energy consumption and the cost of maintenance. In contrast, the savings in overall energy consumption by the use of wood as a building material, in terms of life cycle analysis and demolition, are undeniable. As a natural material, timber requires less energy to produce, to transform, to assemble, and to supply sustainably than other building materials. The public has come to expect this technologically innovative process. Today, public and private clients alike demand novel solutions in terms of both sustainable, high-quality architectural design and construction methods. Architects and planners ought to meet this demand by initiating and guiding innovative processes like the ones addressed here. Furthermore, solutions for so-called "non-standardized" architectural forms that are sustainable and economic should be found.

Sustainability

With the onset of climate change, the concept of sustainability has finally become a central issue for our twenty-first-century society. In this context, the research presented here focuses on the following question: by expanding the applications of wood as a construction material, could its use in the construction of public buildings be boosted?

Regardless of its typology or function, a building always consists of a myriad of small elements. Wood and wood-based materials are made up of the assembly of smaller parts. Solid wood, timber beams, plywood, laminated timber, and laminated veneer timber panels are all produced from the amalgamation of smaller-scale parts. For this reason, the technology of the junction connections should also be considered in the synthesis of these materials in a building. The variety of existing timber materials and the considerable versatility in their application should determine the manufacturing and prefabrication methods.

The aim of the research is to find solutions to a number of questions. We are interested in discovering new construction solutions that can be easily incorporated into hardware stores' offerings, to enable the affordable realization of unconventional architecture. One of the most important ways to reduce construction costs is to use a digital design tool. The development of specific, application-oriented digital tools would thus appear to be imperative. Our tools will help at the interface of architecture/civil engineering, mechanical/geometric design, and form-finding/parametric digital prefabrication, in specific, project-related steps.

Fig. 1 Three-layer panels

The chair of wooden structures, IBOIS/EPFL

Within the framework of the Department of Timber Structures, IBOIS has initiated various research areas that explore the relationship between engineering and architectural design. IBOIS is part of the Civil Engineering Institute ENAC/EPFL, but it is also affiliated with the Department of Architecture, where an architecture studio is made available to engineering master's students. Thus, collaboration between architects and engineers is encouraged, providing the environment for a wider scientific community within architecture schools Europe-wide. The research results presented here focus on construction and the challenges of realizing complex shapes and free forms. What is the relationship between basic research and applied research? What is the connection between pure research and applied research? Or between curiosity-driven research and problem-oriented research? And finally: how can the scientific research in architecture be reconciled with the artistic dimension of research, in order to bring them into harmony with one another?[1]

IBOIS provides a place to innovate, where the fascinating inductive-experimental approach is combined with the clarity of deductive-scientific methods. This is undertaken with the aim of creating new forms and types of structures—particularly timber structures. In addition to its sustainable qualities, timber also has exceptional mechanical properties, which can be utilized in specific structural forms.

For centuries, timber construction has been governed by the use of linear elements connected to truss systems. However, in contrast to steel and reinforced concrete—the dominant building materials of the nineteenth and twentieth centuries—engineers have done little to develop the use of timber as a building material. Now, thanks to the availability of digital tools, applications of this material can be expanded significantly; new geometries can be created; and innovative construction materials and methods can be developed. In short, we can undertake an innovative exploration of structural engineering with regard to timber. Here, the current ability to use structural planes plays a vital role—for example, large-scale multilayered boards, plywood, or laminated veneer lumber panels.

Innate material qualities of timber

Even today, one can see timber construction as a perpetuation of traditional forms and methodologies. The majority of buildings continue to utilize traditional structures or building methods, such as timber-framing or truss systems. However, due to the availability of new timber-based materials, there is a need to invent new construction methodologies. As opposed to steel or reinforced concrete, building with timber relies to a greater degree on the development of cutting types and junctions, and the understanding of the relationship between them. Thus, for example, when designing a timber building, it is imperative to consider the junctions, which are integral to the structure, and to specify the entire structure, including its joints.

It is essential for engineers who specialize in timber to provide comprehensive details that can be integrated into the overall structure when formulating their structural models. In this way, these engineers take on the role of designers. It becomes clear that timber construction calls for interdisciplinary collaboration between engineers, architects, and contractors right from the beginning of the planning process. In particular, the choice of panel types should play a central role in determining the form and typology of the selected support system.

Characteristic properties	Small Specimen	Constructional Element	Difference to the small specimen
Bending Strength (mean value N/mm²)	68	37	46%
Tensile Strength parallel to fibers (mean value N/mm²)	80	30	63%
Compression Strength parallel to fibers (mean value N/mm²)	40	32	20%

Table 1

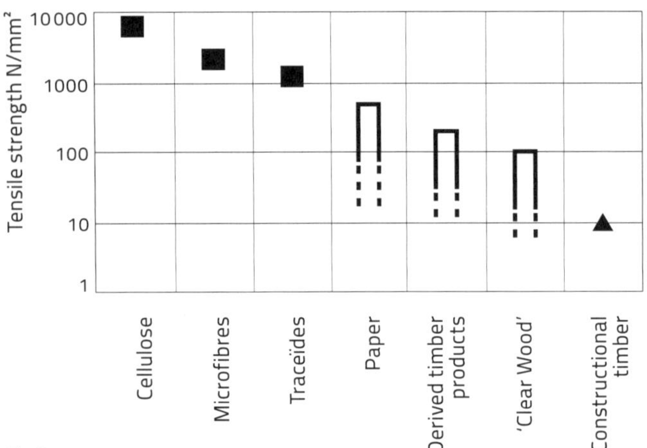

Fig. 2

Table 1 Resistance values for small test elements from spruce

Fig. 2 Development of resistance values for cellulose (f_y = 100 N/mm²) to wood (f_y = 10 N/mm²)

Specifically, the following observations about timber can be listed:

The question of scale

The size of the timber member used determines its strength and hence the range of its applications. Comparatively, a small "clear wood" test block performs far better than a standard-sized beam.

The surprisingly high values of small test blocks (for example, spruce) were categorized by Peter Niemz[2] in table 1. Therefore, great losses in the assessment process can be determined, since in this case considerably lower values were reported.

The anatomy of wood

Figure 3 shows the three principal axes of wood fibers using the example of a cut tree trunk: longitudinal, tangential, and radial alignment of the fibers. In practice, radial and tangential alignments are hardly distinguishable from one another, and an average value is generally taken. The three axes are positioned in a Cartesian axial system. This corresponds to historically applied geometries that define the principles of material strength. A Cartesian axis system describes an isotropic material in an efficient manner, but this is less true for an anisotropic material, such as timber. The longitudinal direction of the fibers, for example, is assumed to be perfectly rectilinear. Though this is partially true, in reality the natural longitudinal orientation of tree growth (i.e., tree fiber) tends to converge conically toward the top of the trunk. More accurate modeling techniques should therefore take into account these specific properties. It would be interesting to develop "tree-specific" mechanical models, such as the scanning of the exact fiber configuration of a particular tree trunk in order to mechanically evaluate a specific application.

The problem of timber anisotropy has already been mentioned, such as in the invention of plywood, where the fibers are confined. If several layers are glued across each other, this results in a homogeneous, or quasi-isotropic, structure.

Systems

A chain breaks as soon as its weakest link fails. In contrast, a system will continue to function even with a broken weakest link. Engineered wood—a dual system of laminated beams, or veneer laminated lumber panels—that consists of several elements glued together can be regarded as a system. These are known as "multilayered systems." The principle can be applied to a specific material as well as to a structural system. A common example is a laminated beam, which consists of several superimposed layers. The resistance value can be adjusted, depending on the number of layers. If the number of layers does not exceed four, then the beam is regarded as a conventional support. If the number of layers exceeds four, then it can be regarded as a system, thereby increasing the resistance of the beam. From the viewpoint of probability theory, the likelihood of failure of a beam decreases with the increase in the number of its layers.

Another pertinent example is a reciprocal system, such as the Zollinger system. If a member of the diamond-shaped configuration of a Zollinger mesh network fails (for example, due to a particularly strong wind load), then the system does not fail as a whole. One also speaks of a "social" support system, where the weakest link is supported by the adjacent members.

These observations lead to the following conclusions:

- It makes sense to produce timber materials where the "system effect" strengthens the total resistance.
- It makes sense to develop structural systems where the interdependency of the elements is maximized.

If these considerations are applied to wood-based materials such as laminated beams, it is foreseeable that, in future applications, these structural systems will also be able to benefit from the system factor.

Traditional carpentry would rarely benefit from a system effect; most of the time, the failure of a roof beam or joist will lead to the collapse of the roof or attic. The same is true for traditional timber connections, where local failure of a connection results in the collapse of the element that is held in place by the connection. The structures presented in the following section consist of a large number of small elements. The importance of developing such support systems, which amplify the mutual dependence of the elements, is reflected in all the structural systems demonstrated below.

A new generation of structures

Can wood perform better than it currently does, or historically has? And could the architectural image of timber buildings be given a more contemporary expression? When one takes a close look at the buildings of Philibert de l'Orme, already envisaged back in the early 1600s, the answers to these questions tend toward the positive.

On the basis of his research, de l'Orme foresaw the use of small-scale wood elements that could be used as an alternative material. He combined this principle with geometric innovations, allowing him to achieve greater spans. Unfortunately, de l'Orme's innovative ideas did not succeed, as they were too labor-intensive: every individual piece and each connection had to be cut individually by hand. These past obstacles to building de l'Orme's structures could be overcome today with industrial fabrication and CNC milling, and the construction could thereby be made affordable. The architectural expression of a networked system can certainly be classified as contemporary.

De l'Orme's findings influenced the French army's military structures. General Armand Rose Emy advocated the use of a large number of small-scale timber elements to cover their arena structures. In this instance, however, the boards were installed horizontally, or rather in a horizontally curved position. Thus the total curvature of the arched structure was achieved by the local bending of each board. This avoided waste; a second advantage was that, as the orientation of the longitudinal fibers of each board coincided with the line of force of the arc, they functioned far better structurally.

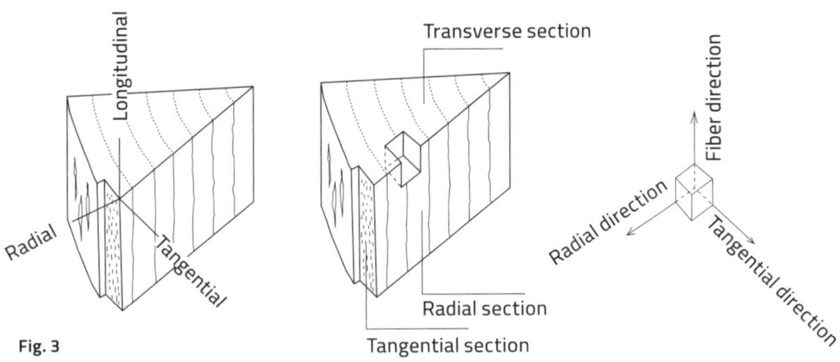

Fig. 3

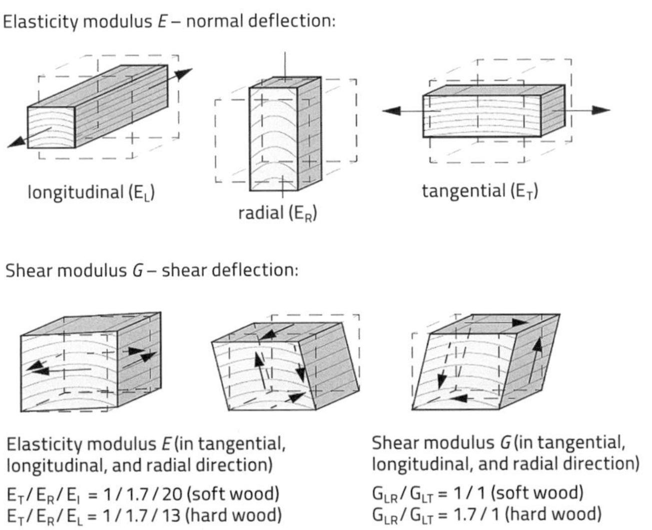

Fig. 4

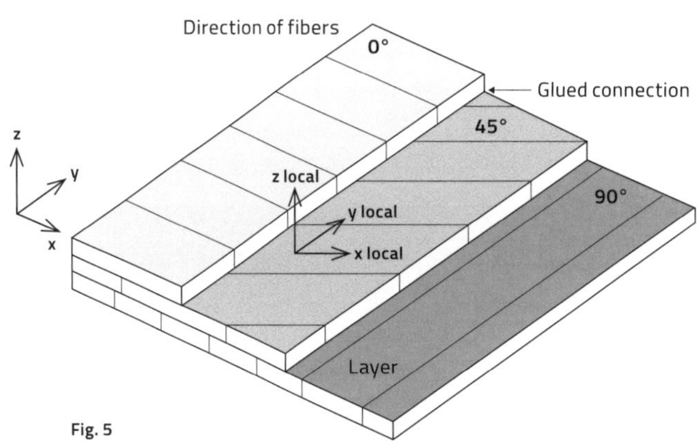

Fig. 5

Fig. 3 The three strands of wood fibers are inserted into a Cartesian axial system.

Fig. 4 Elastic and tangential modules vary greatly.

Fig. 5 The principle of cross-laminated timber

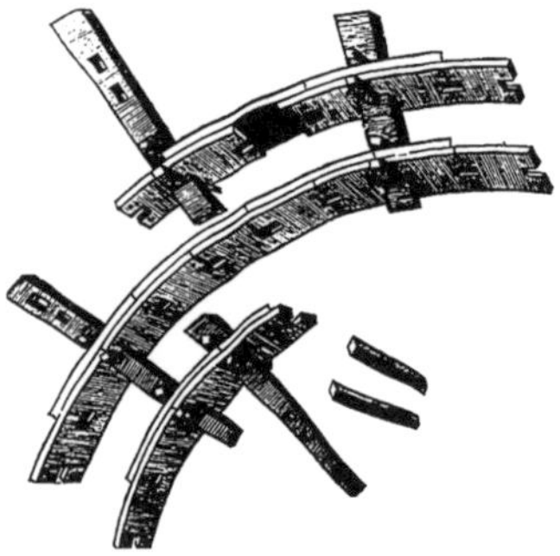

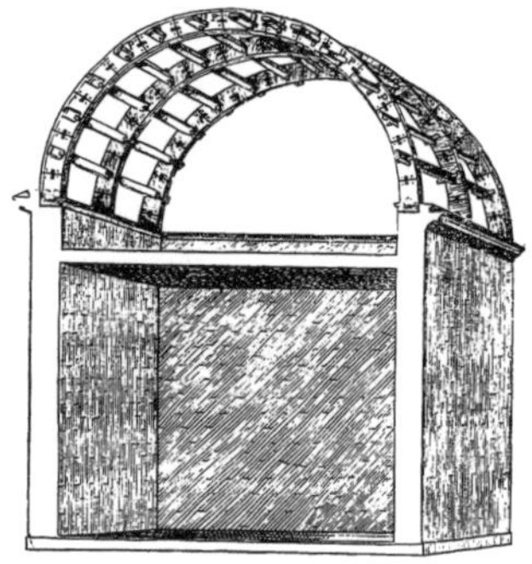

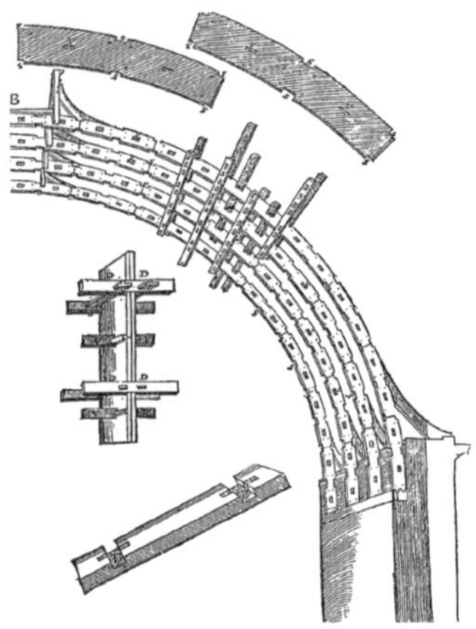

Fig. 6 Philibert de l'Orme: The panels define a part of the curve and are installed vertically. Thus, the edge of each panel describes a section of the arc. A significant proportion of a panel's wood fiber therefore runs into a void, or into the edge of the panel, as it does not follow the primary direction of the curve, as this would require the structural behavior of an arc.

Fig. 6

Fig. 7

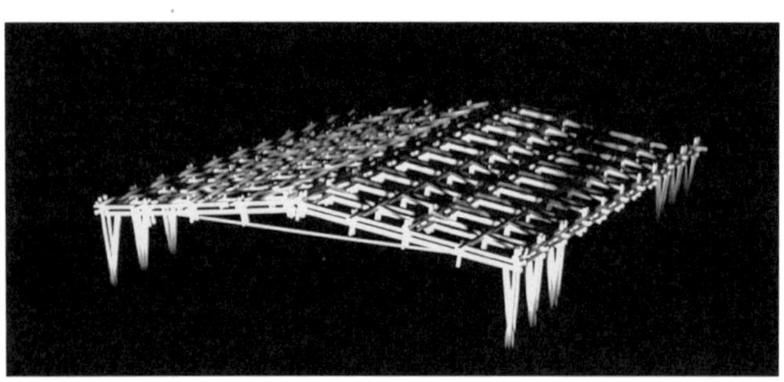

Fig. 8

Fig. 7 Multihalle Mannheim: General view of the shell structure with a span of 60 m. Four rib layers are connected in one node.

Fig. 8 Competition project for an industrial hall in Botrange, Belgium

The much-celebrated Mannheim Multihalle by Otto Mutschler (1975) is an extraordinary example of a spatial structure composed of networked elements. The double-layered network structure consists of actively bent timber slats with a square cross-section, which follow the thrust line and absorb the normal forces well.

The third and final example of an upgraded structural system is the design for an industrial hall in Botrange, Belgium. The supporting structure consists of simple boards, which are multi-layered and crossed over one another to form a mesh network. The boards pass through the nodal points and are connected only with vertical pins inserted laterally. A spatial structure is thus created out of a complex combination of conjoined small parts. The local rigidity of the nodes can be increased by inserting an additional bolt, thereby increasing the overall rigidity of the system. Timber-frame construction and post-and-beam structures were, and remain, widely used systems in timber. With these composite systems, junctions are added individually on-site. In addition, semi-prefabricated floor elements and wall structures are now available, which can also be incorporated.

As a result, an attempt should be made to create made-to-measure prefabricated systems incorporating connection technologies and precise prefabricated elements. Due to their specific shape or geometry, these custom-made fixtures could only be installed in a specific location and in a unique position within the overall system. Errors that often occur on-site could thereby be avoided.

The manner in which building sites are organized today corresponds, in many ways, to nineteenth-century models. The necessity for a foreman who reads and understands construction plans, and then connects them with the delivered components, needs to be replaced by a stronger and different kind of planning. Access connectivity systems, predetermined assembly sequences, and integral mechanical connection techniques should determine the site-work schedule. For this reason, we are interested in geometric algorithms, subdivision processes, planarization processes, connective sequences, automated milling technologies, tool development, mechanical test trials, and the execution of manual as well as robotic joining processes. IBOIS's areas of interactive research have been summarized in the diagram below. Folding systems, discretized free-form structures, woven and actively bent structures, and mechanically induced structural systems will be presented. Special attention will be paid to the connections.

Notes

1 Thanks to Professor Pierre-Alain Croset for his critical notes in this area.

2 Keunecke, D. and P. Niemz. "Axial stiffness and selected structural properties of yew and spruce micro-tensile specimens." *Wood Research*, 53, 1–14, 2008.

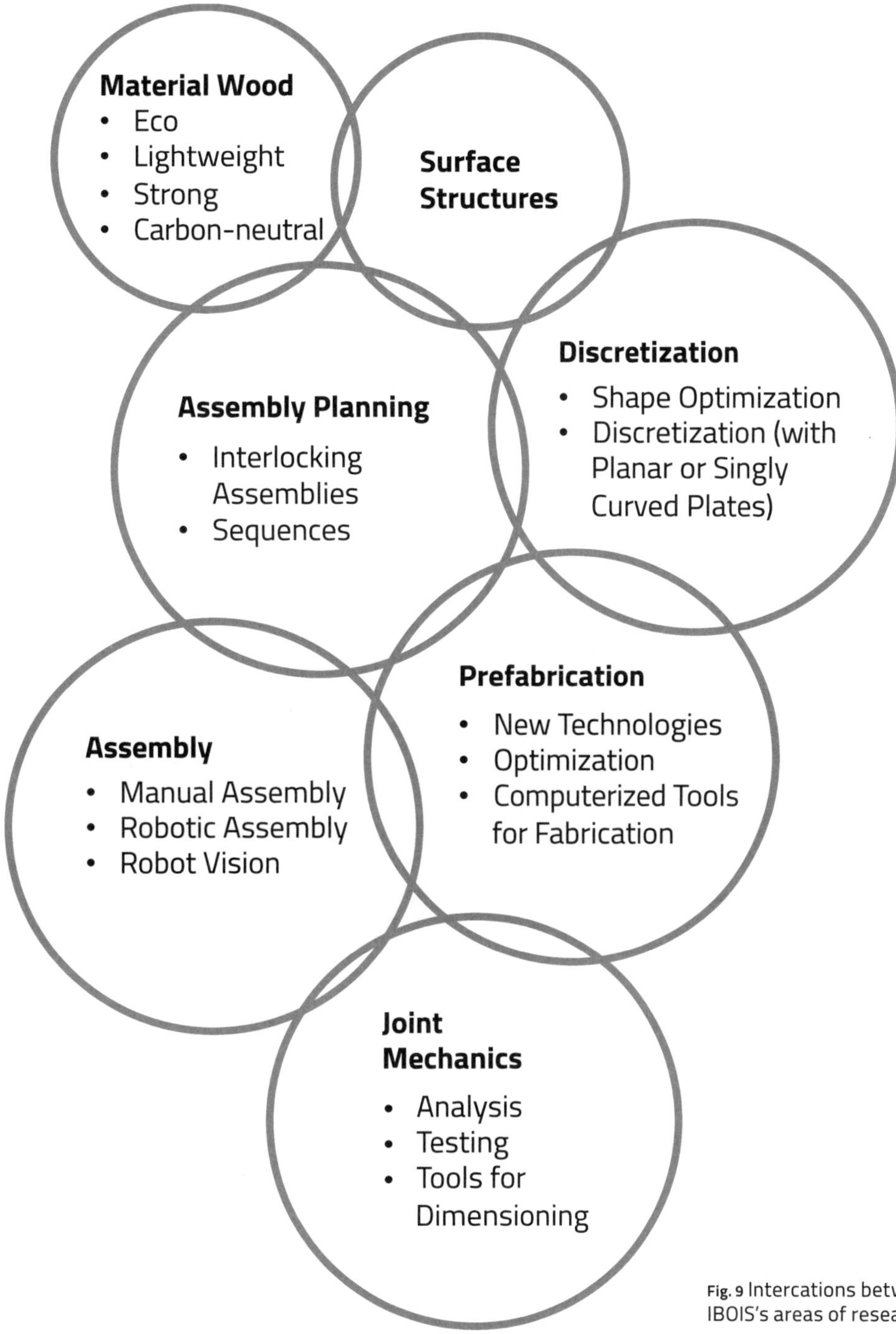

Fig. 9 Intercations between IBOIS's areas of research

1 Folded plate structures

1.1

Folded plates made from timber panels

Yves Weinand

Folded plate structures are fascinating constructions, where form and structure are inextricably linked to one another. Folded panels are also able to assume many different forms. By folding timber, these structures acquire a high level of rigidity, despite the fact that the material strength remains reduced in relation to the span. Furthermore, this material allows filigree structures that are also economical.

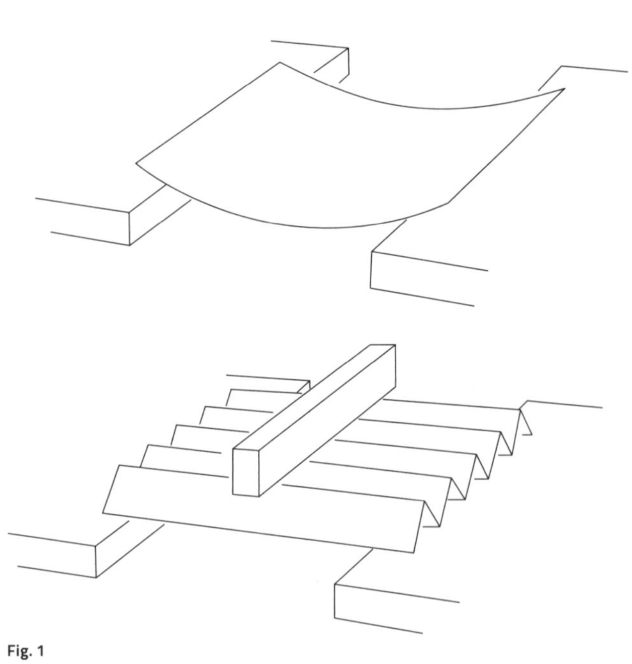

Fig. 1

Thus far, structures made of folded wooden panels could not be implemented. Formwork structures made of reinforced concrete were already being designed and executed in the 1930s. At that time, the construction of identical, prefabricated prismatic supports was significant, as it minimized the cost of timber shuttering. Structures made completely from folded timber panels were being developed in the 1970s. However, this concept was not pursued any further.

In this essay, we will look at possible ways that folded timber panel structures might establish themselves in the marketplace. To illustrate the concept, three projects will be presented to highlight the possible development of these structures. In each project, form-related, mechanical, and production-related aspects are highlighted. As already mentioned, folding structures have the ability to torsion form and structure. Experimenting with relatively soft, large surfaces or panels is made possible by combining the panels along their edges, in "wrinkles" or folds. The geometric manipulation of the folds and the "freezing" of a certain arbitrarily chosen

Fig. 2

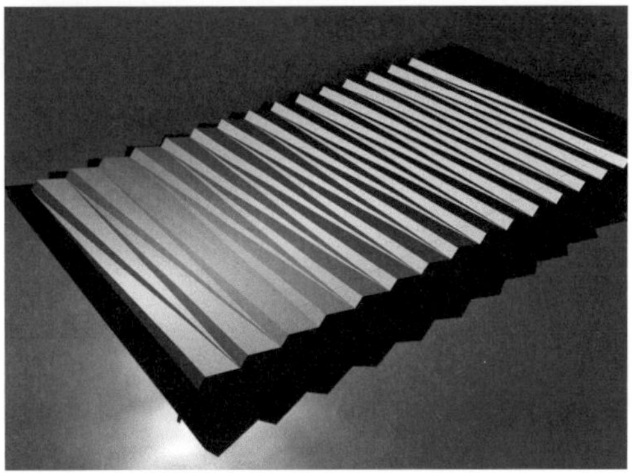

Fig. 3 a

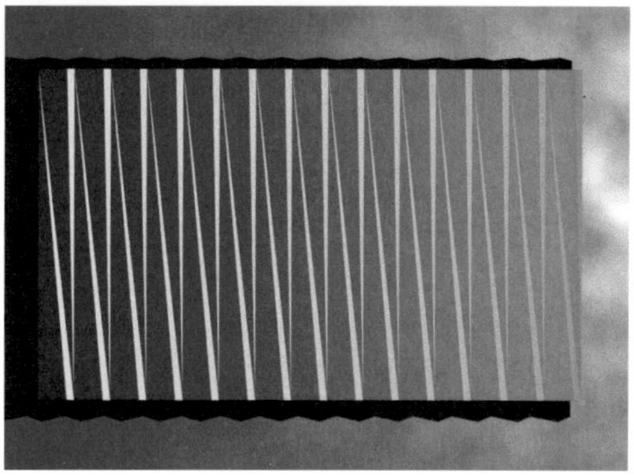

Fig. 3 b

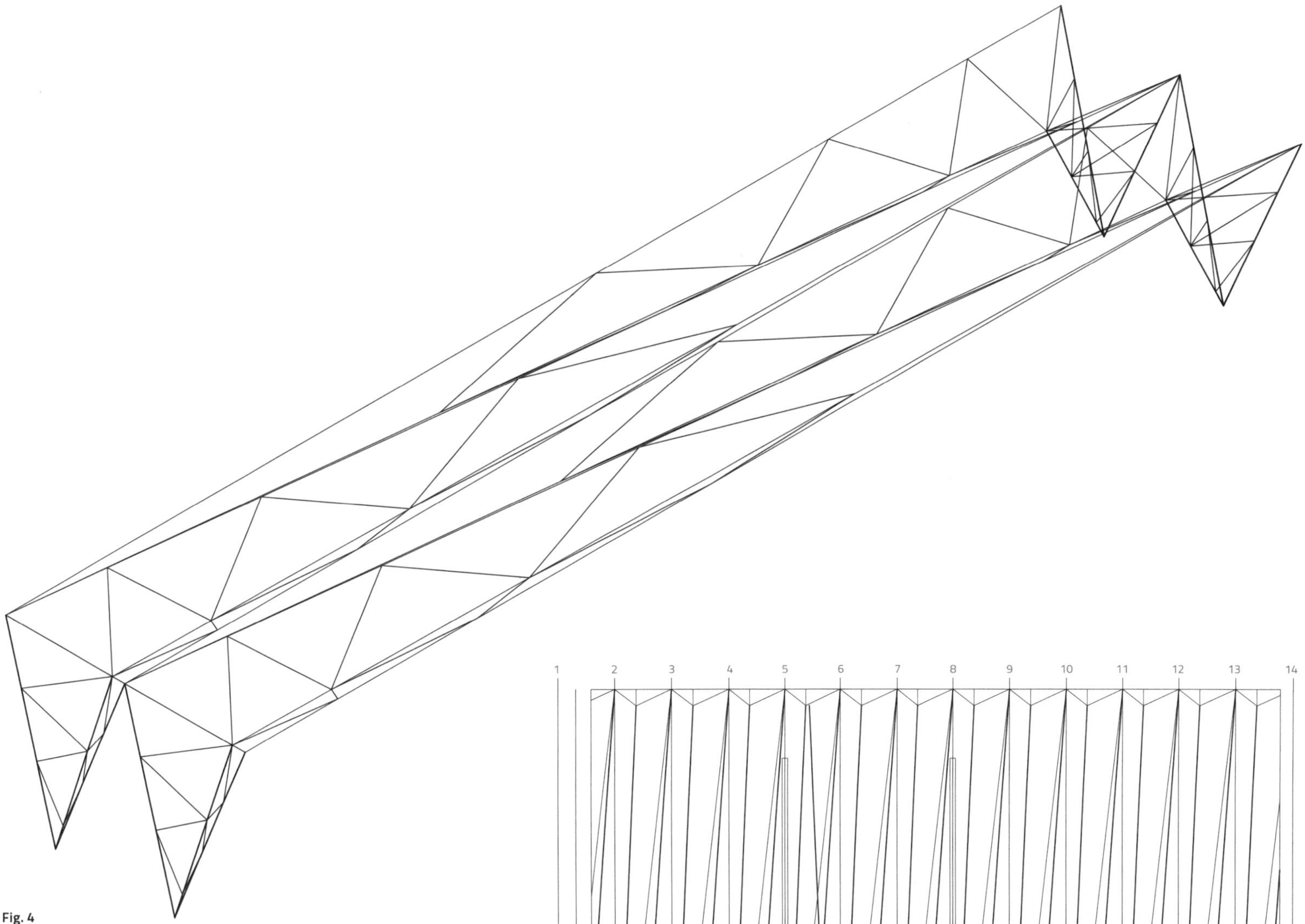

Fig. 4

Fig. 1 Folding principle: rigidity is achieved by folding.

Fig. 2 Experimental folded plane

Fig. 3 a and b Axonometric representation of a folded structure: elevation and plan

Fig. 4 Axonometric representation of the basic elements of the original geometric configuration

Fig. 5 Floor plan of the base element of the original geometric configuration

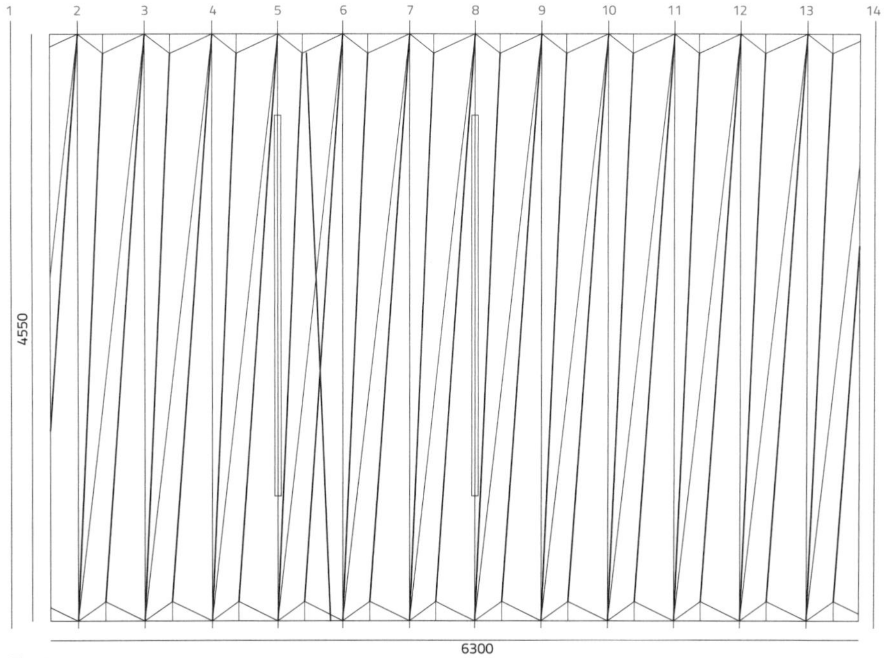

Fig. 5

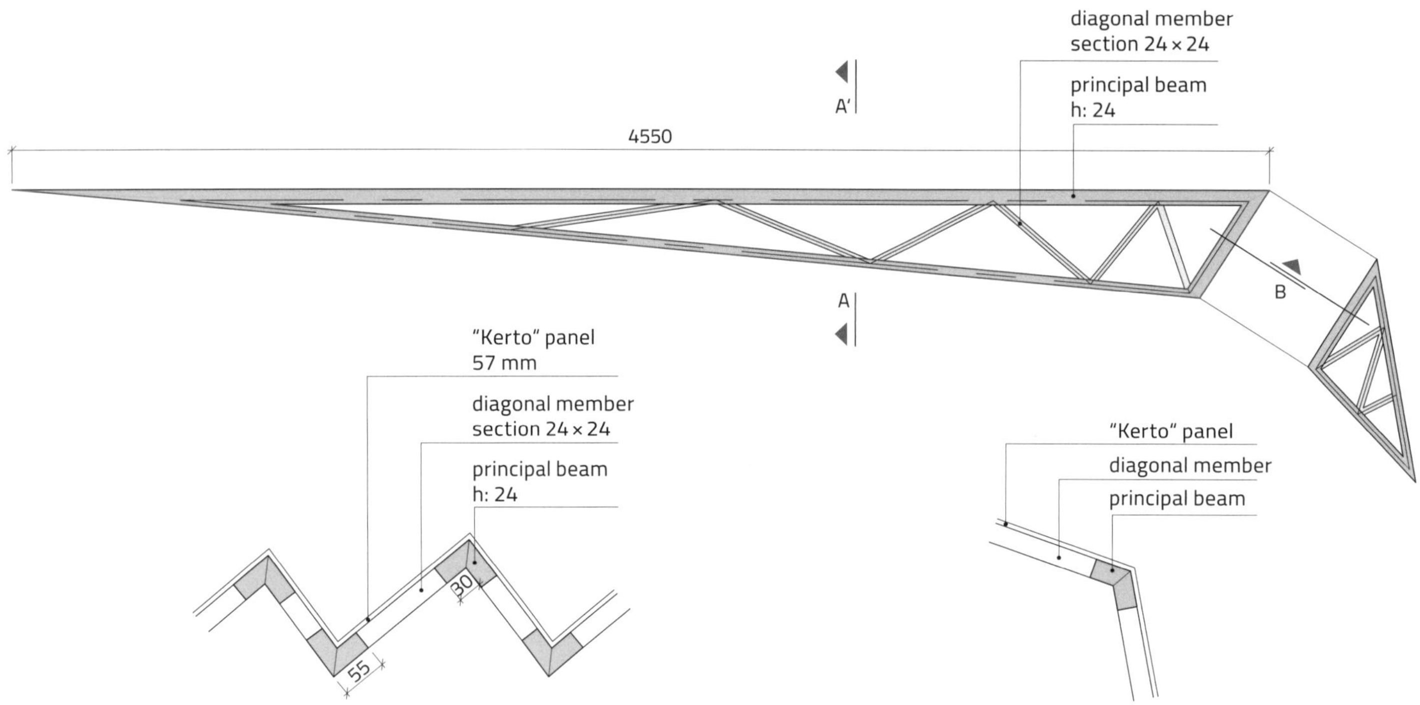

Fig. 6 Solution of the support system: the panel is transformed into a truss, clad in far thinner panels.

geometric position during the folding process opens up the potential for the structural optimization of such structures.

The first example to be analyzed is the roof of the sports hall "des Îles" in the Swiss lakeside town of Yverdon.
This building was initially planned as a folded-panel structure. In preliminary studies, the geometric consequence of having a folding roof with a span of over 45 m was investigated. If one calculates the scenario as illustrated in fig. 3 a—the geometry as sole self-supporting membrane made of laminated veneer lumber panels—then this results in a required panel thickness of around 14 cm. Panels of this thickness are, however, rather expensive.

The challenge of this type of construction is not in the cutting or the installation of the panels, but rather in the execution of their connections. For this reason, alternative construction options were proposed that aimed to reduce the panel thickness and find optimal locations for the connections. One of these alternative solutions is illustrated in fig. 6.

Laminated beams form a triangular truss, which is covered with panels. Thus, the truss is visible from within the hall.
On the exterior, the panels provide the necessary insulation and roofing surface. The connection is made possible here by cutting the edge supports. The peripheral laminated beams running along the edges have to be angled, cut, and then bolted together. The overall geometry of the fold creates a geometric challenge along its "folded joint," which in this case is solved by the cut. The structural consequence of this is that the initial structural strength gained from surface areas is, thanks to the edge rigidity, effectively altered into a truss support structure. Due to the superimposition of linear elements and their triangular stiffeners onto the folded geometry, the unified appearance of the folded structure is unfortunately lost.

Even for this altered folded structure for the hall in Yverdon, the high tenders given by contractors were an indication of the skepticism regarding its economic viability. Ultimately, the structural efficiency of the folded panel was rejected in favor of a frame system. However, additional rigidity still does occur spatially above the folded roof structure and the wall truss.

In summary, the following is ascertained:

- With cost saving in mind and the aim of reducing the thickness of the panels, a panel thickness of 14 cm was determined.
- Due to this reduction, a strengthening of the slenderer panels is necessary. The superposition of the original "natural" folding geometry over that of the edge support and its stiffeners diminishes the conceptual strength of the folding geometry.
- The folded structure is effectively changed from a structural supporting surface to a frame structure by shifting the main components of the shear forces along the edge slats.

Fortunately, our subsequent attempt to build a folded structure succeeded, with our execution of the Chapel of Saint Loup. However, it is important to note that the span of only 9.5 m was far less than in Yverdon, where the span was approximately 45 m. For the chapel, 60-mm-thick, three-layer panels span the chapel space, achieving a reduction of 1/158. The folded geometry relates to the roof but also to the walls, as the static efficiency of folding in the roof area is employed in a similar form in the wall surfaces.

The tool developed by Hans Ulrich Buri at IBOIS generates the overall geometry, as well as determining the cutting and bending of the panels. The montage images display a self-supporting framework that functions efficiently as a single fold. Our idea was to leave the supporting structure—i.e., the panels—visible, at least on the inner side. The panels are therefore connected to one another on the exterior by means of a 2-mm-thick metal plate.

The transfer of forces from one plate to the next can thus take place by linear supports. As a consequence of the cut sizes, five detail types were identified. The nail plates remain identical, but the number of nails per plate varies from one type to the next. The torsion peaks that occur at the frame corners can be avoided. These bending moments can be distributed along the edges and do not all have to be absorbed at the frame corner.
In the most extreme structural state that might occur in the case of the chapel, bending moments at the frame corners of about 10 kNm/m can be reduced by half.

The proposed detail thus corresponds to the surface structural nature of this supporting folded structure. Unlike traditional timber construction, where the prismatic sections are connected by pins of some sort (nails, screws, bolts), that weaken the cross-section, a folded structure, such as the one at Saint Loup, allows an optimal exploitation of the panel thickness as structural support.

In summary, the following can be ascertained:

- 40-mm-thick panels for the walls and 60-mm-thick panels for the roof were found to be optimal sizes for an efficient and material-saving system in this category range.
- With the selected connection method, weakening of the panel cross-section is avoided.
- The connection method suits the surface supporting nature of the structure.

It should be noted that the generation of the Chapel of Saint Loup's overall geometry is rather arbitrary, as the lateral surface of the chapel corresponds to an unfolded sheet of paper. In theory, there are no offcuts. An incline for rainwater runoff was generated by the folds.
The obvious weak point of the overall structure is the open edge—the first fold located above the entrance, where the largest span occurs.

The history of architecture reveals how the understanding of structural folding has led to a wide variety of effective solutions.

The understanding of the mechanical rigidity of the folds in the overall geometry described here does not result in only a single feasible geometry. Rather, numerous geometric options were demonstrated by Hans Ulrich Buri's work.

But ultimately, transverse and longitudinally generated profiles remain purely geometric elements. In this respect, the tool proposed by Hans Ulrich Buri remains a formal, architectural, and constructive tool. It also allows for a mechanical optimization that, however, remains intuitive. In the case study (Fig. 24) it is shown how a given folded geometry can be optimized from a mechanical

Fig. 7 Finite-element model of a single-folded, three-layer panel structure: support conditions and edge connections

Fig. 8 Finite-element model of a single-folded, three-layer panel structure: load cases; a) permanent weight, b) dead load, c) snow, d) wind

Fig. 9 Finite-element model of a single-folded, three-layer panel structure: deformation and membrane forces

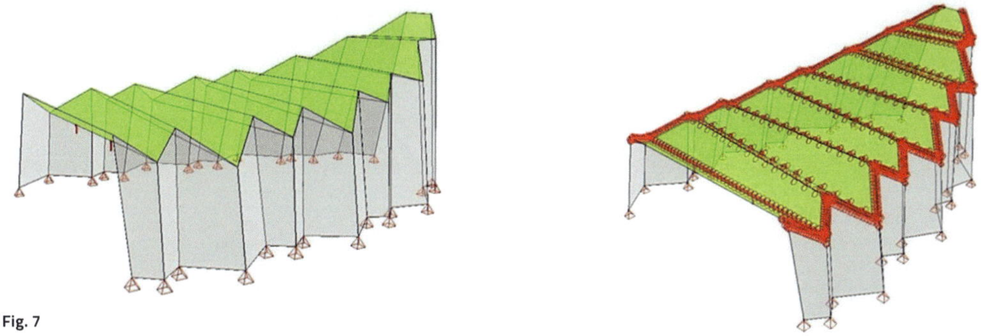

Fig. 7

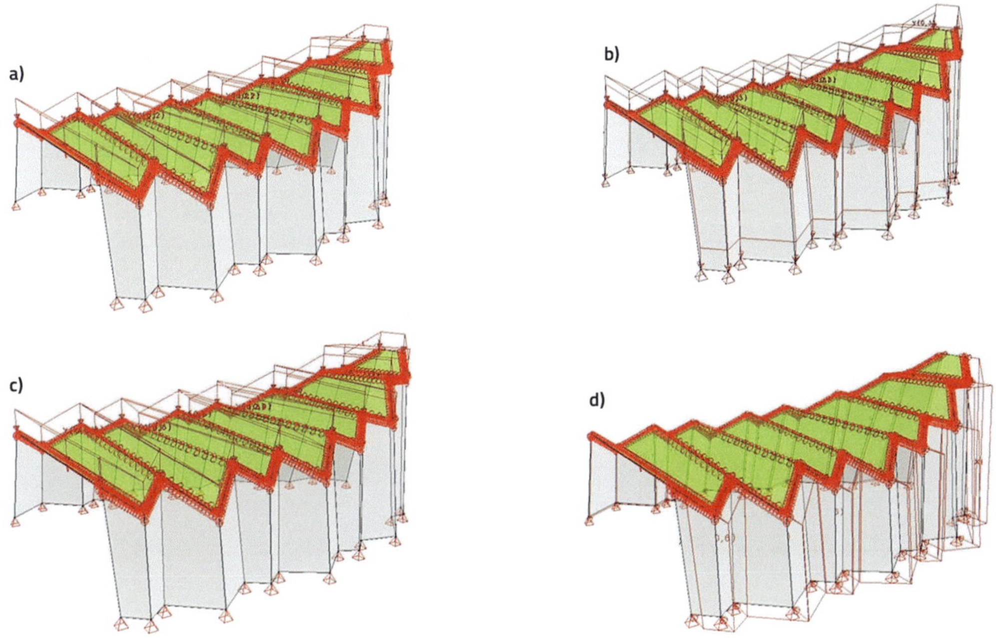

Fig. 8

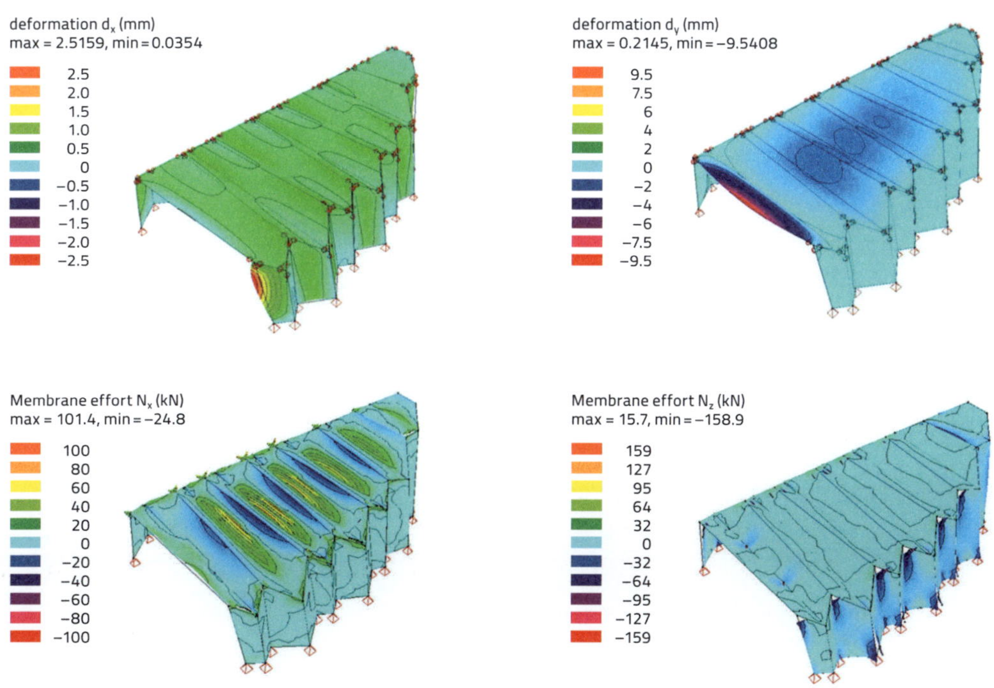

Fig. 9

Fig. 10 a

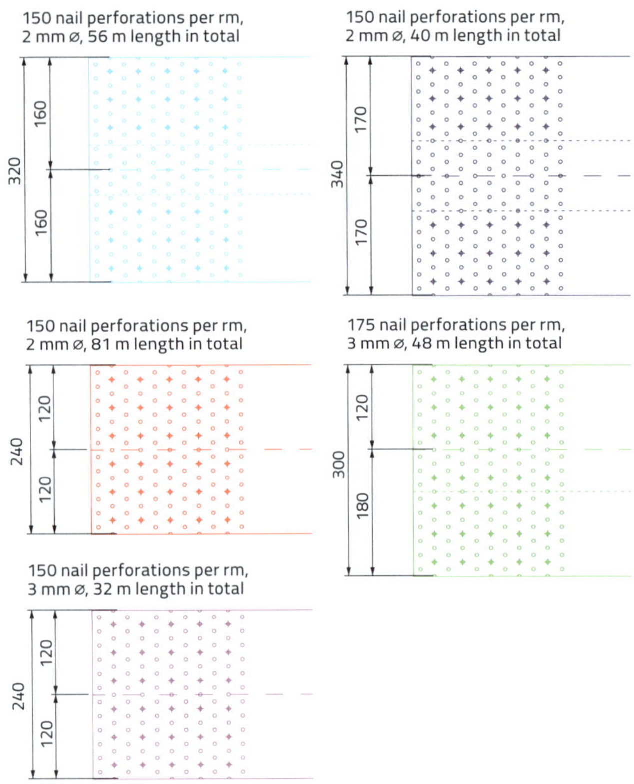

Fig. 10 b

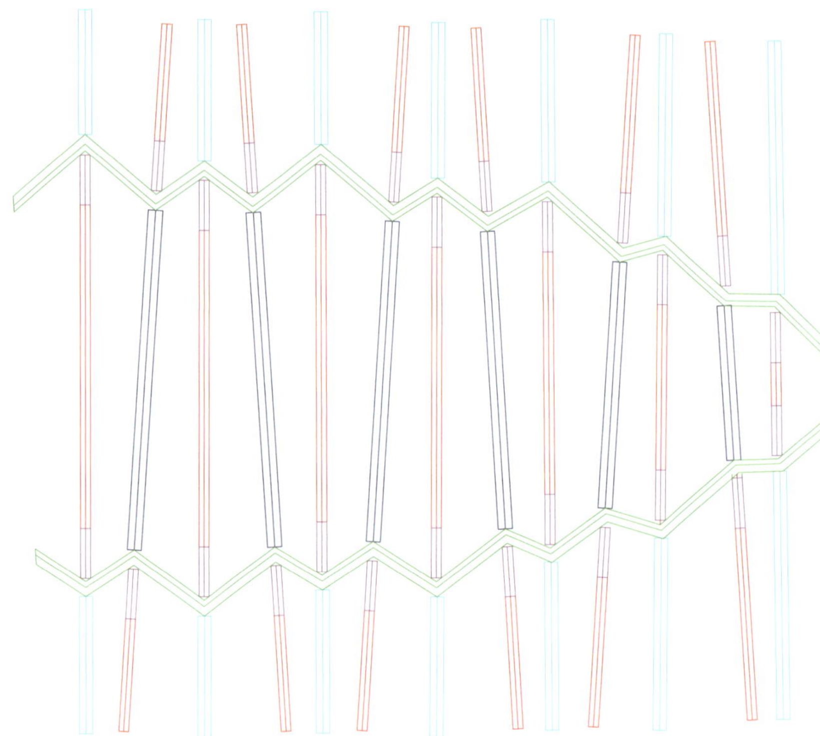

Fig. 10 c

Fig. 10 a, b, and c Detail of a panel connection: 2-mm nail plates are divided into five detail types, where the number of nails varies. The plates themselves were previously folded.

Fig. 11 Detail of the plate connection after completion

Fig. 12: Plate detail at the base

Fig. 13 Detailed view of the roof edge during construction

Fig. 14 Fabrication drawings for each panel

Fig. 15 Axonometric view of the chapel and the same geometry unfolded

Fig. 11

Fig. 12

Fig. 13

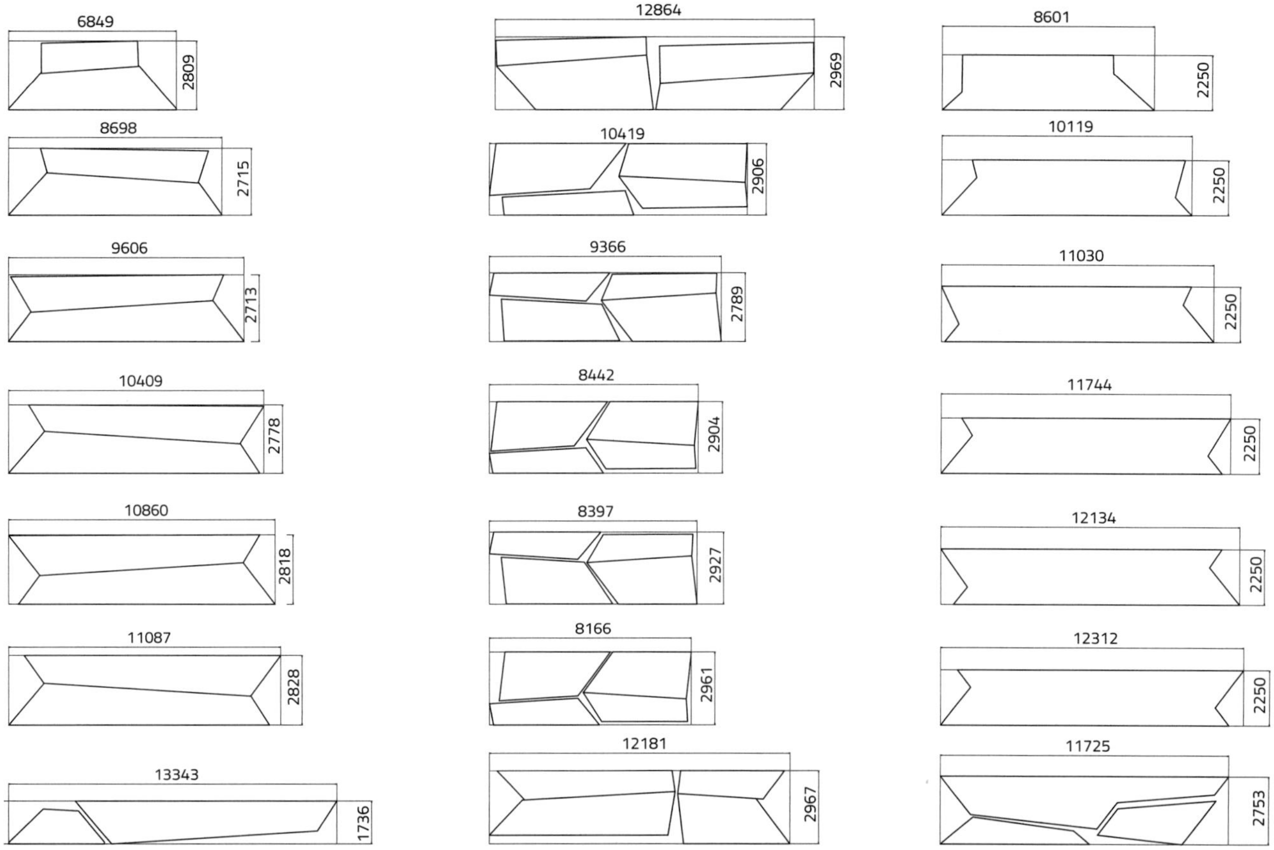
6849
2809
8698
2715
9606
2713
10409
2778
10860
2818
11087
2828
13343
1736
12864
2969
10419
2906
9366
2789
8442
2904
8397
2927
8166
2961
12181
2967
8601
2250
10119
2250
11030
2250
11744
2250
12134
2250
12312
2250
11725
2753

Fig. 14

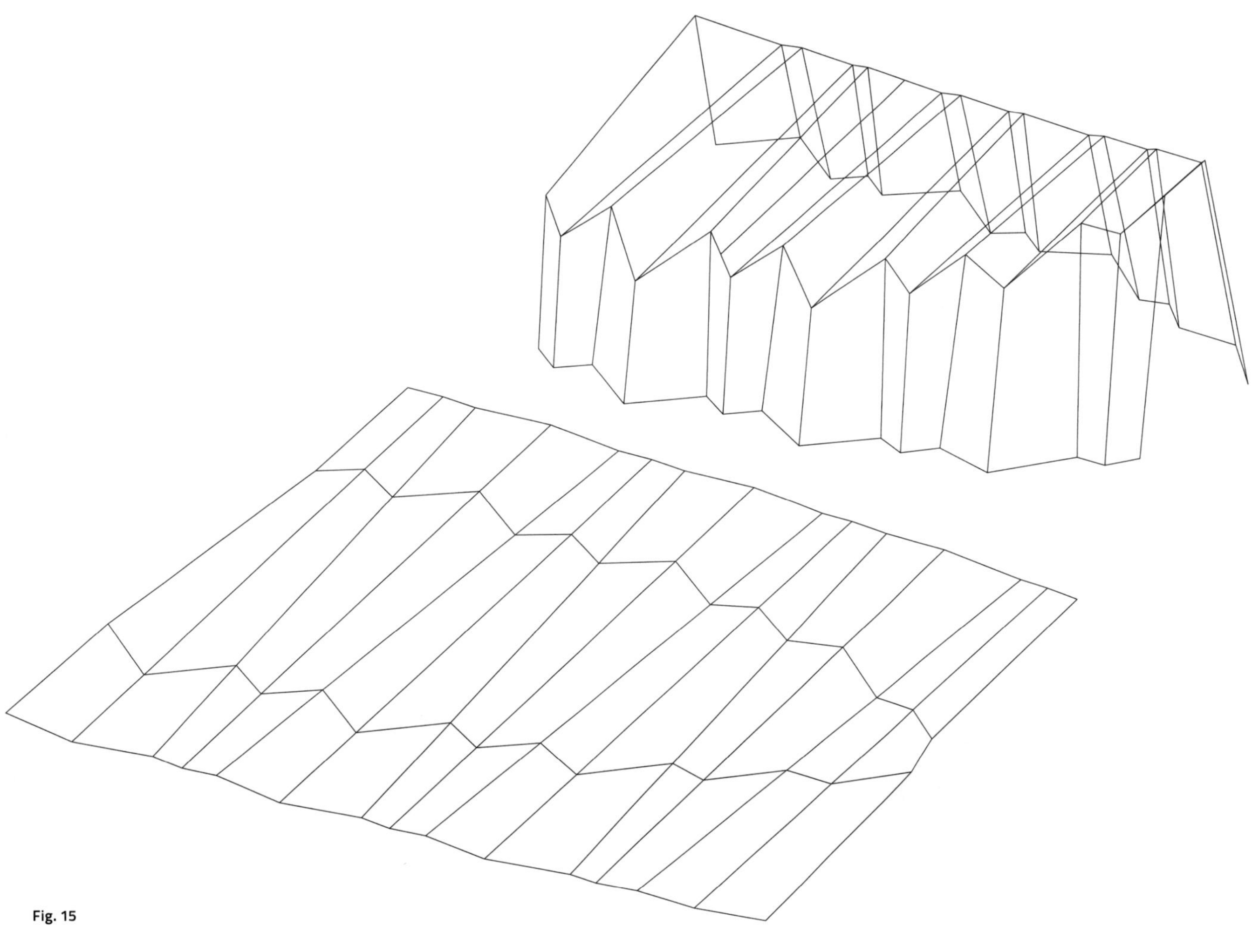

Fig. 15

Fig. 16

Fig. 17

Fig. 18

Fig. 16 Installation of the panels. Each section is self-supporting.

Fig. 17 The chapel during construction

Fig. 18 Mounting of the roof membrane and the battens

Fig. 19 a and b Completed chapel

Fig. 19 a

Fig. 19 b

perspective. By increasing the height of the folds, or by increasing the frequency of the pleats at the edge of the shell, structural performance can be enhanced.

In summary, the geometric adjustments that contribute to overall rigidity:

- By increasing the fold depth, the moment of inertia is increased.
- If the frequency of the pleats is increased toward the edges of the shell, then additional horizontal rigidity can be generated in these edge regions by simply adapting the geometry. This adaptation is convincing as, unlike conventional shell construction solutions, where the edge has to be reinforced, here it is possible to enhance the structural capacity by simply taking advantage of the folded geometry, thereby avoiding the need for an extra element in a different, non-timber material.

Fig. 20

Fig. 21

Fig. 20 The trapezoidal cross-section also corresponds to a fold.

Fig. 21 The scale of the folds can be adapted to the specific building—here, the roof of Graz city hall.

Fig. 22 a and b Generation of possible folding geometries

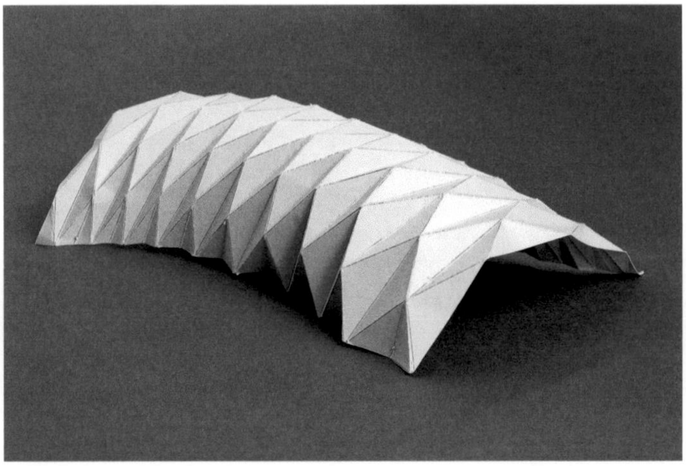

Fig. 22 a

Fig. 22 b

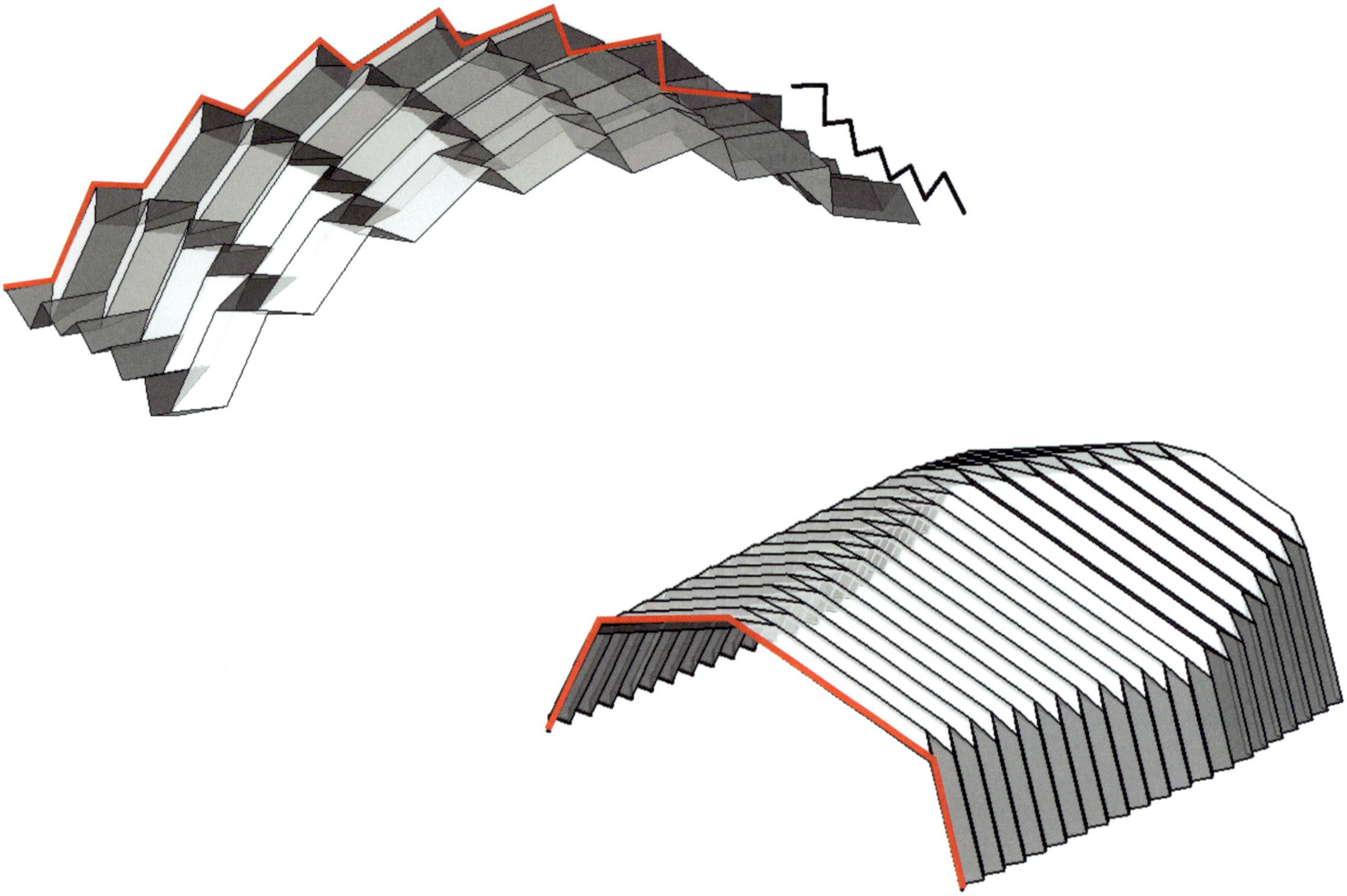

Fig. 23 Verbiers Concert Hall: Transverse and longitudinal profiles enable the creation of a variety of folding geometries.

- The danger of local buckling of the panels can be mitigated by reducing the surface of the panels and increasing the fold frequency. (Since the fold edges are substantially more rigid than the panels, they would buckle before the fold edge fails due to instability.)
- The reduction of the panel surfaces also offers an economic advantage, as the price of wood panels depends on their size: the smaller the panel, the lower the price per square meter, and the lower the transport costs.

Thus far, the presented tool has been aimed specifically at the engineering community. We hope to arouse their interest in the overall form and the mechanical optimization. In addition, the above-mentioned geometric manipulations improve the rigidity of the shell, to accommodate both horizontal and vertical forces.

Fig. 24 Geometric manipulation of folds

Fig. 25 Static improvements achieved: deformations can be reduced by one-third.

Fig. 26 Prototype

Fig. 27 Typical reverse-fold geometry. Transverse and longitudinally generated profiles enable the creation of various folding geometries.

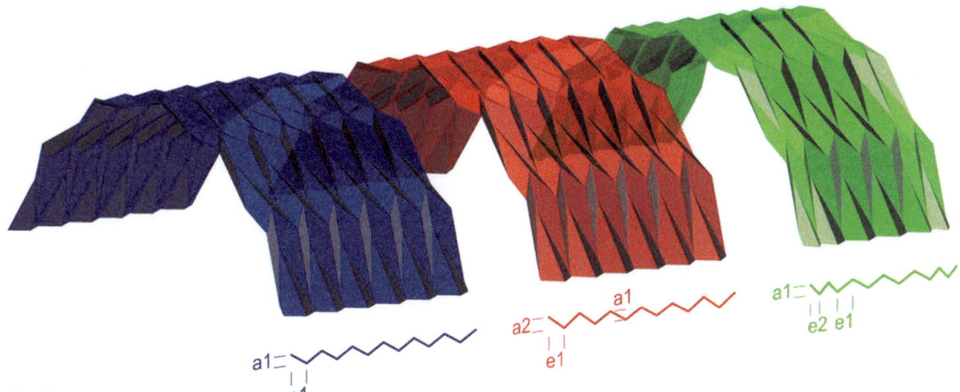

Fig. 24

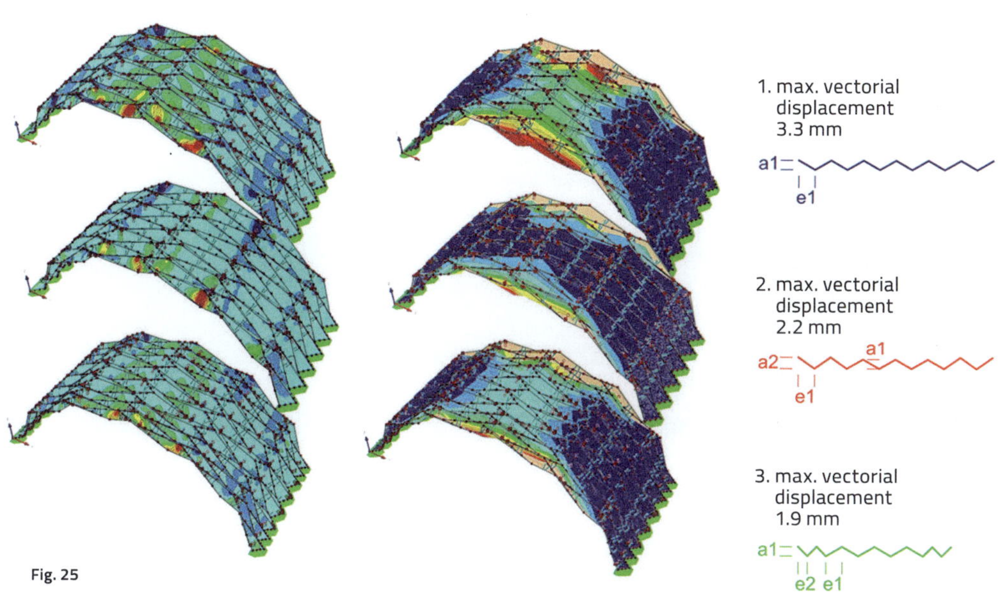

Fig. 25

Fig. 26

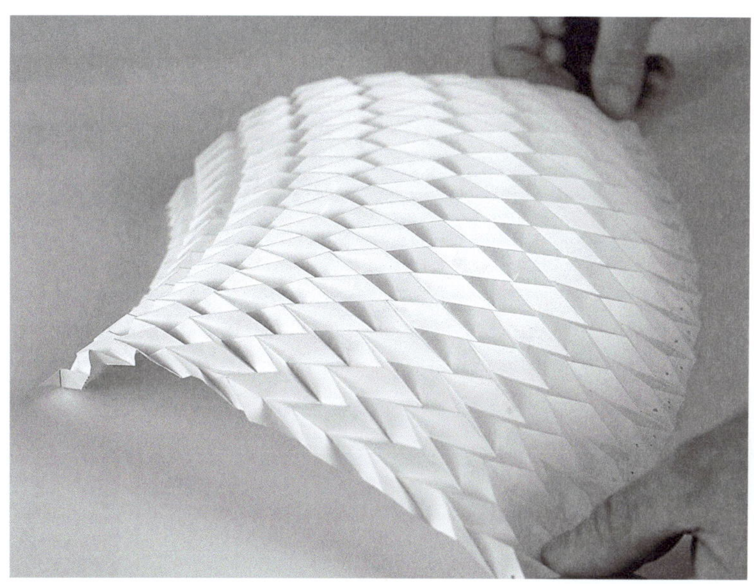

Fig. 27

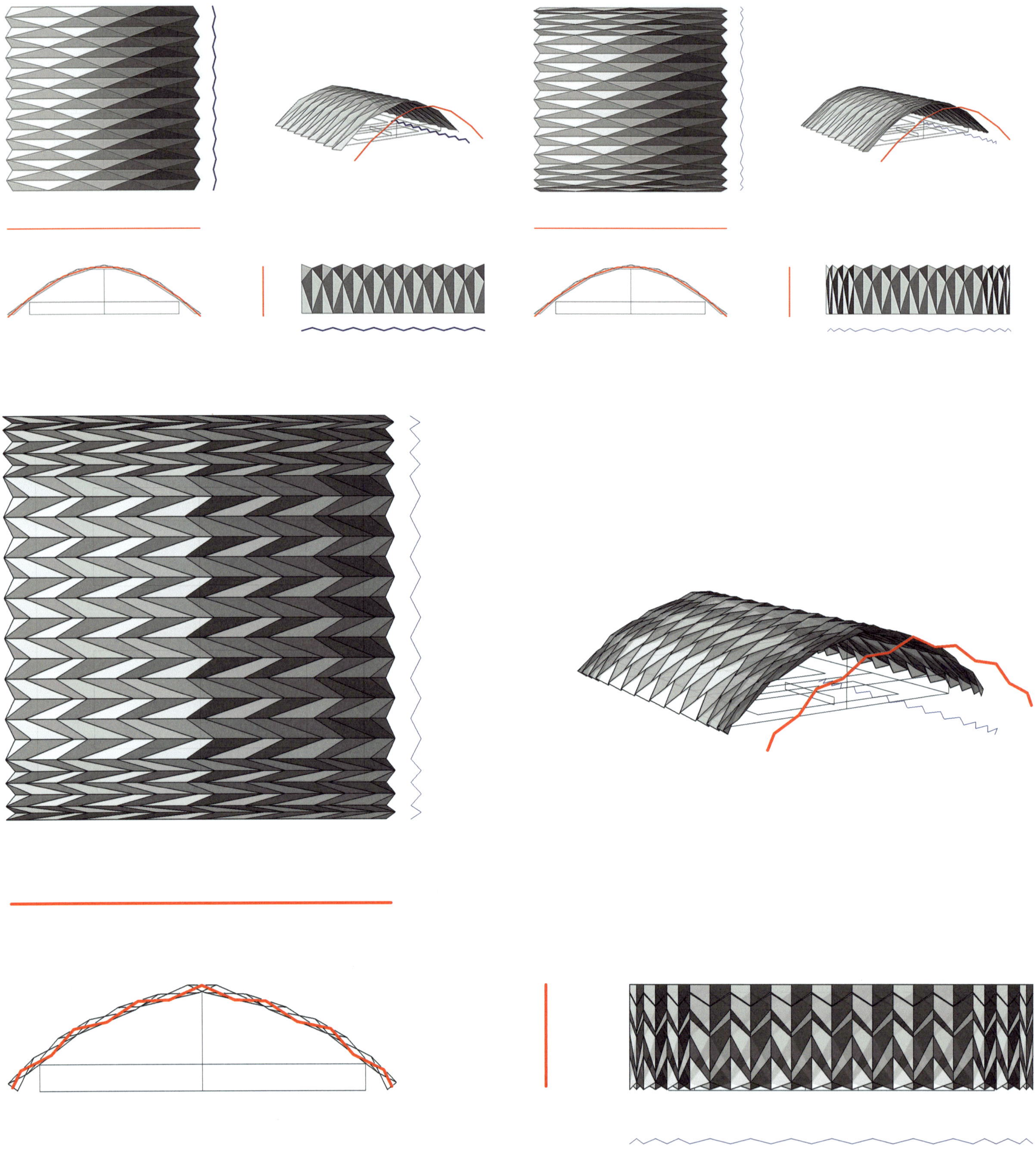

Fig. 28 The folding geometry of a tennis court hall is optimized. The folds are positioned close to the edges of the hall.

The design of a prototype was generated with the following geometric characteristics:

A given area integrates the following surfaces:

- A continuous fold with a V-shaped cross-section
- A complete fold with the longitudinal profile of a triangle (inverse fold)
- Diamond fold
- Herringbone fold

The comparative parameters are as follows:

- Vertical deflection in the center
- Vertical deflection at the edge
- Absolute tension under its own weight
- Structural behavior under asymmetrical wind loads

Here, a potential mechanical optimization of a factor of 5 can be determined.

As with previous projects, careful consideration was given to the means of connection. Although the connections chosen for the Chapel of Saint Loup led to a coherent solution, the chosen panels did not form an integral part of the geometry. Bolted connections that do not comply with regulations were used for initial trials.

The distances stipulated by regulations could not initially be complied with, as the idea was to design with relatively thin panels. As a result, various different options for positioning of the panels and panel thicknesses were explored. All test configurations are feasible and have the same advantages provided by the panel connections as detailed in the chapel described above.

Nevertheless, these connections remain costly. Furthermore, the traditional allocation of cost groups for timber volume, connections, and assembly costs are unaltered. Therefore, the critical consideration lies not only in the design of the overall geometry from a formal, topological, and mechanical perspective (as is shown in Andrea Stitic's work) but also in the integration of the connection details. In a third project, the ideas developed in the hall in Yverdon and the Chapel of Saint Loup, along with the integration of a new connection, are unified and executed. A folded system with small sections was chosen for Vidy's new theater. By doubling the number of folds, two 40-mm-thick panels are positioned 25 cm apart, thus forming two shells.

The principle behind the connection of the two shells is illustrated in fig. 34. Both surfaces of the inner shells of the folds project through to the outer shell. This results in a stiffening of the fold, a kind of rigid corner. Thus, conventional purlins or timber beams can be omitted.

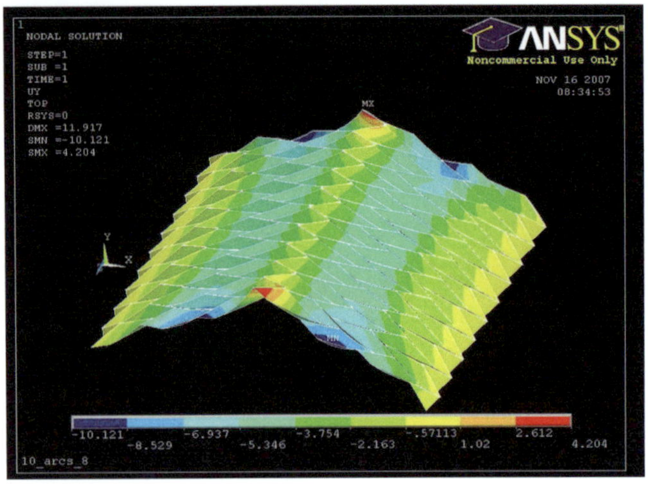

Fig. 29

Fig. 30 a

Fig. 30 b

Fig. 30 c

Fig. 30 d

Fig. 29 Finite-element model of a folded structure

Fig. 30 a, b, c, and d Execution of prototypes, which also led to technical test experiments.

Fig. 31 Variations of folds: a sketch design for a concert hall for the Verbier Festival

Fig. 32 A double-folded structure is generated. Visualization of the virtual parametric model

Fig. 33 Testing a prototype. The double-shell structure consists of 8-mm-thick panels situated 10 cm apart.

An innovative timber connection is created, derived directly from the existing geometry.

The strengthening of the panels along the folds, or the reinforcement of the edges, also decreases the risk of the panels bending. The trapezoidal surface is reduced by the reinforced edges. In addition, all the panel connections were designed as timber–timber connections.

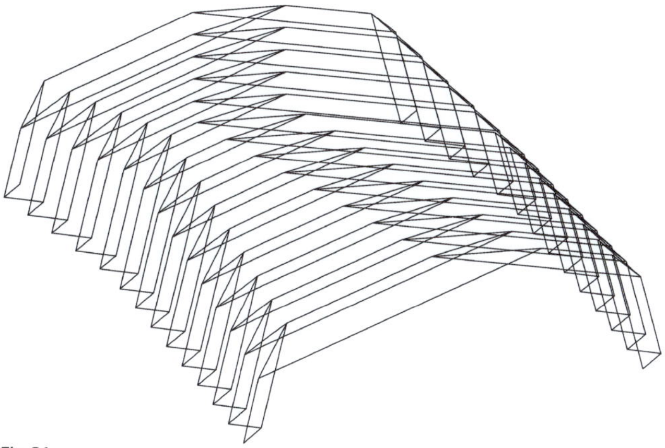

Fig. 31

Fig. 32

Fig. 33

Fig. 34

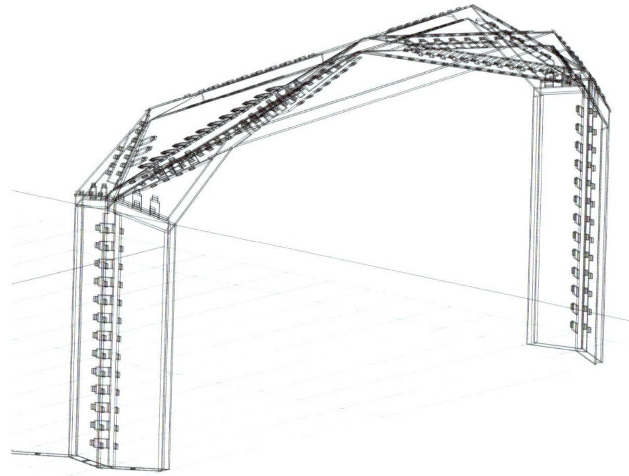

Fig. 35

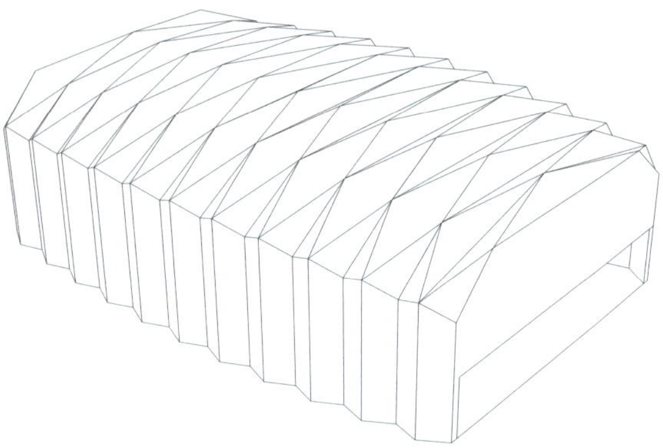

Fig. 36

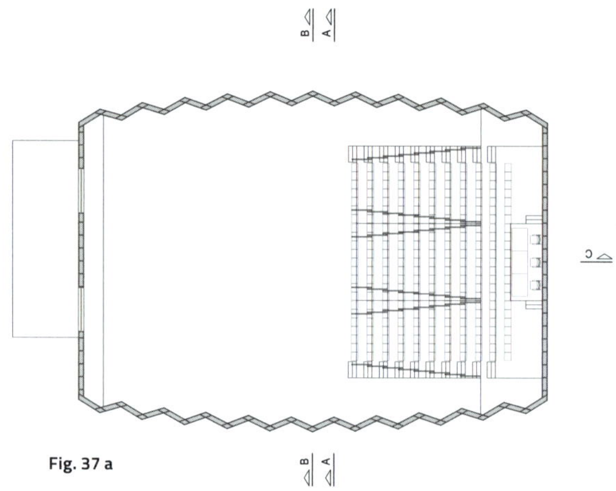

Fig. 37 a

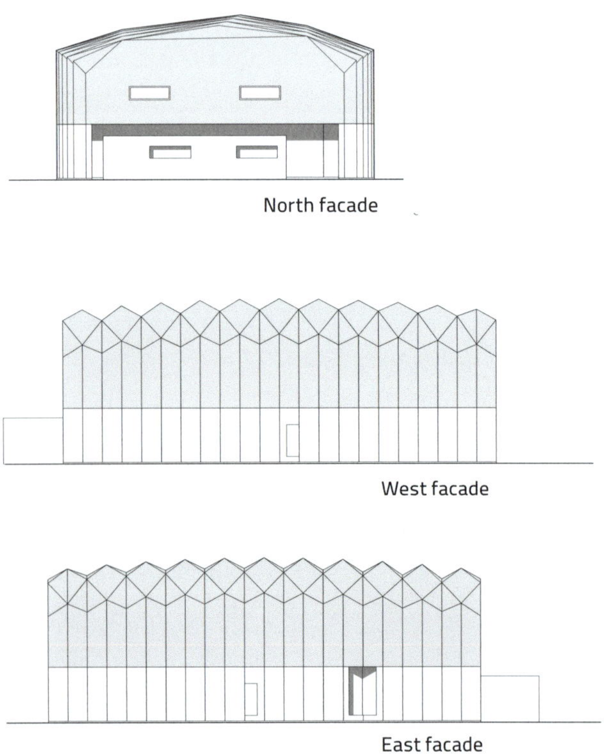

Fig. 37 b

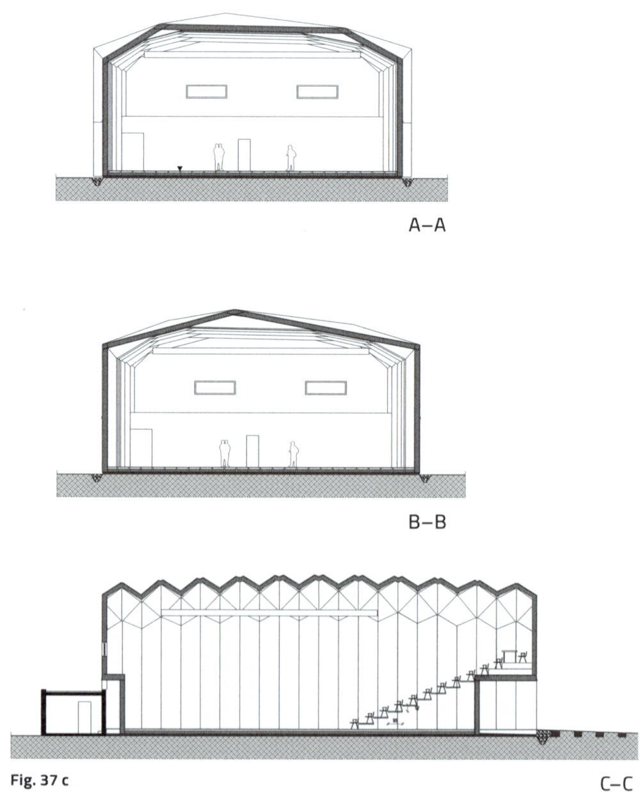

Fig. 37 c

Fig. 34 Principle of the corner junction in each fold. Both layers of shell project to the outer shell and are connected.

Fig. 35 Woven patterns are created along the edges, as the panels are connected by pins throughout.

Fig. 36 Axonometric drawing and plan of folded geometry, Vidy Theater

Fig. 37 a, b, and c Ground-floor plan, elevations, cross-sections, Vidy Theater

Fig. 38 Vidy Theater, Lausanne: presentation images

Fig. 38

1.2

Structural optimization of timber folded-plate structures

Andrea Stitic and Yves Weinand

This paper analyzes the potential of different possible folded-form topologies for generating folded-surface structures made from timber. The advantage of such structures lies primarily in the realm of ecology and sustainability. By utilizing an integrated method of construction that fulfills both a supporting and a covering function, extremely lightweight structures can be achieved. Also, a greater degree of prefabrication is possible leading to a reduction in the overall cost. Timber folded-surface structures consist of a large number of discrete, thin plane elements, connected together to form an overall folded surface. Proper edgewise connection details are needed in order to ensure an efficient load-bearing system. For structures made from timber products, this presents an enormous challenge, requiring the use of state-of-the art joining techniques. Thus, the use of folded timber plates in civil engineering applications has been very limited to date. However, new technical solutions have recently been proposed for efficient edgewise joining of thin timber panels. This paper focuses on integrated mechanical attachment techniques that utilize digital prefabrication to integrate connectors through panel geometry. Taking into consideration material, fabrication, and connection detail constraints, various topologies are examined for the considered application. Furthermore, the structural behavior of folded systems is studied, and three feasible forms are compared by means of finite element analysis. Finally, observations are made on a case study of a built prototype structure, and the structural potential of the proposed systems is outlined.

Keywords ***folded-surface structures, folded form topology, integral mechanical attachments, folded plates structural behavior***

1 Introduction

Architectural and technical applications of origami-inspired structures (the resulting forms need not necessarily be developable) employ the structural potential of the folding principle to use less material and increase structural efficiency. By placing the material farther away from the axis of flexure, i.e. folding, the moment of inertia is increased, inherently leading to greater structural stiffness. This paper focuses on folded structures of engineered panels made from timber. Common terminology for describing these structures, which utilize the benefits of folding, include: "folded plates," "folded slabs," and "corrugated structures." However, the structures being considered are composed of multiple inclined-plane structural surfaces, joined together to form a globally folded form, while their load-bearing behavior combines both a slab and a plate mechanism (**Fig. 1**). In this instance, the terminology "folded-surface structures" is preferred.[1]

Several structures made from timber engineered panels have been realized, as described by Hans Ulrich Buri[2] and Regine Schineis[3]. In these examples, the width of the structure was spanned with a single element. However, due to manufacturing and transportation constraints, timber elements are only available in limited sizes. Consequently, in order to cover longer distances, efficient connection details between adjacent plane elements are needed along the span of the structure. A folded-surface structure made from prefabricated cross-laminated timber panels and assembled with screwed miter joints was proposed.[4] The folded-form geometry chosen was based on folded rhombus elements. After examining its load-bearing performance, it was concluded that the screwed miter joint connections were not sufficiently resistant to withstand the resulting transverse bending moments, and that the joints for large panel assemblies such as these would have to be improved.

The issue of designing adequate thin panel edgewise joining details presents a major challenge in timber engineering, as it is difficult to address by using standard timber panel joining techniques. However, the performance of folded-surface structures depends on these linear edge-to-edge panel connections, and they are considered a key design component. Since on-site gluing is not possible due to a lack of uniform, stable conditions for curing of the adhesive, joints are usually achieved by using metal fasteners. Nevertheless, according to current regulations, the minimum distance from the screw to the panel edge is set to $4d$ (d = diameter of the screw) which restricts the minimum value of the panel thickness, depending on the size of the fastener.[5] Consequently, the final panel thickness is usually not dimensioned according to the structural requirements regarding the load-bearing capacity, but more in line with the minimum requirements imposed by the connection detail. The recent rediscovery of integral mechanical attachments has provided an innovative method for edgewise jointing of timber panels.[6,7] In this paper, we focus on an integrated mechanical attachment technique developed by Robeller et al.[8] that utilizes digital prefabrication to integrate connectors through panel geometry. The main advantage to using these form-fitting joints is that, in addition to their load-bearing function (connector feature), they also integrate features for fast and precise positioning of thin elements (locator feature). Moreover, these joints do not impose any constraints on the panel thickness.

In this paper we present various known folded form topologies and examine their structural potential for application in timber folded-surface structures. Subsequently, feasible forms are derived, taking into account the material and the chosen connection detail fabrication, as well as assembly-related constraints. The structural behavior of timber folded plates is studied and the chosen forms are compared by means of finite element analysis. Finally, a case study of a built prototype structure is presented, along with important observations regarding the material, fabrication, and element assembly.

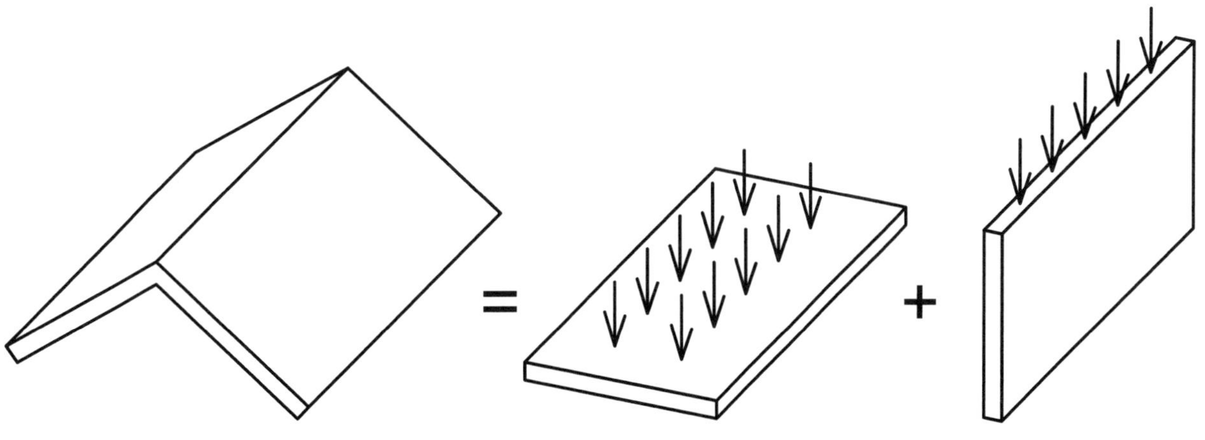

Fig. 1 Load-bearing behavior of a single fold, combination of the slab and plate effect

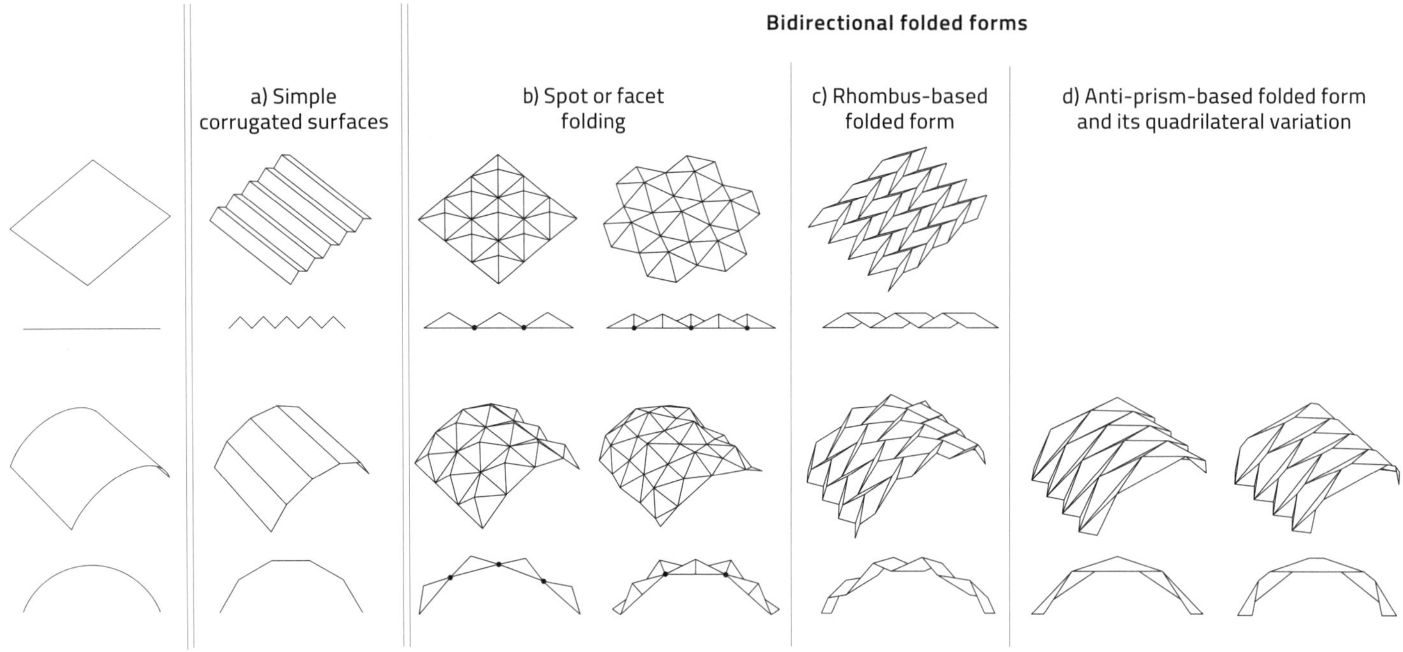

Fig. 2 Classification of folded-surface structures; perspective and transverse cross-section view

2 Topology of the folded form

Fig. 2 shows the various folded-form topologies that are classified by their ability to discretize flat and singly-curved surfaces. In order to describe these forms, we will use the terminology derived from computer graphics where a surface is represented by a polygon mesh.[9] The topology of such a mesh is defined using a set of vertices and additional information on how they are connected. This connectivity further determines the bounding edges of the form's surfaces. The spatial arrangement of vertices, as well as their connectivity, can be regular or irregular, creating a form that is composed of either identical or diverse base polygon geometry.

The first group **(Fig. 2a)** contains simple corrugated surfaces. They are composed of quadrilateral faces with bounding edges that form a succession of either parallel or oblique lines. Folded forms such as these are commonly used in practice. However, due to size constraints on timber panel elements, the covered surface area of such structures is limited. In order to overcome this constraint, topologies where vertices and their connections form a spatial grid are used. By doing so, bidirectional folded forms are obtained **(Fig. 2b, c, d)**. Such forms consist of multiple elements in two distinct directions of the structure.

The first bidirectional folded form considered **(Fig. 2b)** includes spot or facet folded forms, consisting of vertices where several faces converge together in one single spot.[10] Forms such as these are obtained by taking a basic polygon and vertically raising its centroid point. This vertex is then connected to the vertices of the initial polygon to form triangular faces. For such forms, only three types of polygons allow regular tessellations of a surface: equilateral triangles, squares, and regular hexagons. Others either require the use of semi-regular tessellations or irregular polygon geometry. In the classification presented, only regular tessellations are shown, as irregular tessellations offer a myriad of different topologies. The second group **(Fig. 2c)** consists of rhombus-based folded surface forms. In origami literature, this form is also known as the *Miura Ori* or herringbone pattern. The third and final bidirectional folded form considered is the form based on the anti-prism. This is composed of isosceles triangle faces and is also known as the *Yoshimura* or diamond pattern. A variation of this form where quadrilateral faces are used is also included **(Fig. 2d)**. This is obtained by duplicating the anti-prism form vertices and introducing an additional connection line between them, resulting in trapezoid faces. Neither of these forms can discretize flat surfaces, and they are limited to curved cross-sections.[11]

The bidirectional folded forms classified above are subsequently compared according to their folding principles, making them more or less feasible for structural application in timber folded-surface structures. All structures are considered to have pinned supports at the sides, while connections between the plates are regarded as line hinges, allowing rotations about the face edge direction. A cutting plane is positioned perpendicular to the longitudinal axis and the obtained transverse cross-section profile is observed **(Fig. 2)**. It should be noted that for quadrilateral-based facet folded forms, both flat and singly curved, a continuous longitudinal hinge line is formed in every second vertex of the transverse cross-section. Even with a rise sufficient to span the ratio of the structure, the number of such hinges should not exceed three (including the ones at side supports).[12] When more than three hinges are used, the system transforms into a mechanism and is no longer stable for more than two quadrilateral polygons per span. Also, the static height of the structure at this section point is very low and equal only to the panel thickness. With regard to other polygonal facet folded forms, the system remains stable irrespective of the number of hinges due to the offset of each transverse string of polygon elements in relation to the previous one. Each individual string is still a mechanism in its own right, but is kept stable by the neighboring string of elements. Furthermore, as the number of polygon sides in facet-folded forms increases, the dihedral angle between the faces surrounding the raised centroid point also increases. It is considered that angles close to 90° function particularly well for folded surface systems, whereas the structure will lack rigidity when the angle is incrementally increased.[13] Moreover, the dihedral angle values in integrally attached timber structures are restricted to angles between 50° and 130° by the connection detail choice (see section 3.1 below). Consequently, as the dihedral angles become very obtuse for five-sided polygons, such forms are regarded as unsuitable for this study.

As a result, only the rhombus-based folded form and the anti-prism-based form, along with its trapezoid variation, were considered worthy of further investigation. These three forms offer a particular structural advantage by having adjacent folds both in a longitudinal and a transverse direction, together providing the necessary structural height along the entire cross-section.

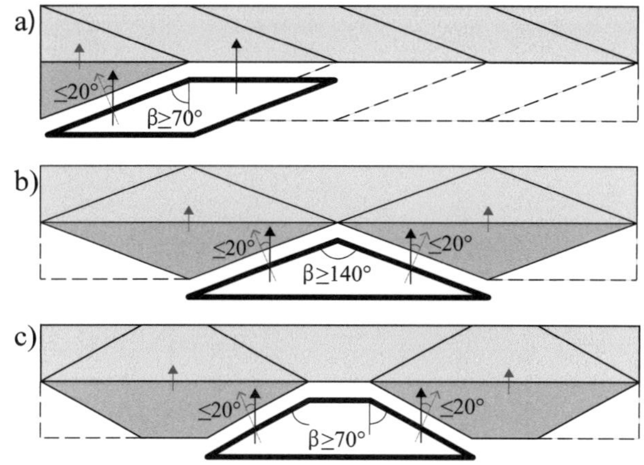

Fig. 3 Minimum value of the apex angle, β, and the maximum value for the assembly direction deviation from the plane normal to the assembled edges for the **a)** rhombus-based, **b)** anti-prism-based and **c)** trapezoid-based folded forms

3 Structural considerations

The chosen bidirectional forms were further examined by comparing their ability to discretize a hollow cylindrical segment with a set of predefined parameters. Its outer radius was set to R_{ext}= 2.5 m and the cord length, i.e. span, to S = 3 m in order to obtain an optimal structure height of h = 0.5 m, as well as a favorable height-to-span ratio of 1 : 6. Additionally, for restricting the maximum static height of the folded system, h_s, the inner radius was set to R_{int} = 2.3 m.

3.1 Joint constraints

The integral attachment technique chosen for connecting the panels imposes a certain number of constraints which need to be taken into consideration when designing the folded surface geometry. One of the main constraint considered is robotic accessibility where, due to the limited tool inclination, the dihedral angle between the connecting panels, φ, is restricted to angles between 50° and 130°. Furthermore, due to the local joint geometry, the orthogonal as well as the non-orthogonal assembly of individual elements also presents specific restrictions concerning the geometry and the number of panel edges that can be connected simultaneously. The simultaneous assembly of multiple edges requires that the panel assembly direction be chosen within the intersection of what are called the "individual edge rotation windows." These "windows" represent all possible assembly directions for each edge. The range of this intersection depends on the number of assembled edges, tool inclination limitations, and the dihedral angle between the plates.[14] In order to ensure that assembly is possible, i.e. that respective rotation windows intersect, the geometry of the panels must be chosen in such a way that the final assembly direction does not deviate from the plane normal to each assembled edge by more than 20° **(Fig. 3)**.

3.2 Geometry of the structures

For each folded form, the base element geometry was kept constant throughout the whole system so that the form remained regular. Apex angles, β, were kept within the limits imposed by assembly constraints and held to a minimum of 70° for the rhombus- and trapezoid-based folded forms, and a minimum of 140° for the anti-prism-based form **(Fig. 3)**. The panel thickness t, widths of elements w, and the dihedral angle φ, were also taken as invariables for all forms. The number of individual elements along a transverse cross-section, which can be used to approximate a set curvature with a defined maximum static height and structure span, was determined according to the following formulae:

$$n = \left\lceil \frac{\psi_{tot}}{\cos^{-1}\left(\frac{R_{ext}-h_s}{R_{ext}}\right)} \right\rceil$$

$$\tilde{n} = \left\lceil \frac{\psi_{tot}}{\cos^{-1}\left(\frac{R_{ext}-h_s}{R_{ext}-(h_s/2)}\right)} \right\rceil \qquad \text{eq. 1}$$

$$n \geq \begin{cases} \tilde{n} & \text{if } \tilde{n} \text{ is even} \\ \tilde{n}+1 & \text{if } \tilde{n} \text{ is odd} \end{cases} \qquad \text{eq. 2}$$

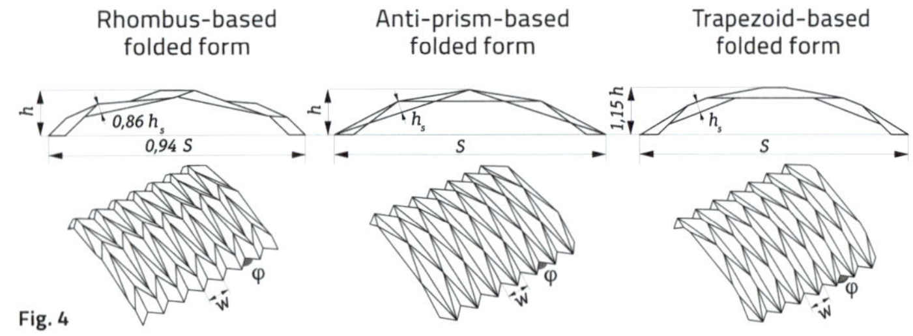

Fig. 4

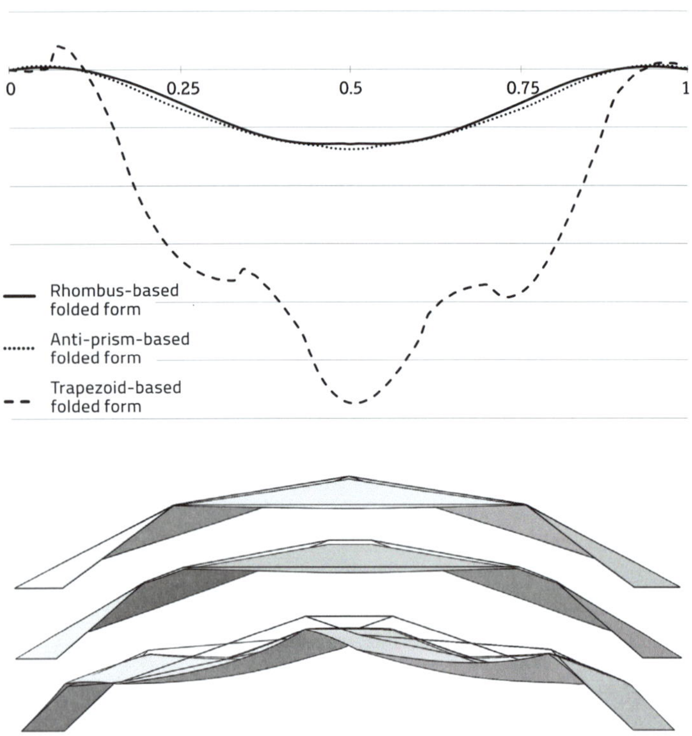

Fig. 5

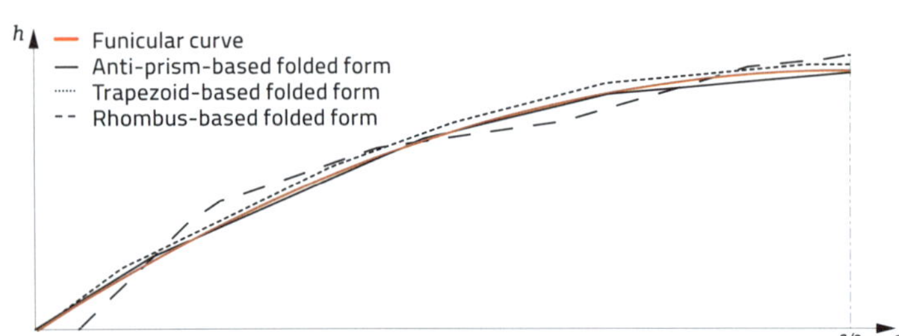

Fig. 6

Fig. 4 Transverse cross-section and perspective view of the three examined structures

Fig. 5 Vertical displacements along a normalized span (above) and an overlay plot of the deformed and undeformed state, uniformly scaled for all structures (below)

Fig. 6 Deviation of folded form central line from the equivalent smooth shell funicular curve

Where ψ_{tot} is the central angle of the circular segment observed, R_{ext} is the outer radius of approximated hollow cylinder, and h_s is the static height. For anti-prism- and trapezoid-based forms, there is a unique result for a required number of whole, uncut triangular elements **(equ. 1)**. On the other hand, for rhombus-based folded forms, the result is defined as the minimum one necessary. Additionally, only an even number of elements is considered, as any odd number results in end segments with different orientations, leading to a height difference between side supports. Therefore, the result is rounded up to the next even number **(equ. 2)**. For the set parameters, three segments for the anti-prism- and trapezoid-based form, and a minimum of six segments for the rhombus-based form were obtained **(Fig. 4)**. It can be noted that approximating the same curvature with a defined height by using the rhombus-based folded form resulted in a slightly smaller span, as well as reduced maximum static height of the system. In the case of the trapezoid-based form, the overall height of the system had to be increased in order to keep the side cross-sectional segments uncut and to avoid unnecessary corrugations at the supports.

3.3 Structural analysis

Bidirectional folded-surface structures carry the load via a combination of extensional and flexural action. The relative proportions of extensional and flexural effects depend on several factors: the overall form of the structure, connection detail stiffness, support conditions, and the loading configuration. Extensional or shell action resists the external transverse loading through the action of in-plane forces. This is a result of the overall shell-like form of the structures, but also the individual position of the plates in relation to the acting force. The structural behavior of an inclined plate such as this is divided into two different mechanisms of resistance and their combinations. Firstly, plate mechanisms, where the component of the acting force directed parallel to the surface is resisted by in-plane forces, and secondly, slab mechanisms, where the component of the acting force directed at right angles to the surface is resisted by bending and twisting moments. The latter mechanism of local plate bending forms the flexural part of the overall structure's load resisting action.

The conversion of out-of-plane external load into in-plane extensional action is explained below. The load normal to the plate is transferred by bending to the edges, where it is resolved into components situated in the planes of the joining plates. When these in-plane forces are not in equilibrium, the resulting force is further transferred to the supporting plates by in-plane shear along certain edges. The size of this in-plane shear depends on a number of factors, such as the uniformity of the dihedral angle between the observed plate and its supporting plates and stiffness of the plates, as well as the edge connection detail and the nature of the load.

The local plate bending behavior also depends on similar factors. However, the values of the edge bending moments that cause edge rotation depend primarily on the rotational stiffness of the connection detail.

The three obtained structures were modeled in finite element analysis software (ABAQUS) and their performance was observed under structural self-weight. It was concluded that the local structural behavior of the system and the inherent global structural stiffness depend to a great degree on the connection details' capacity to transfer the occurring shear forces and bending moments between the adjacent plates. It was decided to model the actual semi-rigid behavior of the joints as being completely rigid. This simplification enabled us to obtain maximum values of the occurring edge forces for each folded form and to compare them with each other. Furthermore, the boundary conditions were modeled as if pinned, allowing rotation, but no movement in all three directions. The plates were modeled as conventional shell elements and their geometry was defined at the reference surface that coincides with each plate's mid surface. In this way, a three-dimensional continuum could be approximated with a two-dimensional theory due to the fact that the plate thickness is small compared to its remaining dimensions. Material properties of simplified linear elastic, orthotropic, 21-mm-thick, structural LVL timber panels were fed into the model with values according to the VTT No. 184.[15]

Fig. 5 shows vertical displacements along a normalized span of the mid transverse cross-section of the observed structures. The overlay plot of the deformed and underformed structure shape, uniformly scaled for all three forms, is shown below. The lowest vertical displacements under a dead load were found for the anti-prism-based structure. The ratios between the values obtained for the other two observed structures and the anti-prism-based form were equal to about 1.5 for the trapezoid-based structure and up to 3.5 for the rhombus-based structure. Even higher discrepancies were found for the values of the bending moments about the individual edges. The ratio between the rhombus-based and the anti-prism-based structure amounted to 5. Values for shear forces were highest for the rhombus-based structure, with a ratio of 3:1 when compared to the value for the anti-prism-based structure.

The ratio between the trapezoid-based and the anti-prism-based structure was equal to about 1:3. A distinct distribution of shear forces was found when comparing the rhombus-based structure with the other two structures. The shear forces appear on all edges of the rhombus-based structure, whereas in the anti-prism-based as well as the trapezoid-based structure they are only present at the edges that are not parallel to the main axis of the structure. The edges are shorter in the trapezoid-based structure than in the anti-prism-based structure. This accounts for the aforementioned increase in shear.

As a result of the analysis presented here, it is concluded that the choice of the folded form has a significant influence on the edge connection load-bearing requirements and that the semi-rigidity of the joint will certainly play an important role in the behavior of such systems in reality. Furthermore, the best structural performance of the folded-surface structure is found when the external load is primarily resisted by shell-like extensional action. This can be enabled by choosing a form which best approximates a corresponding hollow cylinder segment, i.e. a smooth shell solution. To that end, a comparison is made between each folded transverse cross-section central line and a funicular curve of the equivalent smooth shell. It can be seen that the central line of the anti-prism form is the one closest to the funicular solution (Fig. 6). The amount of flexural action generated in order to resist the load depends on how much the folded form deviates from the funicular. A substantial difference observed between the rhombus-based form and the other two structures is a result of their fundamental topological distinction. The rhombus-based form consists of quadrilateral faces with vertices alternately lying on the outer and inner surface of the hollow cylinder, i.e. the distance between these two surfaces is equal to the maximum static height of the system. On the other hand, the anti-prism-based and the trapezoid-based forms consist of faces where all vertices are situated on the outer surface. This topological difference makes the structural height of the latter two forms more uniformly distributed along the transverse cross-section, consequently providing a higher overall rigidity.

4 Observations on a prototype structure

A prototype structure was constructed and tested in order to explore the behavior of timber folded-surface structures under load. Initial findings concerning the fabrication and assembly of its constituting elements were collated. Integral mechanical attachments were used as connection details, and their geometry was designed according to certain extant examples.[15,16,17] Regarding the material, two types of engineered wood products were considered: Kerto-Q structural grade laminated veneer lumber (LVL) and cross-laminated timber (CLT) panels. It was noticed that the more homogeneous and mechanically sturdy peeled-veneer laminates offer particular advantages over CLT panels. First, considerably thinner cross-sections are possible, making lighter-weight structures possible. Second, an important shortcoming of CLT was recognized while milling the panel edges. As CLT panels consist of several layers of longitudinal timber planks that are not glued to each other on the sides, depending on the angle of the joint with respect to the individual layer plank orientation, there is a risk that considerable pieces of the joint could simply chip off after manufacture. Another issue when using integral mechanical attachments was recognized early on in the

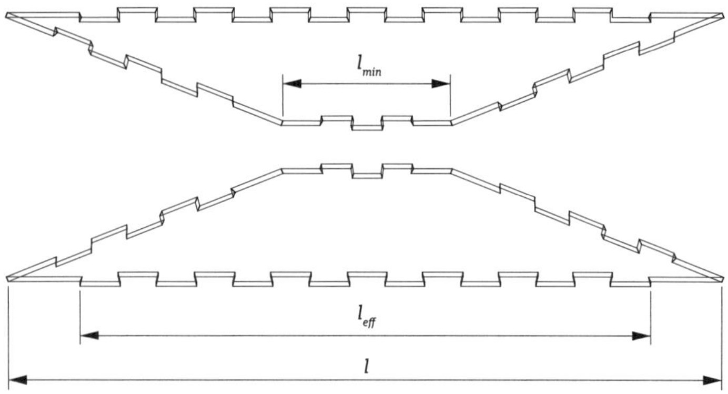

Fig. 7

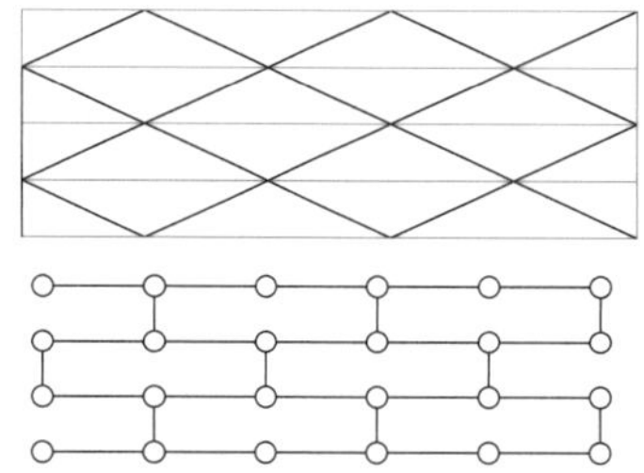

Fig. 8 a

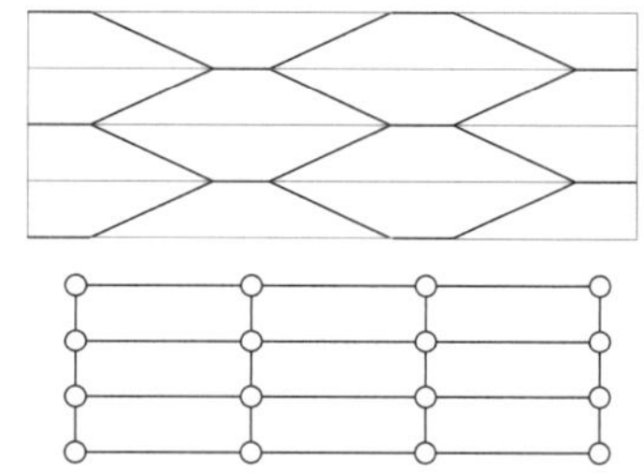

Fig. 8 b

prototype design process. Due to specific joint geometry, the effective connecting length of each edge is always shorter than its total length. This can pose a problem when trying to achieve an efficient connection of a number of plane elements which converge at one vertex. In order to reduce the vertex valence number, the prototype was designed using the isosceles trapezoid-based folded form (vertex valence = 4) rather than the anti-prism-based one (vertex valence = 6).

The minimum edge length of the trapezoid element was restricted in order to secure a minimum connection between the panels and to provide at least one pin on each adjoining plate edge **(Fig. 7)**. As a result, an additional benefit of the trapezoid-based folded form was recognized. In such a form, the edge-to-edge connectivity is realized between every element along the span and the neighboring transverse string element **(Fig. 8b)**. In contrast to this is the isosceles triangle solution where a connection is only achieved in every second element **(Fig. 8a)**. Additionally, by using isosceles trapezoids, a wider span can be realized with the same number of elements along a transverse cross-section.

Using the RhinoPython application programming interface, a computational tool was developed that instantly generates both the geometry of the individual components and the machine G-code required for fabrication. Exploiting this geometric freedom, we have tested our computational tool by designing a double-curved folded-surface prototype with alternating convex-concave transverse curvature **(Fig. 9)**. The built structure spanned three meters and was constructed with 21-mm-thick Kerto-Q structural grade LVL panels (7-layer, I-III-I). The structure's total weight amounted to only 192 kg. Boundary conditions that restrain displacements of the supports in the transverse direction were applied on both sides. This was achieved by fabricating form-fitting lateral timber supports that made it possible to straighten the edges in order to position two large

Fig. 7 Connection detail in a two-panel assembly

Fig. 8 Edge-to-edge connectivity between the faces of **a)** an anti-prism-based folded form and **b)** the isosceles trapezoid version

Fig. 9 Timber folded-surface prototype

Fig. 10 Load-displacement curve of the load test

Fig. 9

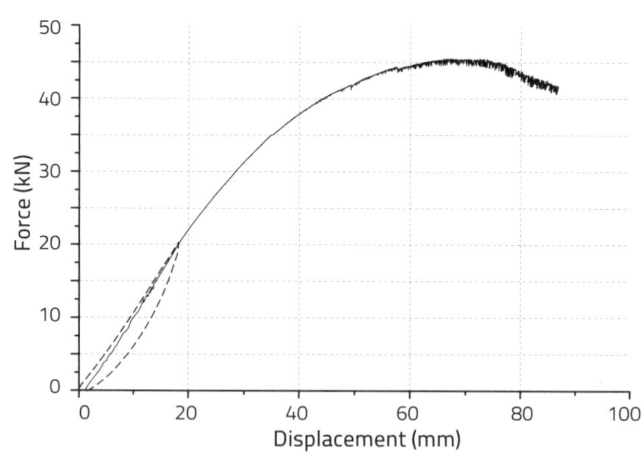

Fig. 10

concrete blocks on each side of the structure. A longitudinal line load was introduced gradually by applying pressure to a leveled steel beam placed on top of the structure. Vertical displacements were measured at the center point of the transverse cross-section. Fig. 10 shows the load-displacement curve obtained from the test. The dashed line represents the first applied load cycle where the maximum load reached up to the approximated proportional limit of the curve.

It can be seen that the load of 20 kN corresponds to a vertical displacement of 18 mm. Next, a load up to the failure of the structure was applied. This displayed a high structural efficiency (ratio of the maximum load over the dead weight of the structure) of the tested prototype, which reached to 23.44 when loaded with 45 kN.

The failure of the structure occurred in the connection detail when the out-of-plane rotation of the edges closest to the applied line load exceeded the joint capacity. This kind of failure mode indicated that the connections used can provide a certain ductile capacity for the folded surface system made from brittle timber members. In order to assess the overall ductility of the structure and the influence of the joint semi-rigidity, the prototype was modeled in the FE analysis software with completely rigid joints, in the same way as previously explained (see section 3.3). The load was applied in the same manner as in the test, to the limit of 20 kN, which resulted in a 2 mm vertical displacement. Such results suggest that the introduction of semi-rigid connections has a significant influence on the overall structural behavior and can potentially provide a beneficial ductile behavior for timber folded-surface structures under load. However, further studies are needed to precisely quantify their effects.

5 Conclusion

The prototype presented here demonstrated the realization of a very lightweight structure with a weight-to-surface-area ratio of only 11.5 kg/m^2. It successfully combined the structural advantages of timber panels with the efficiency of folded plates, while conforming to the constraints of integrated mechanical attachments. Furthermore, it demonstrated the high structural capacity of the chosen folded form and validated the fabrication and assembly methods used for its construction. However, it should be noted that further research is required regarding the physical connections used. Their mechanical behavior and load-bearing capacity still need to be confirmed. Nevertheless, a great structural potential is recognized in the proposed timber panel structural systems, and the forms discussed within this paper are considered to be of significant influence in establishing these structures on a building scale.

References

1 Sedlak, V. "Folded surface structures." *Architectural science review,* 1978, 21(3), 58–59.

2 Buri, H.U. *Origami-Folded Plate Structures,* PhD Thesis, EPF Lausanne, 2010.

3 Schineis, R. "Gefalteter Klangkörper Musikprobensaal Thannhausen (Thannhausen Rehearsal Room)," in: *10. Internationales Holzbau Forum,* Forum-Holzbau, CH-Biel, Garmisch-Partenkirchen, 2004.

4 See note 2 above.

5 EN 1995-1-1. *Eurocode 5: design of timber structures—part 1-1: general-common rules and rules for buildings.* European Committee for Standardization, Brussels, 2004.

6 La Magna, R. et al. "From nature to fabrication: Biomimetic design principles for the production of complex spatial structures," *International Journal of Space Structures,* 2013, 28(1), 27–40.

7 Robeller, C., A. Stitic, P. Mayencourt, and Y. Weinand. "Interlocking folded plate, integrated mechanical attachment for structural timber panels," in: Block, P., J. Knippers, N. J. Mitra, and W. Wang (eds.) *Advances in architectural geometry 2014: Proceedings of the AAG'14 Conference,* Springer, London, 2014, 281-294.

8 Ibid.

9 Botsch, M. et al. *Polygon Mesh Processing.* AK Peters, 2010.

10 Trautz, M. and R. Herkrath, "The application of folded plate principles on spatial structures with regular, irregular and free-form geometries," in: Domingo, A. and C. Lazaro (eds.) *International Association for Shell and Spatial Structures: Proceedings of the IASS Symposium on Evolution and Trends in Design, Analysis and Construction of Shell and Spatial Structures,* Valencia, 2009.

11 Frostick, P. "Antiprism Based Form Possibilities for Folded Surface Structures." *Architectural science review,* 1978, 21(3), 59–67.

12 Huybers, P. *See-through structuring: A method of construction for large span plastic roofs.* PhD Thesis, TU Delft, 1972.

13 Bechthold, M. *Innovative Surface Structures: Technology and Applications.* Taylor and Francis, 2008.

14 Mattoni, G. *Folded Plate Structure—Design and Analysis of Woodworking Joints for Structural Timber Panels.* MA Thesis, ENCP and EPFL, 2014.

15 "VTT certificate for structural laminated veneer lumber," No. 184–03, 24 March 2009, Certification body (S017, EN 45011) accredited by FINAS.

16 See note 6 above.

17 See note 14 above.

18 Roche, S., C. Robeller, L. Humbert, and Y.Weinand, "On the semi-rigidity of dovetail joint for the joinery of LVL panels." *European Journal of Wood and Wood Products,* 2015.

Reprinted from *International Journal of Space Structures,* Vol. 30, Num. 2, 2015.

1.3

"It's important to progress from research to a product"

Interview between Ueli Brauen (Brauen & Wälchli Architectes, Lausanne) and Yves Weinand
April 21, 2015

Yves Weinand: Let's go back to the history of IBOIS, which you are familiar with. When I became the director of the laboratory, we initially began by placing an emphasis on architecture and form creation. The first theses of Hans Ulrich Buri[1] and Ivo Stotz[2] dealt with the creation of overall forms. Hans Ulrich Buri describes folded structures in his "Origami-Folded plate structures" thesis; he created a tool that allows them to be generated while respecting a certain number of geometric constraints. It's intuitive work, inspired by the Japanese art of folding. This very graphic work illustrates the wide range of geometries that can be used to create pleated structures. But the approach remains purely geometric, as mechanical or structural optimization aspects are not addressed.
Then with the thesis of the architect, Christopher Robeller,[3] we were interested in wood-wood assemblies. The folded structures we had already studied were very rigid. However, during tests on rupture, it is the assembly or connections between panels that are the first to give way. So long as we seek to work with thin panels, we are not able to respect existing norms, since minimum distances between screw axis and panel's edge need to be respected.[4] Therefore, the question of joints is of great interest to us.
In the context of her thesis, civil engineer Andrea Stitic was interested in the topology of form. Her research involves the integration of joints or connections, while simultaneously considering manufacturing constraints. She was able to develop calculation models, that take into account the spring rigidity of wood-wood assemblages. She's interested in the topology of form, which is unusual for an engineer. It's truly an architecture/civil engineering undertaking.
In terms of innovation, the engineer can contribute a lot to the appreciation of a global form.
This brings me to my first question: What do you consider to be the potential of folded structures? Do you think they could be used for types of buildings other than pavilions?

Ueli Brauen: I would like to start with your research, and how you came to work on folded structures. You have an engineering and architecture background. Hans Ulrich Buri undertook this first research on folded structures as an architect. When you take an element like the fold—which is familiar to everybody—as an object of research, it's very interesting from an architectural point of view, because it opens up new horizons in the field. At a given time, the engineer intervenes and undertakes research at a structural and mathematical level, to grasp the forms and determine their structural potential.
It does not surprise me that people consider you to be "unusual," because there are plenty of specialists, who are very competent in their domains, but the great challenge resides in the capacity to see links with other domains. What is really interesting in your research is that it combines the culture of architecture with that of engineering. The drive for architecture is very strong in your work. When you work with a form, it must be pure. This is also true for assemblages. You are interested in wood-wood assemblages that don't need any other binding material. That is your sense of aesthetics and elegance speaking. Your research interest would be different if you were trained solely as an engineer.
For me, it's clear that it will take some time before you are able to bring together static and functional requirements, with the requirements of form (whether it's for folds or assemblages). You will either manage to find a structural form for a broad function that is convincing thanks to its form, or you will find a form that is structurally reliable. For the latter, price will be a determining factor. A system can be interesting at one scale, but no longer at another. Each structure has its optimal scale. You dream of large folded structures, such as the Saint-Loup Chapel. This increase in scale, aligned with a need for rigidity, will be a difficult challenge to address.

Y. W.: We are not talking about mega-structures here, but structures with a range of 40 to 45 meters. A folded structure would be strong enough for the scale of a sports arena, such as the one in Yverdon.[5] We have noted—and this was also the case in the context of the Isles Sports

Center in Yverdon—that when the thickness of the wood becomes too significant, the costs sky-rocket. We therefore reconsidered the project and designed a hybrid structure: a lattice beam reinforced by a continuous panel. It's for this reason that we would like to resolve the problem of assemblages and, in doing so, that of cost. Even in 2006/07, at the time of the Yverdon project, our research was not sufficiently advanced for that.

U. B.: And today, do you have a less expensive solution that could work?

Y. W.: It's a nuanced answer. We are still unable to convince enterprises to open up entirely to production automation and to thereby reduce the costs of production for folded structures. However, as long as wood is increasingly applied to wooden constructions, more and more enterprises are becoming specialized in wood manufacturing and the situation could change.
More commonly, companies charge for the cut depending on the number of panels, rather than accounting for "machine-hours." In considering the cost of the panel and the cutting time, then yes, I think production costs could be reduced. So, to answer your question, I have a less expensive solution, so long as you are producing in-house. Unfortunately, to date we have not yet found an enterprise that would be willing to produce at a lower cost.

U. B.: Yes, because they are protecting their market. They have no need to go any further, since demand is there in any case. Glued-laminated is efficient and automated. I think we need to gain the interest of a company that would develop the folding and assemblages with you, and would have a return on investment in the future. But I also think that wood-wood assemblages could be very interesting for residential buildings, for both economic and ecological reasons (no glue, no additional elements or materials). By putting together prefabricated elements, assembly and dismantling would be facilitated. There's tremendous potential.

Y. W.: I'm fascinated by vernacular architecture. I'm participating in a drone airport project in Africa, which will include two systems: one medical, the other commercial. The constructed surface on stilts represents approximately 15 by 50 meters. It will consist of covering areas for drones, which will be stored beneath these structures.
I suggested providing a range of tool data to people on location, who could then use products derived from wood for an in situ production.
I have always liked very sharp angles, patterns reproduced in a very repetitive fashion, which can often be found in African carpets. By transmitting our technology, adaptable on location, for cutting in Africa, the architectural expression could find a local reference. It's not about imposing a modernist attitude here, but transmitting a tool that allows for adapting or specifying architectural solutions based on their cultural context, while taking into account the essence of the available wood, and the local architectural expression. We could create sections of columns with acute angles, of about 15 meters in height, by using connections that would be simply clipped.
This would imply the use of an automated cutting machine (CNC) on location.
Today, from an architectural point of view, the elaboration of construction detailing has become difficult because of the multiplicity of construction layers. The connection is no longer magnified, as it was in the past. I have the impression that wood-wood joint development, like what we are doing, could be a return to basics.

U. B.: Attention to detail still exists today, but in specific buildings, where we invest money for cultural or symbolic reasons. But in most architectural applications, the technical detail must be effective, in that it must be economical and sustainable. In those cases, the aesthetic cost is of little importance. We have much to learn from artisans today. Attention to detail is important. In terms of performance, costs, and sustainability, it must be optimal. I understand that, as an architect, you have a desire for form, which comes through in the elegance of assembly. But in Africa, using a drone and making it work is far more important than the building. In that precise case, I would not talk about the idea of the form as a cultural value, because it adds something excessive that leads to the impression of a cost overrun. In any case, the form is a question of adaptation over time. The Roman aqueduct is a 100% utilitarian building with the purpose of transporting water. Today, most people have an emotional reaction to these forms, as the arch has become something familiar to us. The mast of a high-tension line has the same structural aesthetic value to me, but nobody is moved by it, because we have yet to adopt its form as a cultural value. Architecture must seem self-evident to a user. To arrive at this self-evidence and touch on the collective culture of the user is important.

Y. W.: Can construction detail once again lead to cultural value? In certain cases, the architectural detail holds particular importance. But generally, in ordinary architecture—as you put it so well—it's rarely the case. For me, the constructive detail has much to contribute to architectural expression. Its elaboration should be a priority concern. This was once the case and has been the case for architecture in Japan, or for architects like Mies van der Rohe.

U. B.: Yes indeed, that would be wonderful. But there are other examples, such as Apple, where everything was derived from one person who had an aesthetic sensibility, an attention to detail and business. It was the first time that we considered adapting machines to humans. Which is to say that the common culture of humans allowed for the instantaneous use of a complicated machine. For me, architecture must have this same sense of certainty; we must find the most obvious entry point but, once we're there, we must immediately understand which direction to follow. I hope that attention to detail will be developed with this same sense of evident functionality.

Y. W.: We recently presented our work in Frankfurt,[6] where we were approached by DETAIL magazine. They wanted us to focus on the theme of construction detail, all the while showcasing and demonstrating the structure in its entirety. By the way, we currently have a book in preparation, in which we present the detail in relation to overall form.

U. B.: I admire the overall work you have been doing for years, with the means at your disposal, the engineers' mathematics and tools. I'm under the impression that at IBOIS you are in a sort of playground, where you enjoy yourselves, and have the tools to research the properties and potential of your "toys."

Y. W.: We always work on very specific programs, but I'm not sure how we could go towards more generalist applications.

U. B.: One must recognize that you are involved in fundamental research, not for profit. It's important to progress from fundamental research to arrive one day at a product, the result of which was not distorted by economic or political constraints. You are free, and so is your research, and it's normal that such things take time. I like collaborating and dreaming with you.
I appreciate your knowledge and your desire to create a pure and elegant architecture, which we see less and less nowadays.

Y. W.: In the end, my intuition is that we should evolve towards the scale of larger buildings in order to be convincing. I'm working actively towards building on a larger scale but, for that, we have to create partnerships with industry. We have the opportunity to perpetuate our research in the context of a NCCR[7]. This allows us to undertake applied research, with the support of the SNF[8], and to collaborate with industrial partners.

References

1 Buri, H. U. "Origami —Folded Plate Structures, thesis n° 4714." EPFL, Lausanne, 2010.

2 Stotz, I. "Iterative geometric design for architecture, thesis n° 4572." EPFL, Lausanne, 2009.

3 Robeller, C. "Integral Mechanical Attachment for Timber Folded Plate Structures, thesis n° 6564." EPFL, Lausanne, 2015.

4 EN 1995-1-1:2004+A1, Eurocode 5 Design of Timber Structures.

5 Isles Sports Center, Yverdon-les-Bains, Architects: Brauen & Wälchli et Yves Weinand; Engineers: Bureau d'études Weinand; Client: Town of Yverdon-les-Bains, Competition: 1991, Project: 2006–09, Implementation dates: 2010–11.

6 Advancing Wood of Architecture, Deutsches Architekturmuseum, Frankfurt, 27.03.2015.

7 National Centre of Competence in Research Digital Fabrication (NCCRs). The National Centres of Competence in Research –NCCRs – are a research instrument of the Swiss National Science Foundation. They promote long-term interdisciplinary research networks in areas of strategic importance for the future of Swiss science, economy and society (www.dfab.ch)

8 FNS/SNF Swiss National Science Foundation (www.snf.ch)

1.4

The Chapel of the Deaconesses of St-Loup at Pompaples, Switzerland

Marielle Savoyat

Design Local Architecture/Danilo Mondada architecture firm and Shel Architecture, Engineering, and Production Design

Client St-Loup Deaconnesses Institute 1318 Pompaples, Switzerland. Contact: Sister Marianne Morel

Completion 2008

Location Pompaples, Switzerland

Fig. 1 Detail

Fig. 2 Structural panels seen from the inside

Following a competition awarded in the summer of 2007 to renovate the buildings occupied by the St-Loup Deaconesses in Pompaples, a wooden chapel with an abundance of folds was realized in 2008. It was intended to be a temporary structure, as renovations to the original chapel in one of the buildings of the deaconess institution would only be accessed again in early 2010. However, the sisters fully appropriated this new innovative space and wanted to keep their temporary timber chapel, even after the renovation period. The idea of building this ephemeral, yet warm and welcoming place of worship—using high-quality, inexpensive material—became an obvious and preferred solution. In collaboration with the Laboratory for Timber Constructions (IBOIS) at the EPFL, Swiss Federal Institute of Technology, Lausanne, an innovative structure based on panels of cross-laminated timber was created, based on research undertaken by Hans Ulrich Buri

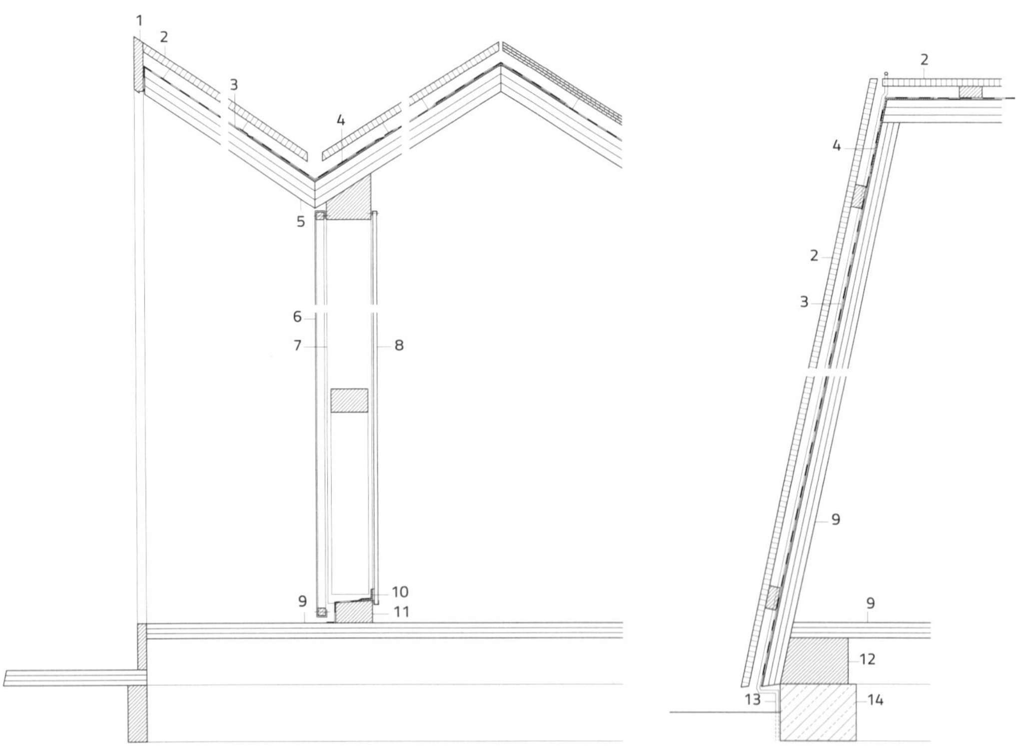

1. 24 mm pine eaves board
2. 20 mm three-ply lam. Sheeting prematurely greyed, screw fixed to 30/60 mm pine distance pieces
3. bituminous sealing layer
4. 2 mm sheet steel bent to shape
5. 60 mm lam. cross-boarded sheeting, inner face untreated
6. fabric covering on 20/20 mm wood framing
7. 60/120 mm pine column
8. 10 mm polycarbonate cellular slab in 10 mm alum. channels
9. 40 mm lam. cross-boarded sheeting
10. 30/30 mm aluminium angle
11. 100/60 mm pine plate
12. timber bearing plate
13. Ø 8 mm copper lightning conductor
14. 200/150 mm concrete blocks

Fig. 1

Fig. 2

Fig. 3

Fig. 3 Sun and shade expression of lateral facade

Fig. 4 View of the entrance with its translucent enclosure

Fig. 5 The building in its countryside setting

Fig. 4

Fig. 5

and Professor Yves Weinand on folded wooden structures, inspired by origami, the Japanese art of folded paper. The audacity of the sisters was mitigated: The cost of the new space was equal to that of renting a provisional structure on the construction site.

IBOIS researcher Hans Ulrich Buri designed a digital tool that generates folds inspired by the art of folded paper. Shel analyzed the structural behavior and established construction plans. Thanks to this invention, the chapel was delivered in the record time of two months. Prior to this, only small-scale prototypes had been produced at the IBOIS.

The fact that a thin surface becomes more rigid when it is folded was applied. The principle made it possible to construct the envelope and chapel structure with large wooden panels, rather than with a traditional post-and-beam frame. This technique was feasible thanks to advances in the wood industry. It is now possible to produce massive surfaces with irregular and differing cutouts for each piece at no additional cost. The panels are cut with a numerically-guided saw, based on 3D drawings. This allows the construction of designs with a high level of complexity. In the case of the St-Loup Deaconesses' chapel, the massive 45-mm-thick plywood vertical panels and 60-mm-thick horizontal panels in spruce, are all different sizes. The panels were simply assembled by metal plates and fixed in place with screws, before being covered on the exterior with a layer of bituminous waterproofing and clad in three-ply wood panels. It is important to note that every second ridge is vertical, thus facilitating on-site assembly. The inverted roof surfaces distribute the water runoff on both sides of the long elevations. The two gabled facades, composed of a textile membrane and skeletal reliefs with diagonal drawings, recall the composition of a stained glass window and allow natural light to enter the space.

The coherence between structure and architecture and the choice of using wood, creates an extraordinary sense of serenity and a profound architectural quality.

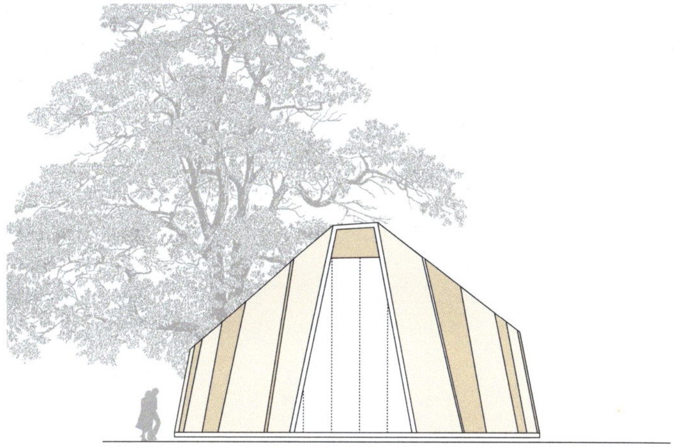

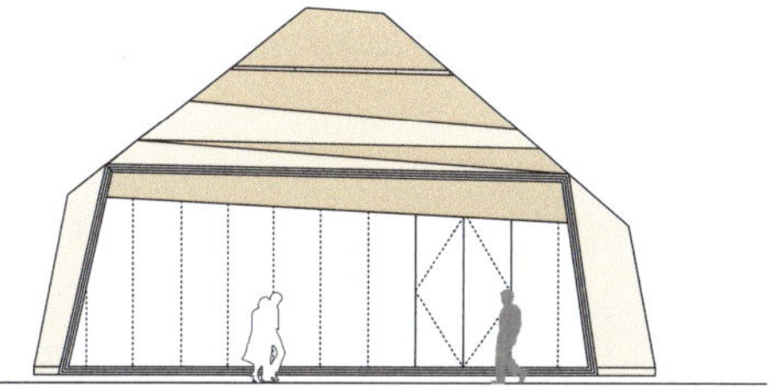

Fig. 7

Fig. 6 View on the altar

Fig. 7 Elevations

Fig. 8 Plan and section

Fig. 6

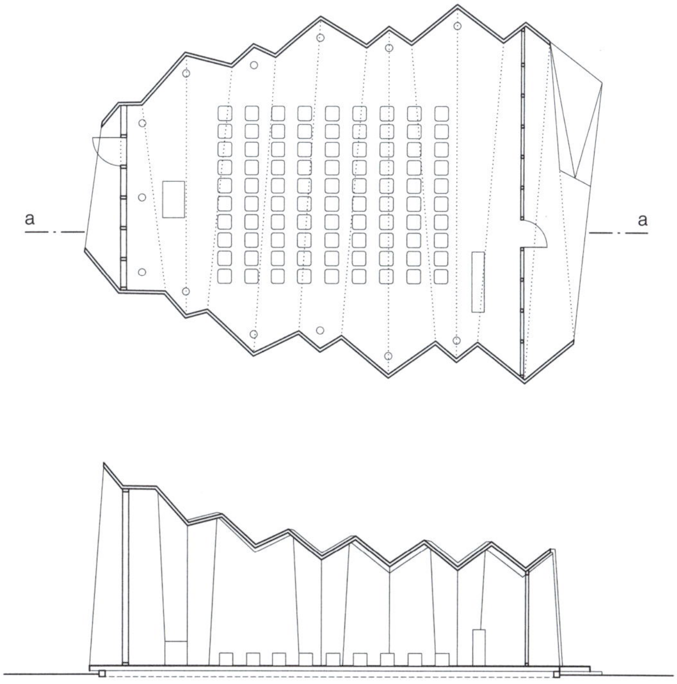

Fig. 8

2 Advanced architectural geometry

2.1

Regular and fractal-shaped, discretized tensile structures

Yves Weinand

In the previous chapter, it was shown how the overall stiffness of a structure can be increased by creating folds. The origami tool undertakes this within the following constraint: the geometry and folds generated can always be reduced to a flat surface. The following project envisages embedding folds into open, free areas.

Shell structures are known to perform well under their own weight and vertical loads, because these forces are dissipated as membrane forces, thereby avoiding any bending in the shell (except at the support points). It is also known, however, that horizontal loads—caused, for example, by wind or earthquakes—produce bending in domes that cannot be absorbed well. Historically, civil engineers have mainly specified standard areas for shell structures. Well-known examples of these are buildings by Félix Candela and Eduardo Torroja. Pier Luigi Nervi also envisaged the use of folds in these standardized areas as, thanks to the folds, asymmetrical and horizontal forces can be better absorbed. Furthermore, light can penetrate into the domes.

The aim of the work presented here was to develop a tool that could generate folds in regulated as well as free-form areas. In addition, our aim was to integrate

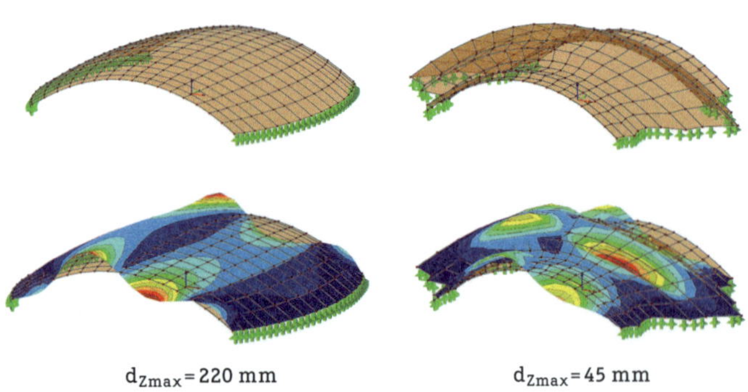

Fig. 1

Fig. 1 Comparison of standard and folded areas. Identical dimensions and support conditions are evaluated for a standard surface and an equivalent surface with folds (under snow load).

Fig. 2 Using the developed tool to create free forms, standard surfaces and folded or fractal surfaces can be generated. The division pattern is also visible.

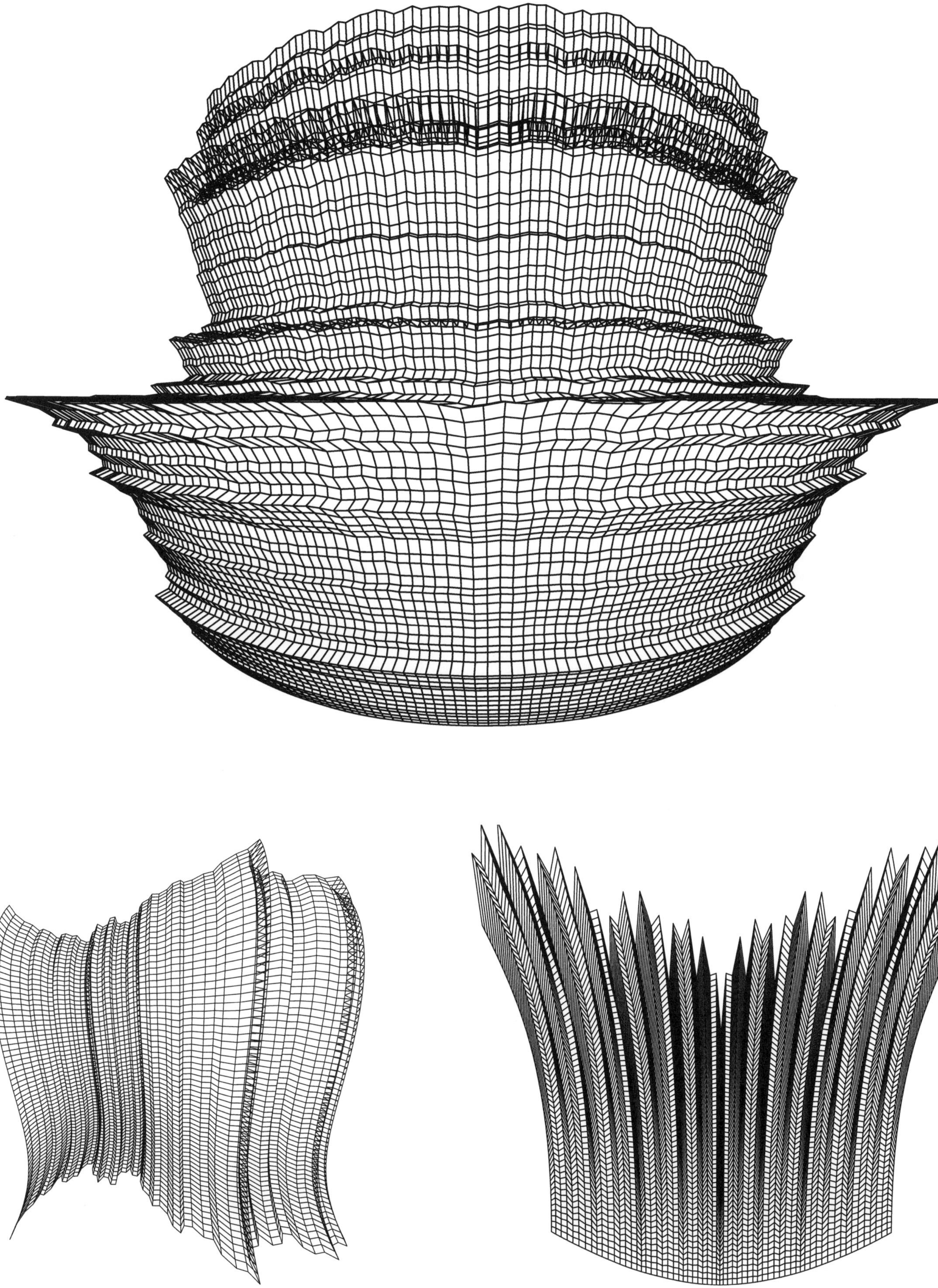

Fig. 2

the subdivision of the surfaces into the form-finding process.

The starting point and the basis of our work are iterative functions that have the ability to retain a selected basic topological form regardless of the transformations applied to the geometry. Thus, it is possible to introduce standardized or fractal curves as transformations. The surface-generated fractals have the advantage of introducing folds into the overall geometry. With the help of weight points, the folds can be strengthened. The subdivision process can be carried out either homogeneously, over the entire surface, or locally, thus supporting the structural requirements.

The developed form-finding tool can thus also be used as a mechanical optimization tool: folds can be moved or increased. The issues regarding the necessary edge reinforcements of domes, raised in the first chapter, can also be solved by the use of folds. The user is free to suggest and generate changes in form, depending on their architectural or engineering-oriented considerations. Thus, the introduction of foreign design elements is rendered unnecessary.

Fig. 1 shows how a standard surface is improved by the subdivision and folding of the overall geometry. The mechanical behavior of both geometries—folded and unfolded—is tested. The rigidity of the folded geometry shows an improvement factor of 5, their deformations thus amounting to a mere 20 percent of those found in a standard, unfolded shell. The partition function also defines the local geometry of the edges or borders of each element, which is also exportable.

The prototype described at the close of this chapter was initiated by Ivo Stotz and Gilles Gouaty, who strove to develop

Fig. 3

Fig. 4

Fig. 3 The tool describes the overall geometry. After subdivision into separate parts, the precise geometric description of each facet is specified.

Fig. 4 The data is then transcribed into machine code, which controls the milling head.

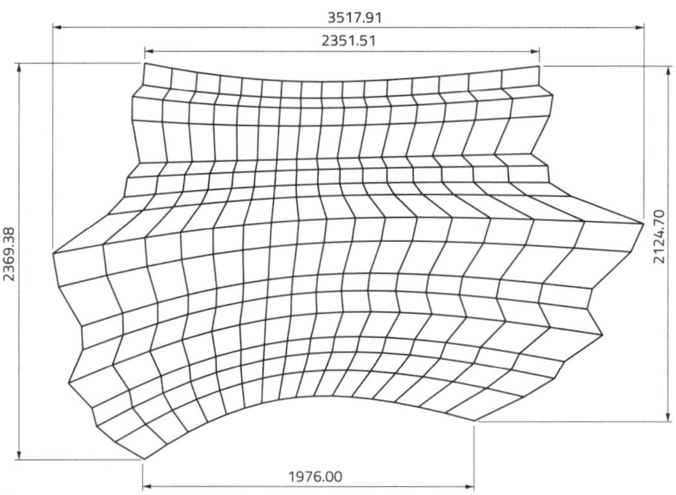

Fig. 5

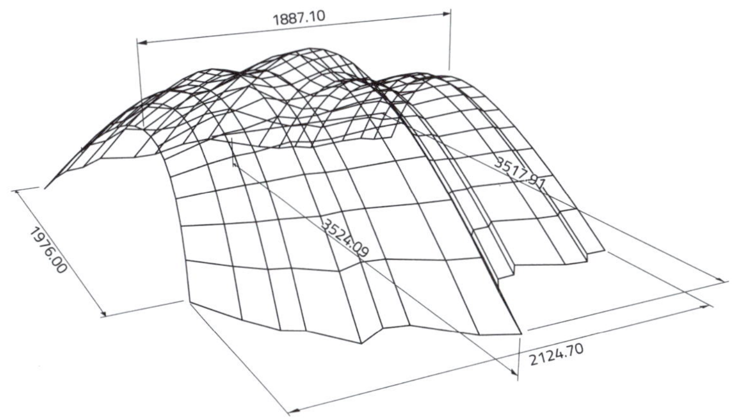

Fig. 6

Fig. 5 Floor plan of the generated shell construction
Fig. 6 Axonometric of the generated shell construction

Fig. 7 A prototype of the generated shell construction

existing structures by the use of discrete elements. The mold was generated by using the tool they described. It remains up to the user to control the criteria for the shape or the parameters that fit the mold.

Which type of connection should be used to join the panels? The digital tool contains the description of the panels' geometry. This description can also be used for the assembly of the connections. The structure must be able to withstand lateral forces, regular forces, and bending moments. The formation of these connections in timber is significant, as this has a considerable effect on construction costs.

Experiments were carried out on four types of connections and mechanically tested. In the initial attempts, only self-tapping, diagonally mounted screws were used. Here, torsion along the edges is hardly prevented. A folded sheet is inserted into the second row in the axis of the plates. Here, a lever arm at half-plate thickness is used.

In the third test series, two folded metal sheets divide the thickness of the panel into thirds, thereby increasing bending rigidity. The fourth series of tests combined plates with screws, thus achieving the highest bending rigidity possible.

The previous example shows a facet-like structure, the execution of which requires specialized connection technology.

Observations of the behavior of the overall geometry have shown that even freely rotatable edges are sufficient for the overall rigidity of the system, though connections remain necessary. However, if one considers the connections to the overall geometry and their complexity, the connections shown here do not solve the problem.

For this reason, a further experiment was undertaken with a different kind of shell structure, this one composed of facets. The connections were directly considered in the initial design phase. If the connections developed in the Stotz/Gouaty prototype were to be added to the example here, then incisions would be made into the panels in order to secure the connections by means of contact zones.

The advantage of the recommended integral connections between the panels is that they do not require any additional measures or connections. The connection is an integral part of the plate and is produced simultaneously with the cut panel. Even if one can assume that this structure looks promising, it is also clear that mechanical improvements are not only possible but are even on the verge of being realized.

Two assessment steps took place with students from the Civil Engineering Department of the ENAC/EPFL. First, the nodes of the structure were analyzed

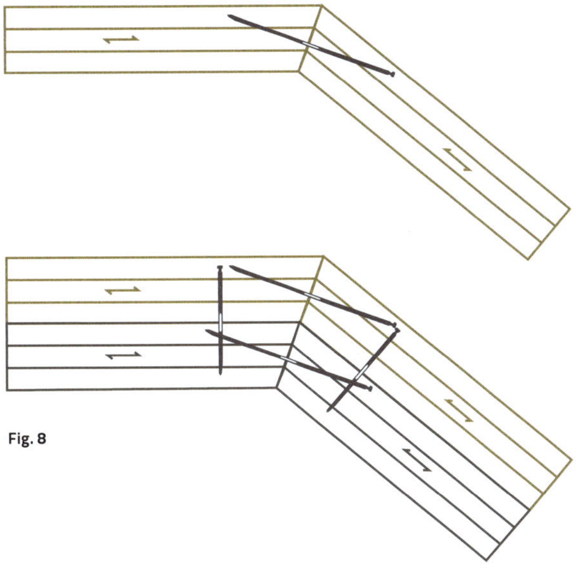

Fig. 8

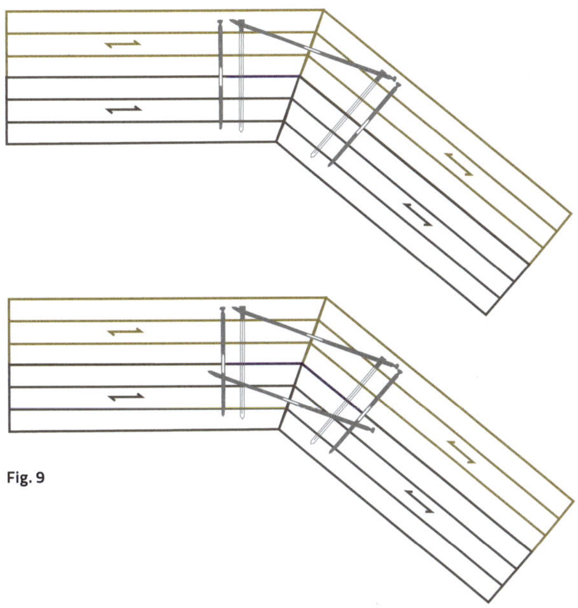

Fig. 9

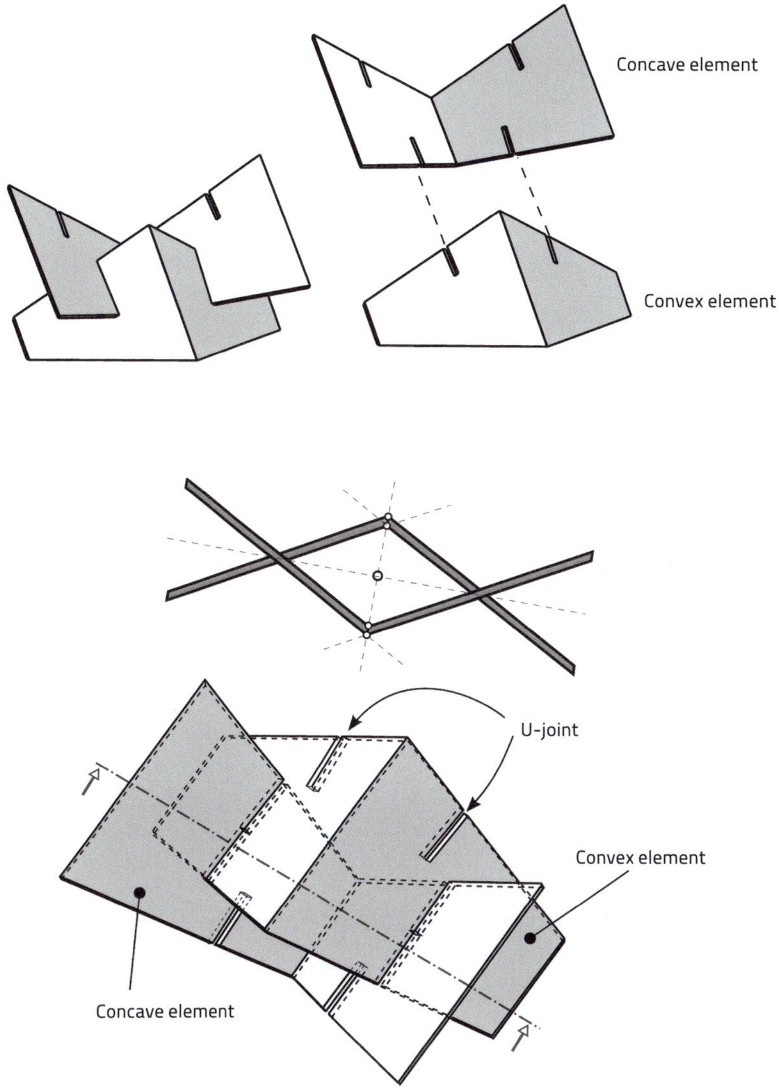

Fig. 10

Fig. 11

Fig. 8 Prototype of a test body in which two 10-cm-thick laminated timber panels are connected with oblique, self-sinking screws. In the lower image, four 10-cm-thick laminated timber panels are connected using such screws. Furthermore, the panels are also screwed together vertically.

Fig. 9 Prototype of a test body. Additional sheets are inserted between the plates.

Fig. 10 Axonometric view of the contact zones

Fig. 11 Detailed view of the shell construction. The panels are interwoven or wedged together.

Fig. 12 Completed arch pavilion

and mechanically calibrated, and the frictional resistances were modeled, measured, and calibrated. As shown in figs. 8 to 12, geometric manipulations of the overall structure were proposed, modeled, and recorded in a secondary step. A satisfactory solution for the connections can only be found after the overall geometry has been revised. The structural optimization described by Sina Nabaei is also affected by a modification of the local connection. The latter is, however, only made possible by the manipulation of the entire geometry. One of the key issues that has concerned us for a long time relates to precisely this point: how can one initiate an optimization process that takes both the global issues of the overall geometry and the local requirements of the connection details equally into account? For this, geometrical, mechanical, and manufacturing-related criteria should be considered.

As the following discussion with Mark Pauly highlights, it is of interest to expand the interaction of edge constraints. If solely geometric constraints play a role with the form-finding tool developed by Stotz and Gouaty, then

mechanical form-finding remains intuitive. Thus, the generation and distribution of folds remains intuitive and a quantitative control is not incorporated. The pavilion shown in fig. 12 introduces a connection technology that will be discussed in the fourth chapter.
In fact, it is already possible to expand the digital form-finding tool, and thereby expand the overall geometric conditions respecting material-related and mechanical constraints. An example of this is presented in the work on mechanical form-finding shown in Chapter 4.

2.2

Iterative geometric design for architecture

Ivo Stotz, Gilles Gouaty, and Yves Weinand

This interdisciplinary research project is presented by a group of architects, mathematicians, and computer scientists who research new methods for the efficient realization of complex architectural forms. The present work investigates methods of iterative geometric design inspired by the work of Michael Fielding Barnsley. Several iteratively constructed geometric figures will be discussed in order to introduce the notion of transformation-driven geometric design. The design method studied allows interaction with the design, forming affine transformations and generating discrete geometries.

Furthermore, the handling of specific constraints is discussed. Geometrical and topological constraints aim to facilitate the production of architectural free-form objects. A surface method based on vector sums is studied, allowing the design of free-form surfaces that are entirely composed of planar quadrilateral elements. The combination of the proposed surface method and transformation driven iterative design provides new form-finding possibilities while fulfilling a number of material and construction constraints. Finally, the findings are tested on a series of applications. The studied test scenarios aim to evaluate the advantages of discrete geometric design in terms of efficient integrated production of free-form architecture.

Keywords *architecture, applied discrete geometry, IFS, timber construction*

1 Introduction

In order to present the geometric design method studied here, the mathematical background must first be clarified. Before explaining the principles of transformation-driven geometric design, a series of historic examples will be examined. This will introduce the reader to the methods of iterative geometric design. The relationship between the mathematical method of geometric surface design and the physically constructed building will be shown by examples in the second part of this presentation.

2 Mathematical background

2.1 Monster curves

The Cantor set (Fig. 1), also called Cantor dust, is named after the German mathematician Georg Cantor. It describes a set of points that lie on a straight line. At the end of the 19th century, this figure attracted the attention of mathematicians because of its apparently contradictory properties. Cantor himself described it as a perfect set, which is not dense anywhere.[1] Further properties, such as self-similarity, compactness, and discontinuity, were studied years later.

The geometric construction of the Cantor set can be explained as follows: Take a straight line segment, divide it into three parts of equal length and remove its middle third; divide again each of the resulting line segments and keep removing their middle thirds. If you repeat this for each of the new line segments, you will end up with the Cantor set.

The von Koch curve is one of the best-known fractal objects and among the first found. In 1904, the Swedish mathematician Helge von Koch described it for the first time.[2] The curve is constructed step by step. Beginning from a straight line, a meandering curve with strange properties is created:

- It does not possess a tangent, which means that it cannot be differentiated.
- The length of any of its sections is always infinite.

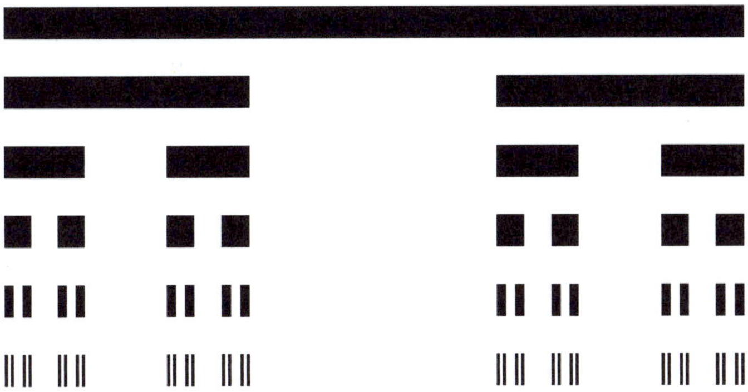

Fig. 1

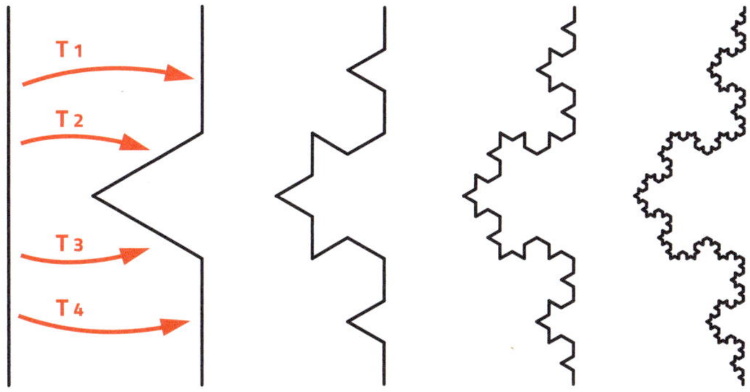

Fig. 2

Fig. 1 Cantor Set

Fig. 2 Von Koch Curve

Fig. 3 Example of a free-form object designed using the presaged surface method

The geometric construction of the von Koch curve is iterative, where each of the construction steps consists of four affine geometric transformations. The primitive is a section of a straight line, which is scaled, rotated, and displaced by each of the transformations $\{T_1...T_4\}$. Four duplicates are generated per construction step and each of these will, in turn, produce four more duplicates in the following construction step **(Fig. 2)**.

2.2 Iterative geometric figures

The peculiar properties of the aforementioned objects led mathematicians to name them "monster curves." In 1981, Barnsley defined a formalism based on Hutchinson's operator[3] that was able to describe objects such as these in a deterministic way.[4] His IFS-method (see section 2.3) consists of a set of contracting functions that are applied iteratively. In our case, a function is an affine geometric transformation. Iterative means that the construction is done step by step. The input of a construction step is the result of the previous step.

What is really new in Barnsley's work is that the resulting geometric figures are not defined by the primitive used, but rather by its transformations. As shown in fig. 4, the construction of a Sierpinski triangle may use a fish as a primitive. Analogous to this, the von Koch curve might be constructed on the basis of the letter "A." The end result remains exactly the same.

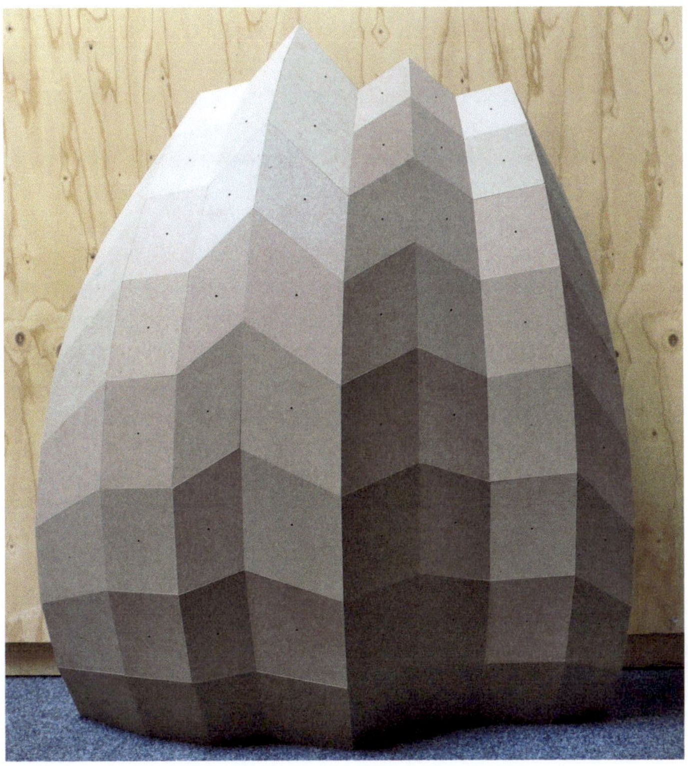

Fig. 3

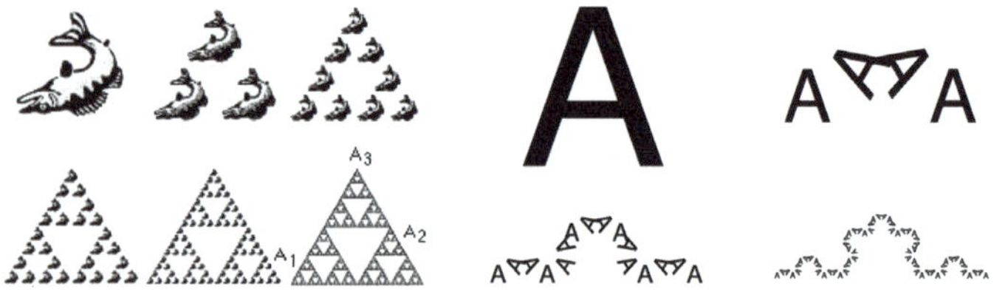

Fig. 4

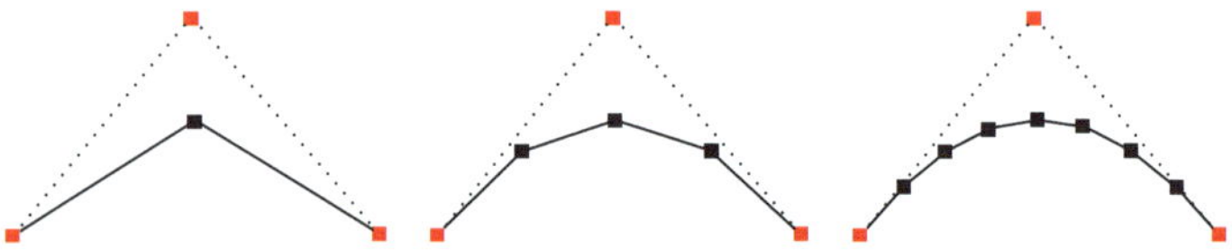

Fig. 5

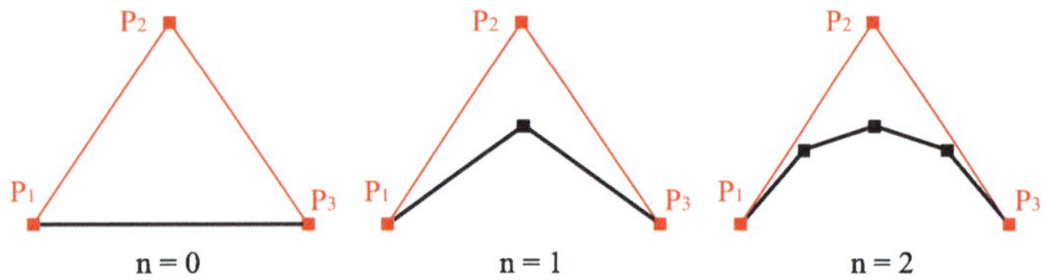

Fig. 6

Fig. 4 Sierpinski triangle and von Koch curve according to Barnsley's IFS-formalism

Fig. 5 Iterative construction of a Bézier curve

Fig. 6 First construction steps of a Bézier curve

The conclusion is that it is theoretically possible to use any form of primitive for the construction of geometric figures such as these. This led to the hypothesis that it is basically possible to use construction elements as primitives. Instead of using fishes like Barnsley did, we would rather choose to use construction elements such as beams or panels.

In order to complete this series of introductory examples, we would like to briefly address the Bézier curve. In 1959, de Casteljau discovered a method for the construction of the curve known today as the Bézier curve. De Casteljau's method[5] is based on iterative construction **(Fig. 5)**, which is extremely similar to the construction of a von Koch curve. The actual Bézier curve was analytically described by Bézier in 1961 as a polynomial function. His pioneering work forms the cornerstone of today's CAD software.

2.3 Iterated function systems (IFS)

The geometric figures of the examples shown at the beginning of this section are all defined by a set of transformations $\{T_i\}$. As Barnsley teaches us, the result is indifferent to the initial object on which the set of transformations is applied. This is true for the limit state, also called attractor A, which is the resulting figure if the set of transformation $\{T_i\}$ is applied an infinite number of times. The attractor A is the unique non-empty compact, such that

$$A = \bigcup_i T_i A$$

The attractor is a theoretical object, which will not be analyzed any further here. We will work on the intermediate construction steps that are given by the following sequence:

$$K_{n+1} = \bigcup_i T_i K_n$$

Where K represents the initial figure, the so-called "germ." K might be any arbitrary object, for example a fish, as shown in fig. 4, or a straight line segment, as used for the construction of a von Koch curve or a Bézier curve **(Figs. 3 and 5)**. Finally, there is the equation relating to the geometric figure K in the construction step $n \to \infty$ to the attractor A:

$$\lim_{n \to \infty} (K_n) = A$$

Within the following, the modeled objects are defined as the projection of an IFS via a projection operator P.

3 Discrete iterative geometric design

The strange properties of the geometric figures discussed in section 2.1 and 2.2 concern the limit state; the attractor. For practical applications, the theoretical object of the attractor is far less relevant than its intermediate construction state K. In the scope of this work, the construction state K has the following properties that are beneficial for the application of free-form geometries in architecture:

- K is computational point by point
- The resulting geometry is always expressed by a finite number of elements

3.1 Transformation-driven geometric design

As stated in section 2.3, the final aspect of the iteratively constructed geometry is defined by transformations. In order to control the resulting figure, the geometric design method has to provide solutions that will allow the manipulation of the transformations used for the construction of the figure.

Let us analyze an example of an iteratively constructed Bézier curve as shown in fig. 6. The figure shows the first construction steps of a Bézier curve with three control points P shown in red.

$$P = (p_1, p_2, p_3) = \begin{pmatrix} p_{1,x} & p_{2,x} & p_{3,x} \\ p_{1,y} & p_{2,y} & p_{3,y} \\ p_{1,z} & p_{2,z} & p_{3,z} \end{pmatrix}$$

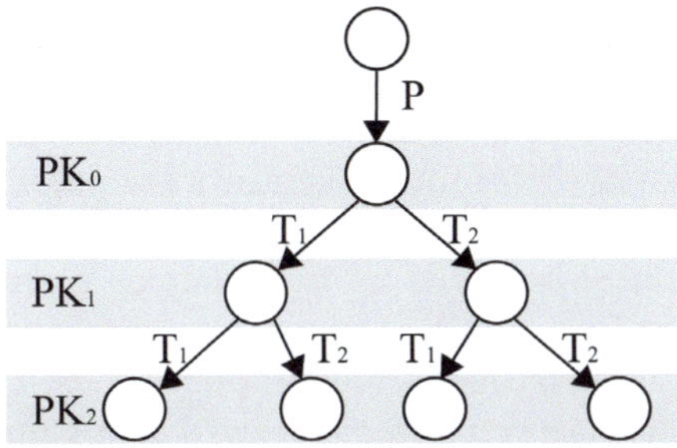

Fig. 7

At the beginning (n=0), the initial figure K consists of the line connecting the end points of the control polygon [p1, p3]. Per construction step, two transformations *{T1, T2}* are applied to each element. The first three steps of the construction sequence are shown below.

$$PK_0 = PK$$
$$PK_1 = PT_1K_0 \cup PT_2K_0 = PT_1K \cup PT_2K$$
$$PK_2 = PT_1K_1 \cup PT_2K_1 = PT_1T_1K \cup PT_1T_2K \cup PT_2T_1K \cup PT_2T_2K$$
$$PK_3 = PT_1K_2 \cup PT_2K_2 = PT_1T_1T_1K \cup \ldots \cup PT_2T_2T_2K$$

Each element of PK_i can be computed by following the construction tree shown in fig. 7.

We use transformation matrices to describe affine geometric transformations. We work with barycentric coordinates, which means that each computed point is based on a combination of the entry points. In the present example of a Bézier curve with three control points, each transformation can be expressed by a single 3 × 3 matrix:

$$T_1 = (c_1, c_2, c_3) = \begin{pmatrix} 1 & 0.5 & 0.25 \\ 0 & 0.5 & 0.5 \\ 0 & 0 & 0.25 \end{pmatrix}$$

$$T_2 = (c_4, c_5, c_6) = \begin{pmatrix} 0.25 & 0 & 0 \\ 0.5 & 0.5 & 0 \\ 0.25 & 0.5 & 1 \end{pmatrix}$$

The transformation matrices shown above are fairly standard. These matrices work with fixed values that always produce the same result: a smooth Bézier curve with three control points.

A closer look at the transformation matrices of the Bézier curve shows that they have certain values in common. For example: The column c_1 of T_1 as well as the last column c_6 of T_2 is fixed. Furthermore, the last column of T_1 is identical to the first column of T_2. These dependencies guarantee that the resulting curve will be continuous. The rest of the values are free and can be modified according to the designer's wishes.

If we generalize the above-mentioned findings, we can say that certain values of the transformation matrices are constrained, whereas others are free. Based on the example of an iteratively constructed continuous curve with two transformations and three control points, the general scheme of the transformation matrices is as follows:

$$T_1 = \begin{pmatrix} 1 & c_{2,1} & c_{3,1} \\ 0 & c_{2,2} & c_{3,2} \\ 0 & c_{2,3} & c_{3,3} \end{pmatrix}, \quad T_2 = \begin{pmatrix} c_{3,1} & c_{5,1} & 0 \\ c_{3,2} & c_{5,2} & 0 \\ c_{3,3} & c_{5,3} & 1 \end{pmatrix}$$

In order to act on the free values of the transformation matrices, we defined a graphic data input method. We introduce the set of points s_i which we call "subdivision points" **(Fig. 8)**. Each point s_i corresponds to the projection via P of columns c_i.

$$S_i = Pc_i$$

By this equation, the values of the columns c_i can be obtained by the position of the subdivision points s, relative to the position of the control points P **(Fig. 9)** using the equation below:

$$c_i = P^{-1}s_i$$

Fig. 7 Construction sequence: Tree representation

Fig. 8 Representation of control and subdivision points

Fig. 9 Acting on the transformation matrices by manipulating the subdivision points (blue)

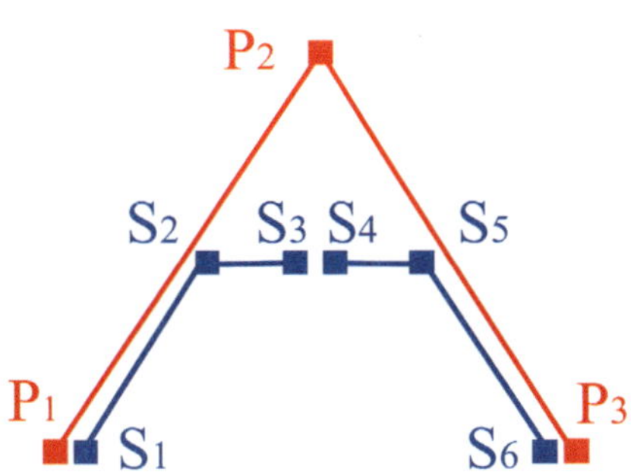

Fig. 8

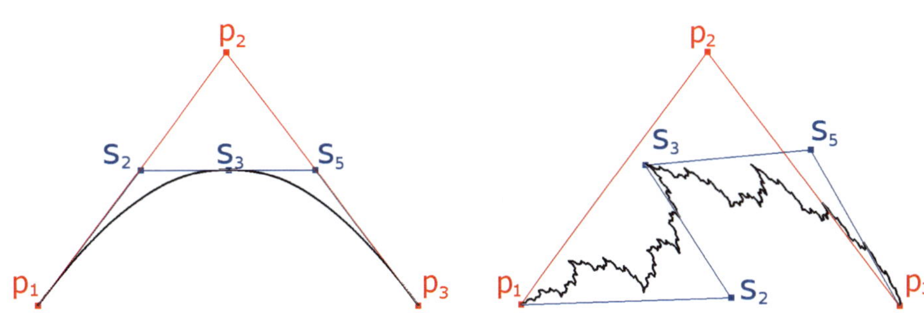

Fig. 9

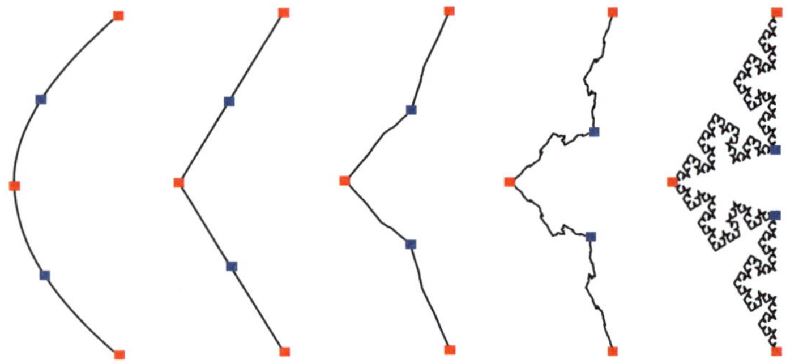

Fig. 10

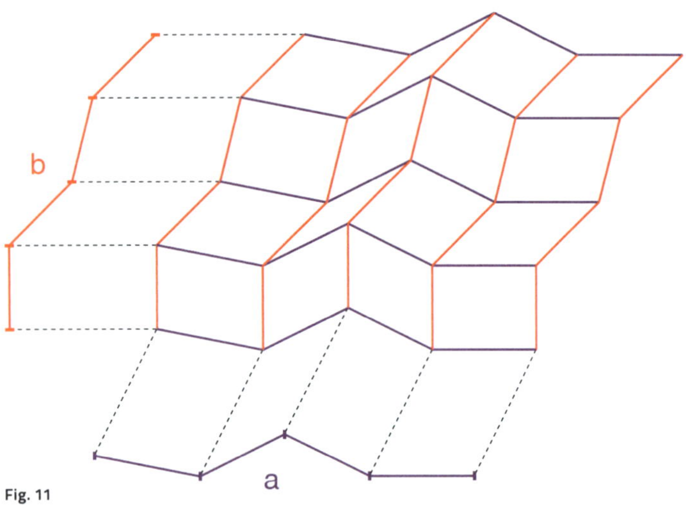

Fig. 11

Fig. 10 IFS-Curve curve design: adjustment of roughness / smoothness

Fig. 11 Surface design by vector sum

3.2 Potential of transformation-driven geometric design

Whether a figure is smooth or rough only depends on the affine geometric transformations. The same curve might be smooth or rough. By changing the subdivision parameters, the smoothness and the roughness respectively can be adjusted, as shown in fig. 10. The input of the subdivision parameters is given by the position of the subdivision points (as we referred to them). Alongside the control points, which are widely known in classical CAD-software, subdivision points augment a variety of design possibilities. They provide a graphic way to manipulate the affine geometric transformations, which are expressed in the user-unfriendly form of n-dimensional matrices. These points work below the surface of the graphic user interface.

3.3 Constrained geometric design

The goal is to develop design strategies that make the design and production of free-form surfaces easier. Therefore, the geometric design should meet certain topological and geometrical constraints. The constraints are mainly dictated by physical and production conditions from the field of construction. Within the following, a few examples of different constraints are presented.

For example, an important point might be that the free-form object will be built out of planar timber panels. According to this, the geometric constraint demands that the virtual 3D-model ought to be completely comprised of planar panels. We will work on this constraint in sections 3.4 and 4.2.

In section 2, we have presented a series of iteratively constructed objects. Not all of them are suitable for physical realization. The Cantor set, for instance, is simply a set of discontinuous line fractions. Since we generally need material continuity (unless designing ornaments or similar forms), we have to verify that the created elements building up the geometric figure are connected with one another. Continuity represents a topological constraint.

In order to avoid complex detailing of the nodes of a wire frame structure, it is advantageous to know the number of bars coinciding in one node. To keep the number of bars per node at 6, we might work with surfaces composed entirely of regular triangular faces. This is a topological constraint.

On the one hand, constraints will make the physical realization of free-form objects easier. On the other hand, they may limit the design possibilities and therefore restrict the form-finding process, and this ought to be avoided as far as possible.

3.4 A constrained surface model

In order to create iterative surfaces that are entirely composed of planar elements, we will work on so-called "vector sums." Generally, classical CAD-software computes NURBS-surfaces by tensor products, which have the unsuitable property of being composed locally of double-curved faces. Great effort is required for their production. The principle of using vector sums, more precisely Minkowski sums[6], for the generation of free-form surfaces has already been studied by Schlaich[7] and Glymph.[8] Surfaces such as these are combinations of two curves. Fig. 11 shows the curves a and b. The vector sum of any two segments of the curves *(a, b)* creates a parallelogram, which is part of the entire surface. The surface is completely composed of parallelograms and therefore it meets the geometrical constraint which requires that all its parts have to be planar.

The discrete curves a and b used for the construction of the surface are represented by two lists of points A and B. The resulting vector sum surface is represented by the quad mesh M:

$$M_{i,j} = M_{0,0} + \overrightarrow{A_0 A_i} + \overrightarrow{B_0 B_i}$$

Note: $M_{0,0}$ is the origin of the quad Mesh.
It may be any arbitrary point in the design space.

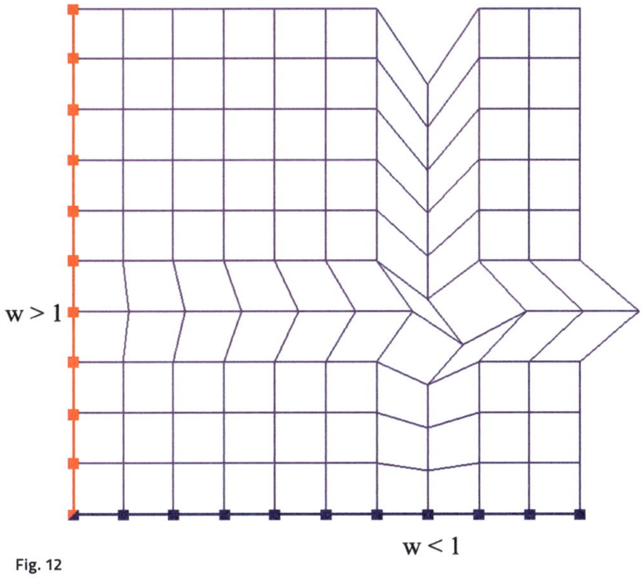

Fig. 12

The design possibilities of vector sums are limited compared to NURBS surfaces. The quad mesh generated by vector sums is composed entirely of parallelograms. In order to augment the design capabilities, Schlaich[7] and Glymph[8] extended the method such that the resulting quad mesh is composed of trapezoids as well as parallelograms.

In order to extend the design capabilities, we employ methods of projective geometry. The IFS-formalism will be extended by the possibility of assigning different weights to its control and subdivision points. Each point will be defined by four coordinates (w, x, y, z). Assigning different weights *w* to the points allows deforming of the quad mesh more or less locally. Thus, the resulting quad mesh is no longer only composed of parallelograms, but largely of convex planar quadrilaterals. To illustrate the effect of point weight editing on vector sums, fig. 12 shows the simple case of a regular quad mesh where two point weights have been edited. Fig. 13 shows the same principle applied to more complex meshes.

Within the scope of projective geometry, the coordinates *(w, x, y, z)* are called homogeneous coordinates. A point *(x, y, z)* in the model space $\mathbf{R}^3$ with a weight *w* has the corresponding homogeneous coordinates *(w, xw, yw, zw)*. Reciprocally, each point of homogeneous coordinates *(w, x, y, z)* has corresponding $\mathbf{R}^3$ coordinates *(x/w, y/w, z/w)*. The path from homogeneous coordinates to $\mathbf{R}^3$ is called projection.

In order to work with IFS-subdivision in homogeneous coordinates, the control points *P* must be defined in homogeneous coordinates. Here, the IFS-subdivision and the computation of the quad mesh M are performed using homogeneous coordinates. Finally, the result obtained is projected back to the model space $\mathbf{R}^3$.

The proposed surface method functions with any pair of discrete curves. However, the use of IFS-curves creates the ability to control the global shape of the surface via its control points. In addition to that, the subdivision points allow the manipulation of the local aspect of the surface: smoothing/roughening.

This surface method offers great design potential. It unifies in one formalism the hitherto separate paradigms of "smooth" and "rough." Furthermore, it verifies a certain number of geometric constraints, allowing the optimization of the production of free-form architecture. This will be shown by examples in the next section.

4 Applications

In order to realize physical buildings out of discrete virtual geometries, the elements that constitute the 3D models are replaced by construction elements. For an iteratively designed curve, the line sections will be replaced by linear construction elements, such as planks or beams. In the case of a discrete surface, we replace its faces by planar construction elements (panels, plates, etc.). The replacement of geometric elements by construction elements poses a certain number of questions, as the geometric figures do not have physical dimensions like thickness. We will first discuss the more demonstrative case of a two-dimensional figure, namely the Bézier curve.

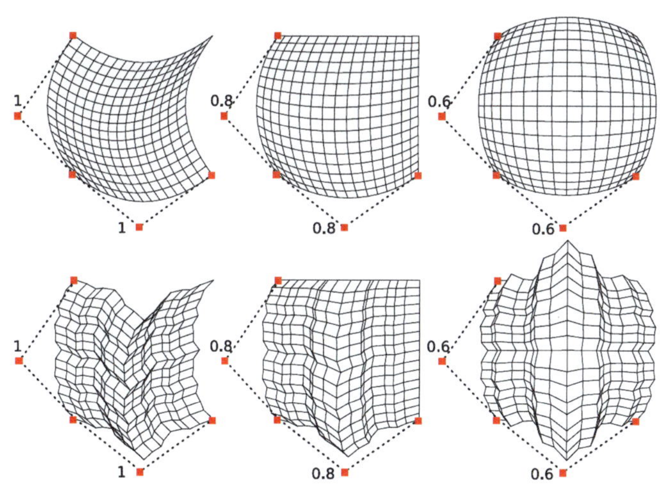

Fig. 13

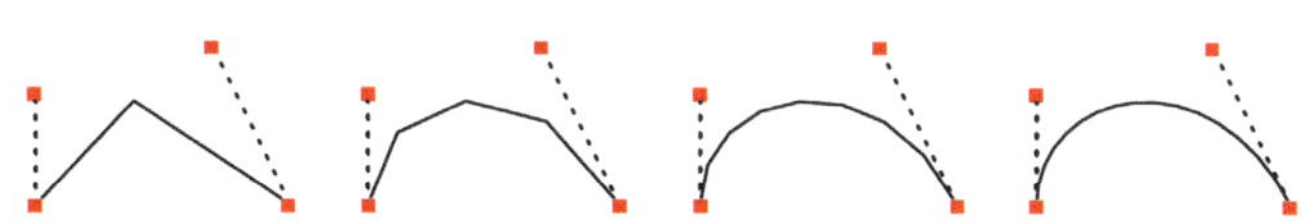

Fig. 14

Fig. 12 Deformation of vector sum meshes by editing the weight of certain points

Fig. 13 Point weight editing of smooth and rough vector sum surfaces

Fig. 14 Iterative Bézier curve

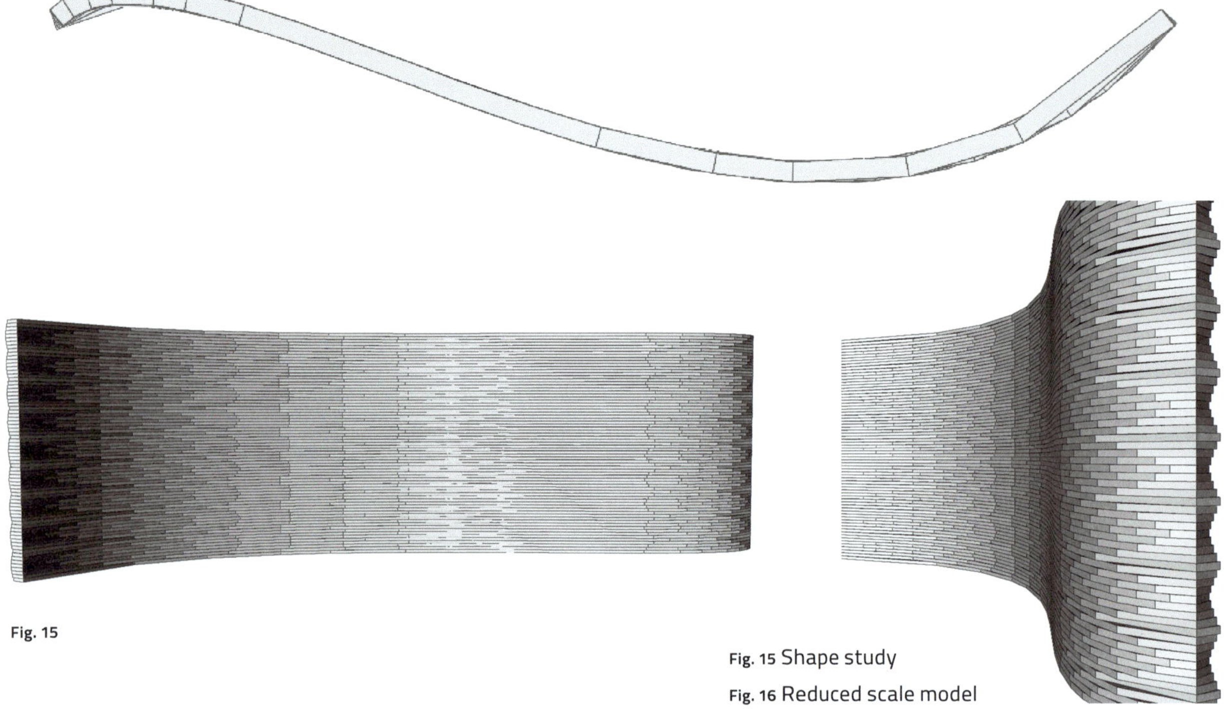

Fig. 15

Fig. 15 Shape study

Fig. 16 Reduced scale model

4.1 Discrete Bézier vault structure

In the following example, we build a vault structure based on an iteratively constructed Bézier curve with four control points (Fig. 14). The straight-line sections that build up the curve will be replaced by untreated timber planks. The vault is composed of a series of arched curves, placed jointlessly alongside one another. The planks are then screwed together in order to create a massive timber vault structure. The shape of the vault's section can be controlled via the control points of the Bézier curve. Fig. 15 shows a form study where the curve's control points have been deformed such that the resulting shape is a meandering element with inflection points. The line segments of the underlying discrete Bézier curve have been replaced by construction elements.

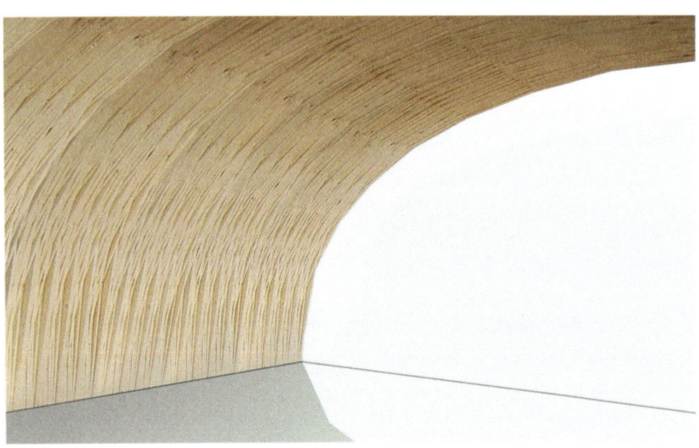

Fig. 16

Once the shape has been defined, the curve will be subdivided into its constituent parts until we obtain a suitable length for the construction elements. On the one hand, the lengths of the elements should not exceed the length of the most common sizes of planks existing on the market. On the other hand, the subdivision should be small enough to obtain a smooth rendering of the curve.

The relevant dimensions needed for the production of the construction elements are directly produced by the geometric figure. The lengths of the planks correspond to the lengths of the curve's line sections. The chamfer angle can also be deduced from the geometric model (bisector angle of two adjacent line segments). The design is therefore limited to two steps:

- Shape control, via the control points
- Subdivision control by choosing the adequate level of iteration.

Fig. 16 shows a reduced scale model of such a massive timber vault structure. The different planks have been cut by a numerically-controlled computerized routing table.

The question of how to subdivide a free-form object into a coherent set of construction elements becomes obsolete as it is directly given by the iterative geometric construction method. A direct link from design to production has been established, which is an important cost and time factor for the production of free-form architecture.

4.2 Shell structure—feasibility test

In this section we will discuss the application of an iteratively constructed free-form surface as a panel construction. The surface method used for the design of the free-form objects has been described in section 3.4 The design used for the realization of the larger prototype **(Fig. 21)** was mainly driven by following parameters: On the one hand, the aim was to create a small, dome-like structure, presenting a smooth arc in its longitudinal section; on the other hand, we designed a rough curve, providing folds to the transverse section of the structure.

In the present example, the faces that compose the surface are replaced by planar timber panels. The choice of the thickness of the timber panel is important, as the virtual 3D surface does not consider any thickness. A volumetric model has to be derived from the surface model. The extrusion process is illustrated by fig. 18. Firstly we generate a parallel offset surface that maintains a constant distance from the initial surface. That distance corresponds to the thickness of the timber panel. Secondly, the bisector planes are calculated, which will be used at a later stage for the chamfer cut of the panels. In this manner, we design free-form objects that are entirely built up of planar constructional elements. Fig. 19 shows an example of an IFS-surface with faces that have been entirely replaced by extruded construction elements.

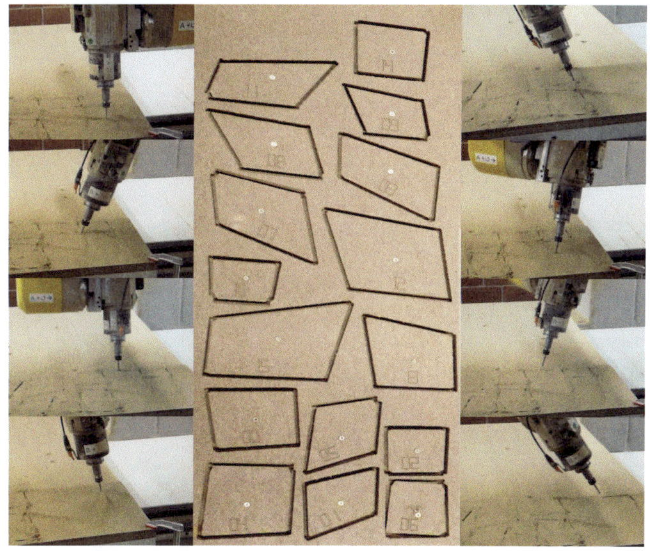

Fig. 17

4.3 Integrated manufacturing

In order to test the established digital production chain, an extract of an iteratively designed free-form surface was first produced by a 5-axis CNC-machine. The procedure to advance from the geometry data of the construction elements to the machine code has mainly been automated. To create the elements of such complex shapes, the following work steps are necessary:

Fig. 17 Integrated manufacturing of the constructional elements

Fig. 18 Parallel offset mesh generation

Fig. 19 Extruded IFS-surface

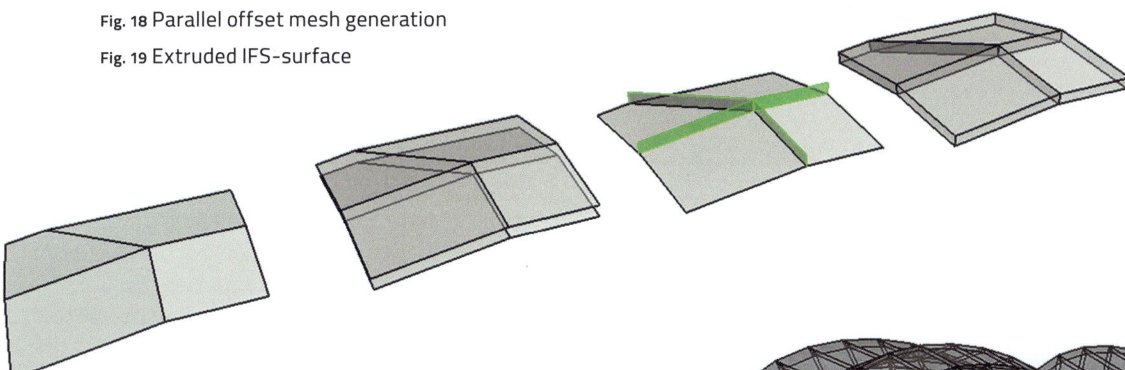

Fig. 18

Fig. 19

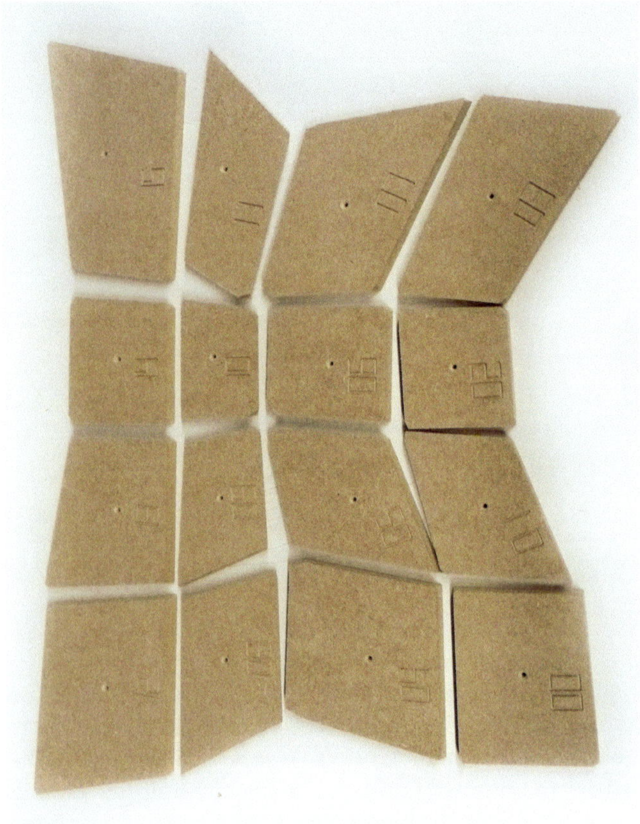

Fig. 20 a

Fig. 20 b

Fig. 20 a and b Assembling of the partial prototype

Fig. 21 a and b Prototype of the generated shell structure

A unique address for each construction element is logistically necessary, in order for the different elements to be assembled in the right place. Each element has to be oriented according to the coordinate system of the CNC-machine, the dimensions of the raw material, and the fiber direction of the plywood panel.

Automatic generation of the machine code for each element—the material properties, the type of machine, and the nature of the cutting tools—is of vital importance for the integrated production of the elements, which all have different sizes and shapes.
Fig. 17 shows a sequence of the machining process. The production of each multi-part plate has been divided into three working steps:

- Piercing of the fixation holes. Each part is screwed on the machining table.
- Engraving of the addresses of each element.
- Contour cut: machining of the actual element.

The assembled manufactured elements give an accurate rendering of the surface designed on the computer screen. This shows that practical realization of iteratively constructed surfaces becomes possible.

After the realization of an early partial prototype (Fig. 20), we continued testing the method on a more complex structure comprising 256 construction elements. The shell structure shown in fig. 21 presents a small vault spanning over 4.5 meters, built from 10-mm-thick spruce plywood panels. Although the realized objects remain relatively small in scale thus far, they allowed us to verify the validity of the proposed design method, since the employed manufacturing techniques are also applicable for 1:1 scale construction elements.

5 Discussion and outlook

The applications presented here show that the design and construction of free-form surfaces using our method requires a relatively small planning effort. Several problems appeared during the manufacturing process due to the extremely low tolerances permitted by the perfectly fitting pieces. Large-scale free-form buildings will probably have higher tolerances, but the logistics and the assembly will probably become more complicated. The efficiency of the method presented is only proved insofar as the processing of the data, design to production, takes only a few moments.

We have shown that the design method which allows the generation of rough and smooth objects could be employed for the design of bearing plate structures using the rigidity of the fold to improve the structural strength of the free-form surfaces.

Fig. 22 illustrates a preliminary investigation into the bearing potential of such folded structures. On its left side, a smooth symmetrical plate structure is

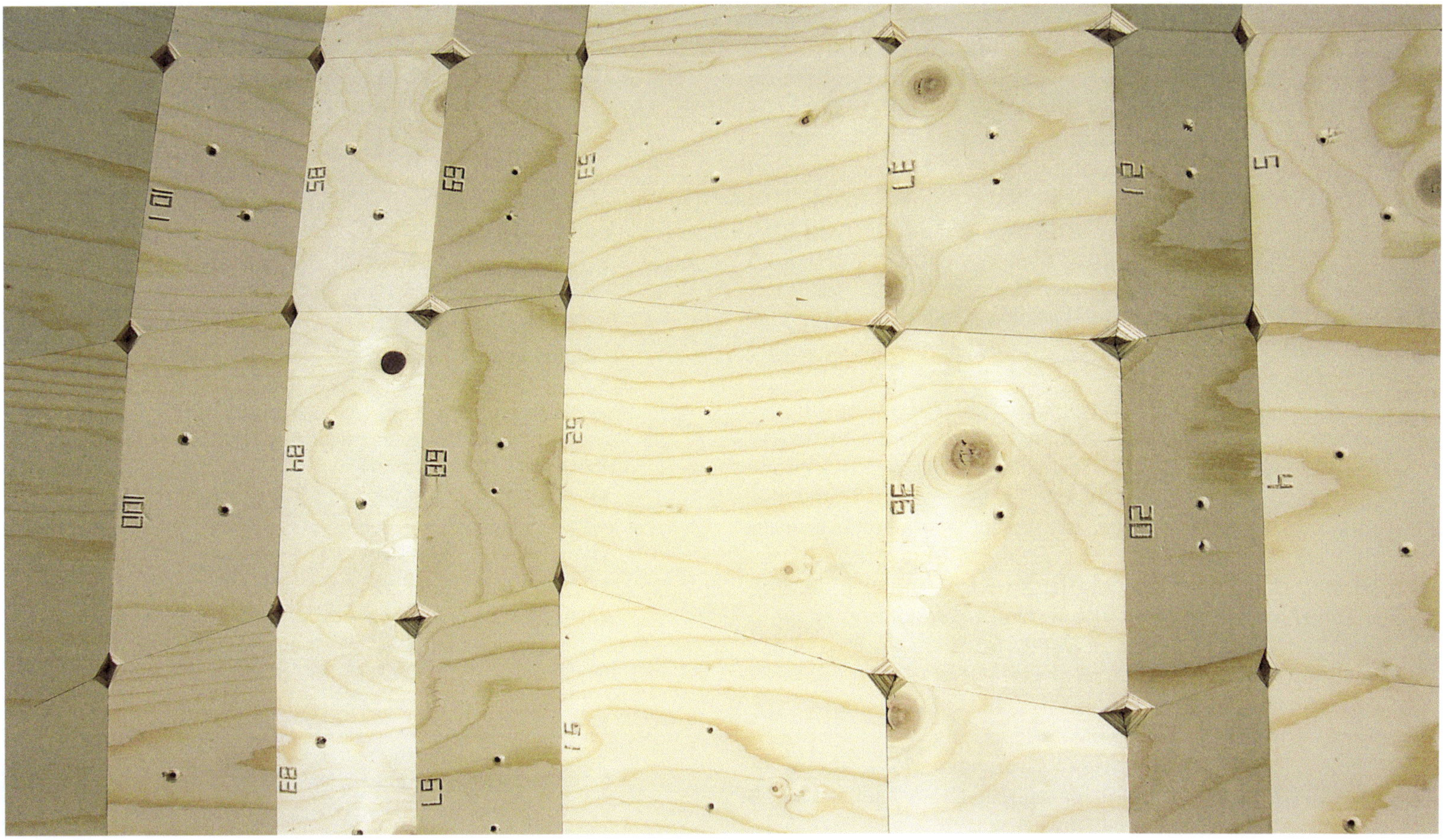
Fig. 21 a

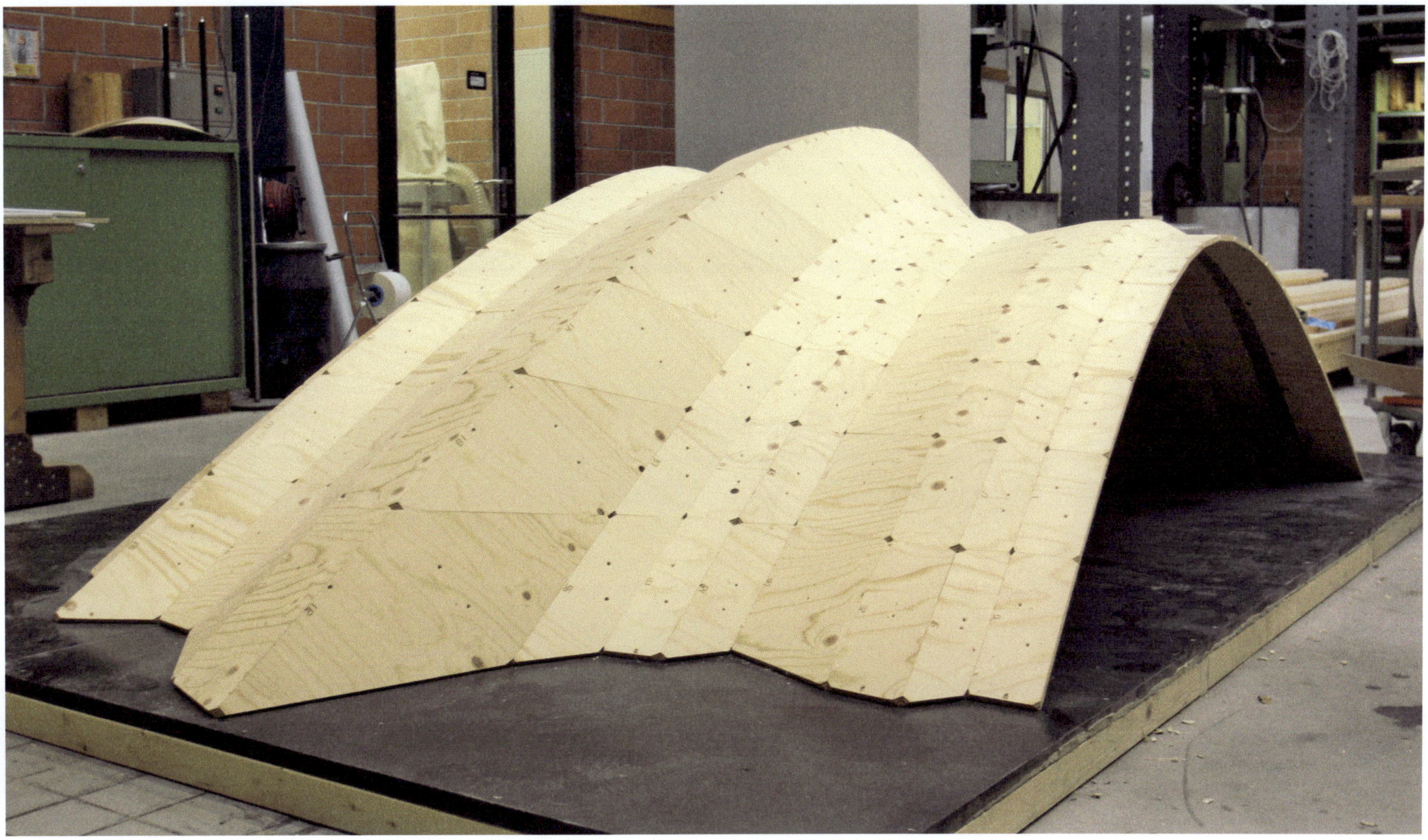
Fig. 21 b

shown. The same model is shown on the right side with additional folds. In order to get a rough idea of the global structural behavior, we imagined a plate structure spanning over 12 meters, built of 20-mm plywood panels (spruce). First, we designed a smooth, dome-like shell structure, where we applied an asymmetrical load of Fz = 1500 N/m²—a typical design value for snow-load in central Switzerland. The maximum deflection occurred along the z-axis, which was at about 266 mm. After adding folds, the FEM-analysis showed a maximum deflection value of 15 mm.

Note: This analysis was assuming hinge joints between the individual construction elements. In reality, the joints may present a certain bending rigidity. Since the construction elements are fixed along all four sides (except the border panels), any possible rotation is greatly limited by the system. For real-scale applications, we believe that adequate detailing will be imperative. Further information about the joints of folded timber plate structures can be found in Buri[9] and Haasis[10]. Buri et al.[9] used 2-mm-thick folded steel plates for joining massive 40-mm-thick timber panels. Haasis et al.[10] studied the bending rigidity of screwed connections.

The above-mentioned considerations about detailing and structural properties of the presented work are possible subjects for future research.

In the future, we will continue the development of larger and more complex objects. The potential of the new design method for free-form surfaces is far from being exhausted. We hope to have further opportunities to test our method on applications such as suspended ceilings, free-form facades, climbing walls, halls, and shell structures.

Acknowledgements

This research has been supported by the Swiss National Fund (200021–112103) and (200020– 120037/1). We would also like to thank our project partner, Dr. Eric Tosan from the LIRIS, Université Lyon I, France, for his support on the development of the mathematical model. Our special thanks goes to Johannes Natterer for the preliminary investigation on the load-bearing capacity of IFS surfaces.

References

1 Cantor, G. "De la puissance des ensembles parfaits de points." *Acta Mathernatica* 1884, p. 381–392.

2 von Koch, H. "Une courbe continué sans tangente, obtenue par une construction géometrique élémentaire." *Arkiv för Matematik I,* 1904, p. 681–704.

3 Hutchinson, J. E. "Fractals and self-similarity." *Indiana University Mathematics Journal 30,* 1981, p. 713–747.

4 Barnsley, M. F. *Fractals Everywhere,* Academic Press. 1988.

5 De Casteljau, P. "Courbes à poles." *INPI,* 1959.

6 Minkowski, H. *Geometrie der Zahlen.* Leipzig, Teubner, 1896.

7 Schlaich, J., H. Schober. "Filigrane Kuppeln. Beispiele, Tendenzen und Entwicklungen." *TEC21,* 2002, vol. 128, no. 12, p. 21–27.

8 Glymph, J., D. Shelden, C. Ceccato, J. Mussel and H. Schober. "A parametric strategy for freeform glass structures using quadrilateral planar facets." *Automation in Construction,* 2004.

9 Buri, H., Y. Weinand. "Gefaltet. Holztragwerke." *TEC21,* 2009, no. 8, p.18–22.

10 Haasis, M., Y. Weinand. "Origami -folded plate structures." *Engineering. 10th World Conference on Timber Engineering.* 2008, Miyazaki, Japan.

Reprinted from *Journal of the International Association for Shell and Spatial Structures*, vol. 50, num. 1, 2009, pp. 11–20.

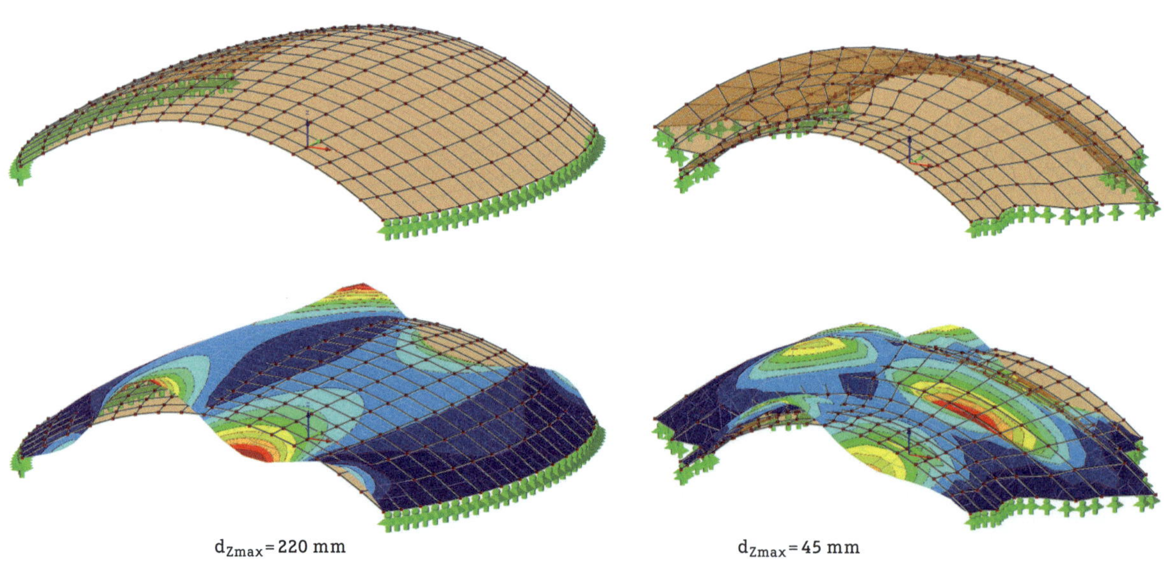

Fig. 22 Comparison of the a smooth and rough IFS-surface under asymmetrical load

2.3

Geometrical description and structural analysis of a modular timber structure

Sina Nabaei, Yves Weinand, and Olivier Baverel

The ambitious goal of the ongoing research at IBOIS, the laboratory of timber constructions at the École Polytechnique Fédérale de Lausanne (EPFL), is to develop the next generation of timber constructions made from innovative timber-derived products, by the application of textile principles on a building scale. The structure presented is a modular composition of timber folded panels, notably demonstrating an example of the application of geometric techniques used to produce modular patterns and lattices to timber structures. Effectively, it is shown that complex spatial structures can be designed using simple connection technology between elements. Moreover, by taking advantage of advanced CAM process, complex planar timber elements are cut in large scale and assembled with high precision, as is the case for the prototype of the structure presented in this paper. The folding concept corresponds to a planar reciprocal frame structure. The basic module consists of two mutually supporting timber folded panels that are joined together consecutively along their slots to build up an arch. The stability of the inter-module connection is provided by contact boundary condition over the sliding joints. The fundamental mechanical properties of the structure are examined using the Finite Element Method considering the nonlinear contact boundary condition. The static behavior is studied under the self-weight load case, as well as the model dynamic response. According to analysis results, and with the aid of a CAD parametric model, structural and geometrical alternatives are proposed to improve the structural performance. A prototype based on this geometric principal has been fabricated and assembled in order to explore the feasibility of the concept at a building scale.

Keywords ***timber spatial structure, reciprocal frame structure, structural system improvement, finite elements analysis, parametric design***

1 Introduction

1.1 IBOIS, the re-interpretation of timber construction

In recent years, the necessity of using renewable and sustainable resources in the building sector has become obvious, and interest in timber as a building material has revived.[1,2,3] Novel timber-derived products, such as massive block panels, have emerged and the use of such products is increasing.[4,5] The ambitious aim of the ongoing research at IBOIS, the laboratory of timber constructions at the École Polytechnique Fédérale de Lausanne is to develop the next generation of timber constructions made from innovative timber-derived products by the application of textile principles on a building scale.[6,7] The unprecedented exploration and study of timber-related structures and their structural analysis is sought within a framework integrating the mechanical and structural principles of textiles. Since timber can be viewed as a fiber-derived product, it follows that the analogy between micro-scale fiber structures and timber-derived wooden structures can be explored at the micro and macro scale. The key to our approach is the underlying notion that timber's fibrous nature, historically perceived as a liability for a construction material, is in fact a positive feature that should be exploited to increase the material's functional and aesthetic value. Its inherent flexibility allows it to be folded into robust, lightweight structures that use material very sparingly. The concise observation of existing textile techniques and fabrication methods, notably here the geometric techniques used to produce modular patterns (as described by Clare Horne[8]), combined with the investigation of the modular structure presented here, is intended to result in the development of a new family of timber constructions based on the logic and principles of textile fabrics.

1.2 Geometry of spatial structures: a survey of "form-element" relationships

Over the past decade, four principal directions dealing with the geometry of spatial structures have been recognized and researched. Fig. 1 shows an example from a related publication on the subject. The various approaches are described and analyzed below.

- *Reciprocal frame structures:* This family of modular spatial structures consists of interlaced linear elements, where the final form is a result of a basic module as well as a connection technology **(Fig. 1a)**.[9,10,11,12]

- *Form-finding practice:* This deals principally with tensile membranes or vaulted masonry construction in order to determine a tensile or compressive geometry. Here, the final form is the direct result of equilibrium and is greatly influenced by the materiality and the boundary condition applied **(Fig. 1b)**.[13,14,15]

- *Topology optimization:* The last stream can be categorized under topology and shape optimization problems. These methods, combined with evolutionary algorithms, often result in free-form spatial objects, which have been generated on the basis of structural criteria. These fall into the first category, where they are geometrically approximated in order to be built.

- *Free-form design:* Here the free form design surface is the input and by means of geometric-mathematical models, the surface is approximated by use of a uniform (or typologically uniform) planar mesh. The mesh generation is clearly constrained by tolerances of the approximation problem as well as the economical sensibility analysis. The supporting structure closely follows this approximated mesh and thus its spatial form is less influenced by the materiality and robust mechanical reflections on the equilibrium and the mechanical governing laws. The materiality and the connection technology clearly come alongside the free-form design surface.[16,17,18]

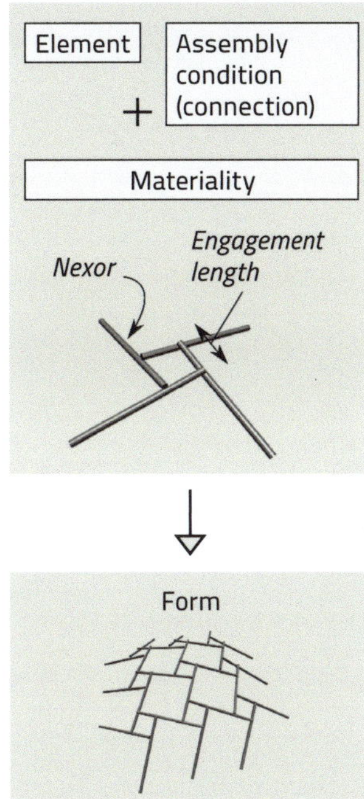

Fig. 1 a

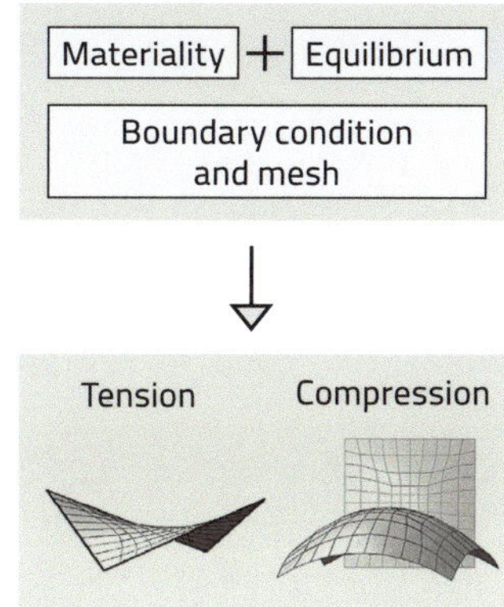

Fig. 1 b

Fig. 1 Forms for spatial structures:
a) Reciprocal frame structures[19]
b) Equilibrium forms either in tension or compression[20]

Fig. 2 Folding concept for the modular structure designed in the Atelier Weinand workshop

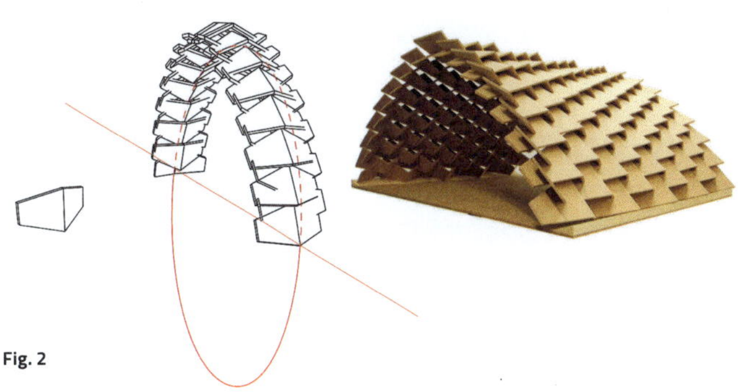

Fig. 2

In fact, when it comes to free-form design, it generally starts with a "complex form" that ends up with multiple typologies of elements in a top-down process and "complex connection technologies" where structural elements follow offsets of the external form. Indeed, we ask the question whether the same free-form practice should be followed in the context of timber construction. Thus it is necessary to take the materiality and the complexity of the connection technology as an important feature here, in order to make a distinction between the four approaches described above. The core idea that will be explored is an investigation of families of spatial structures, where the focus is on materiality and related connection technology, rather than on an irregular surface approximation. According to this approach, the final form is a result of the geometry of connected members as well as the employed connection technology. In the context of timber engineering, we focus on use of CAM facilities to cut complex geometries from thin timber panels and slender beams, while the connections are kept simple, as is appropriate when using timber.

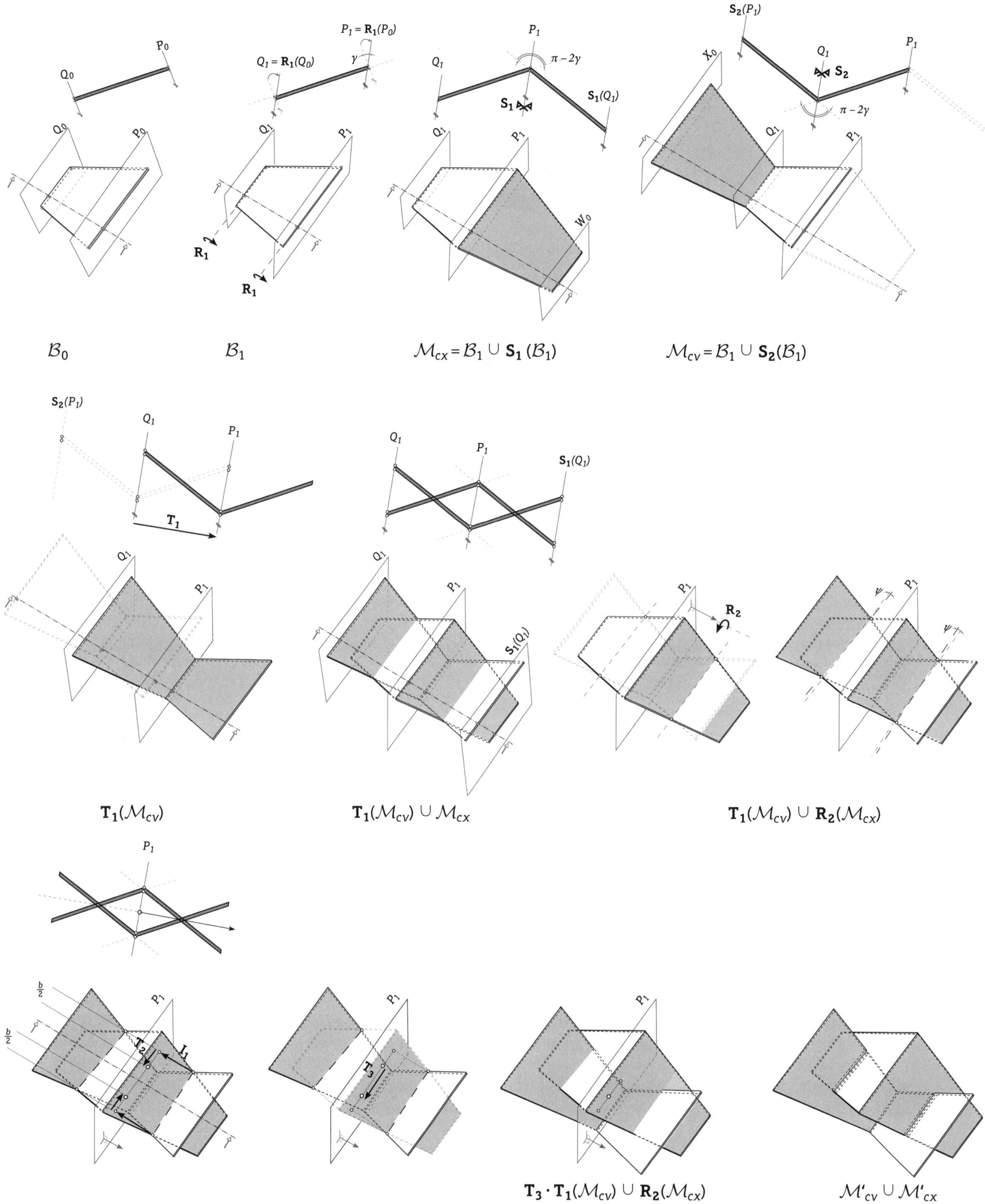

Fig. 3 Geometric transformation of basic trapezoidal panel ($\mathcal{B}_0$) into two slotted V-form modules ($\mathcal{M}_{cv} \cup \mathcal{M}_{cx}$)

The modular structure proposed in this paper is derived from the exploration of form-element relationships. It is not only treated geometrically, but also from a mechanical point of view it is demonstrated that its structural behavior is understood. Moreover, a prototype is realized to complete the investigation.

2 Presentation of structure

2.1 The folding concept

The folding concept presented in this paper was initially examined during an architectural workshop titled the "Atelier Weinand at IBOIS EPFL," focusing on the discrete architectural geometry under the supervision of Yyes Weinand. A V-form basic module is fabricated connecting two timber panels and is then spatially multiplied using consecutive spatial rotations to form an arch (Fig. 2).

The structure can be divided into four principal typologies, each consisting of two mirror-image timber panels joined together as a V-form module through the bisector plane by means of two hardwood dowels. They are placed in the middle as shear keys, and two oval head screws are inserted close to the borders in order to avoid relative translation and rotation. Fasteners are inserted in the direction of the normal to the plane of reflection. These modules are then slid consecutively along their U-shaped incisions to form an arch. The inter-panel stability is provided by roto-rigidity of the slide connection and axial contact of the reciprocal panels.

2.2 Geometric and parametric decomposition of global forms

A trapezoidal plate is introduced as the generating element of the form, denoted as B1 and referred to as the "base panel." In general, this single panel is transformed by means of two classes of operators in order to give shape to the global form. The first operations are the set of Boolean operators representing the manufacturing, connecting, and assembling processes. The second type of operators, referred to as Geometrical operators, introduce rigid structural movements and consist of congruent maps employed to place the object in the space. Among Boolean operations, union, intersection, and remove are used. The geometrical operators we may utilize include rotation around a space vector, reflection against a plane, inversion against a spatial point, or translation in the direction of a space vector as examples of simple isomorphisms.

The geometric transformation permitted to determine the slot cuts between two consecutive modules is illustrated in fig. 3. The manuscript letters (e.g. "$\mathcal{B}$," "$\mathcal{M}$") stand for geometric objects, while bold capital letters (e.g. **S**, **R**, **T**, etc.) symbolize isomorphisms, respectively the symmetry, rotation, and translations. Boolean operations are represented by their mathematical symbols of union, etc.

Fig. 4 Prototype realization at EPFL

By repeating the same geometric transformation, a double cut module is obtained. This "base module," as shown in fig. 4, is the base unit of the modular structure.

As a result of the implementation of the geometric transformation described above in a CAD setting, a parametric model of the modular structure is obtained where the geometry is controlled by means of a set of meaningful scalars.

2.3 Prototype realization

A prototype of this structure has been realized at EPFL in order to test the structural feasibility of the concept as well as to investigate the architectural quality. Two photos illustrating the project are shown in fig. 4. The project included development of relative NC codes for 5-axis machining, as well as designing constructive details and fabrication of the structure for exhibition. All V-form folded modules have been manufactured from 21-mm-thick, three-layer, cross-laminated panels and cut by means of CNC machines on the EPFL campus.

3 Structural analysis

Here the objective is to understand the structural system considering the geometric non-linearity by the aid of appropriate numerical models and to improve it. The diagnostics about the structural system consists of the linear static analysis under the self-weight and the modal dynamic analysis, in order to examine the rigidity. Based on these observations, improvements are proposed in section 4.

3.1 Modeling hypothesis and local boundary condition analysis

According to the adapted modeling approach, the thickness of panels becomes relevant in the reality of modeling, enabling the modeling of the slide connection as it is defined by the geometric configuration. For each slide connection, five normal contacts are defined between two pairs of surfaces as master and slave surfaces. Contact property is considered to be frictionless. The contact boundary condition has to be satisfied along slide joints underlying finite deformations. A 10-node modified quadratic tetrahedron element is chosen to mesh the continuum model, referred to as C310DM ABAQUS® solid element.

The timber is considered to be an elastic, homogeneous material throughout the entire thickness. The Young Modulus, $E = 8000$ *Mpa*, Poisson's ratio, $\upsilon = 0.3$, and material density, $\rho = 500$ kg/m^3 are calculated from documentation disposed by the industrial provider of cross-laminated panels.[21]

3.2 Structural analysis of a single arch

Results for the global deformation field and the von Mises stress driven from a static nonlinear analysis of a single arch under its self-weight load case are shown in fig. 6. It can be seen that the geometric configuration of the slotted-together modules leads to a concentrated distribution of stress at the location of slide cuts. This happens primarily due to the bending behavior of the structure.

Moreover, a modal dynamic analysis for the isolated modular arch has been constructed to have an initial estimation of structural rigidity in lateral and transverse loading conditions by comparing natural frequency values. The first global mode is lateral and has a relatively small natural frequency of 0.59 Hz, compared to practical guidelines, which advise a minimum natural frequency ranging between 1 and 4 Hz.[22]

4 Propositions for structural system improvement

Based on our findings in 3.2, in this section we proceed with two goals: first, to obtain a more uniform stress distribution in panels and to reduce the stress concentration in U-joints, and second, to increase structural rigidity, which is measured by means of the natural frequency of the first global mode.

4.1 Toward a truss system

While already having a geometric superposition concept, one immediate observation would be to change the current beam-like system to a more truss-like system. Two main directions are tracked, as follows:

- Addition of intermediate elements: this could be realized with the aid of additional intermediate elements which are carefully inserted at the mid-plan of the arch to connect consecutive CX-CX and CV-CV modules to each other.
- Opening the U-joint: in the initial geometric configuration, each module's stability is provided by the locking effect of panels across the U-joints. In a truss system with additional intermediate panels, fixed between them and fixed to the slid modules, it would be possible to increase the angle of U-joints. This would help to reduce the concentrated stresses **(Fig. 6a)**. Applying these two main modifications to the original structure, the maximum von Mises stress under the self-weight load case is reduced from 14.1 Mpa for the original configuration, to 1.66 Mpa, keeping the same order of magnitude of maximum total deformation. Furthermore, the stress leads to intermediate elements rather than panels and consequently the concentrated pattern of stress on panels has been resolved. The natural frequency for the first global mode of the structure is increased to 0.98 Hz.

4.2 Increasing panel interlocking effect

Consider a CX module, chosen deliberately from the arch. The region of this module is indicated in equation 1 and fig. 7a. According to the geometric principle shown in fig. 4, CX_1 is connected to CV_1 and CV_2 across two U-joints. The concept is to increase the length of these current U-joints to make CX_1 meet CV_{-1}. **(Fig. 7b)** The entire two-intersection cubes are then removed from CV_{-1} resulting in two extra U-joints on the external part of the panels. Consequently, CV_1, currently connected with CX_0 and CX_1, intersects with CX_{-1}. Removing the two new intersection cubes from CV_{-1} provides two extra U-joint connections, in this instance situated in the internal part of the CV module.

If we resume, the general idea is to keep the cut-pattern for the CX module unchanged, although for the CV module there would be four extra U-form cuts: two internal and two external **(Fig. 7c)**.

To achieve this objective, the original geometric concept has been implemented within a parametric, computer-aided design interface. The important parameters that determine the geometry of each typology of modules in the original design have been identified. Next, the geometric configuration for the montage of the Base modules is set with respect to the height and total span of the original structure.

$$CV_{-2} - CX_{-1} - CV_{-1} - CX_0 - CV_1 - CX_1 - CV_2 \qquad \text{eq. 1}$$

Increasing the notch length, while keeping the total span and height of the structure constant, will increase the number of modules. The original design consisted of 33 slid modules. In fact, by increasing the length of joint by 103 mm to create two more U-joints between panels, 53 modules of nearly the same size are needed to achieve the same height and span as the original design. Therefore, it follows that the interlocked version will be 53/33~1.7 times heavier than the original one. By increasing the number of slide joints and distributing their position across the entire length of the panel, we expect a more uniform load transfer between modules. Indeed, the results for von Mises stress from an elastic, nonlinear analysis, confirm this idea. The maximum von Mises stress for

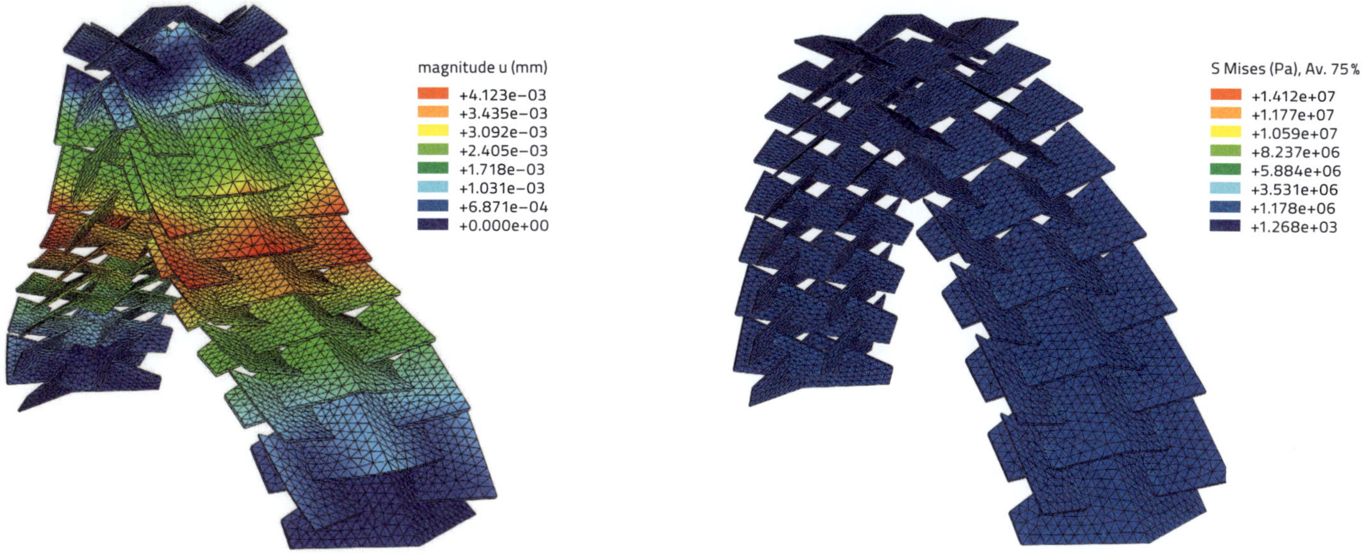

Fig. 5 Global deformation field and von Mises stress in a single, isolated arc model

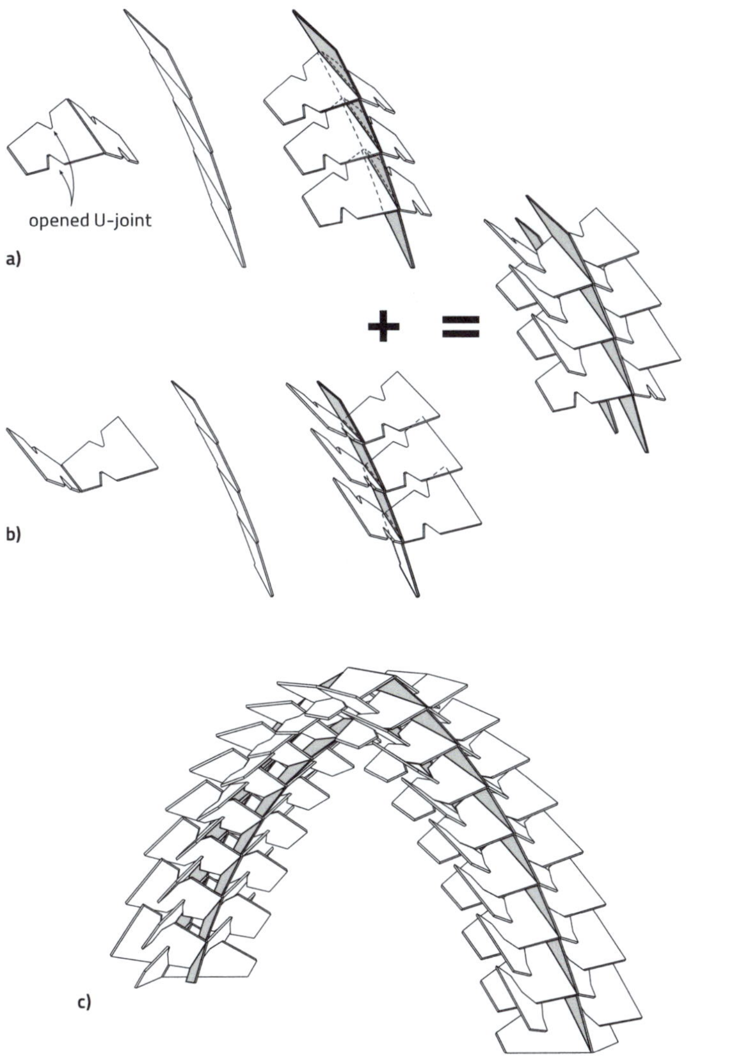

Fig. 6 Toward a truss behavior **a)** CX module modification and intermediate panels **b)** CV module modification and intermediate panels **c)** Isolated arch reinforced with intermediate panels on upper and lower areas

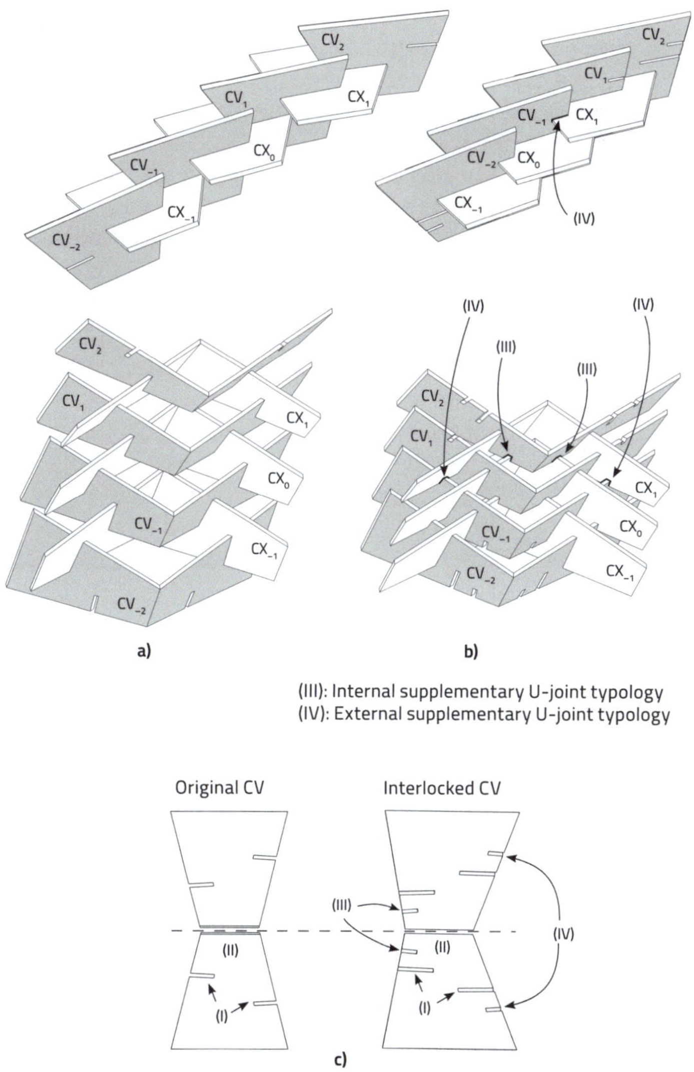

Fig. 7 Increasing panel interlocking effect: two geometric configurations **a)** Original **b)** Interlocked **c)** Comparing geometric modification brought to CV module as well as its connection typology

self-weight load case, reduces to 1.15 Mpa with a maximum total deformation of 1.3 mm, which is still acceptable. This is true even though the interlocked configuration is 1.7 times heavier than the original one. The main gain is in the structural rigidity, where the minimum natural frequency, calculated from a modal dynamic analysis, is estimated to be 5.99 Hz. Using the values of natural frequency (f) and total mass (m) for the original configuration (marked with subscript 0 in equation 2) and the improved interlocked version (marked with subscript 2), one may compare the relative equivalent structural stiffness (k) between these configurations as represented in (Fig. 3), concluding that the new slide locks stiffen the original structure by 165 times.

$$\frac{k_2}{k_0} = \frac{m_2}{m_0}\left(\frac{f_2}{f_0}\right)^2 \cong 165 \qquad \text{eq. 2}$$

5 Conclusion

A modular structural concept consisting of folded planar timber elements is presented in this article. It is shown that the form of the spatial structure is determined by the geometry of its base module and the connection geometry between panels. It represents a novel family of spatial structures that can be interpreted as planar reciprocal frames. The parametric implementation of the concept is analyzed and it is established that this parametric model can be used to propose structural improvements, notably here by increasing the interlocking effect of connections. Further research will look at a generalization of this work based on mathematical exploration of modular structural shapes suitable for timber construction.

Acknowledgements

The authors would like to express their gratitude to Olivier Baverel from Navier laboratory of ENPC for his important contribution to this study.

References

1 Thun, M., *DETAIL Zeitschrift für Architektur + Baudetail,* 2010. 50(10): 982–988.

2 Weinand, Y. "Innovative timber constructions." *Journal of the International Association for Shell and Spatial Structures,* 2009. 50(161): 111–120.

3 Herzog, T., J. Natterer and M. Volz. *Timber Construction Manual.* DETAIL ed. 2000: Birkhäuser Architecture.

4 Dunky, M. and P. Niemz. *Holzwerkstoffe und Leime: Technologie und Einflussfaktoren.* 2002: Springer-Verlag Berlin Heidelberg.

5 Buri, H. and Y. Weinand. "Übersicht Massivholzplatten." *39. Fortbildungskurs Schweizerische Arbeitsgemeinschaft für Holzforschung.* 2007, SAH Schweizerische Arbeitsgemeinschaft für Holzforschung: Weinfelden. 63–84.

6 See note 4 above.

7 Weinand, Y. and M. Hudert. "Timberfabric: Applying textile principles on a building scale." *Architectural Design,* 2010. 80(4): 102–107.

8 Horne, C.E. *Geometric symmetry in patterns and tilings.* 2000: Woodhead Publishing.

9 Baverel, O., H. Nooshin, and Y. Kuroiwa. "Configuration processing of nexorades using genetic algorithms." *Journal of the International Association for Shell and Spatial Structures,* 2004. 45(145): 99–108.

10 Popovic Larsen, O., *Reciprocal Frame Architecture.* 2008: Architectural Press.

11 Douthe, C. and O. Baverel. "Design of nexorades or reciprocal frame systems with the dynamic relaxation method." *Computers and Structures,* 2009. 87(21-22): 1296–1307.

12 Baverel, O. and H. Nooshin. "Nexorades based on regular polyhedra." *Nexus Network Journal,* 2007. 9(2): 281–298.

13 Bletzinger, K.-U. et al. "Optimal shapes of mechanically motivated surfaces." *Computer Methods in Applied Mechanics and Engineering,* 2010. 199(5-8): 324–333.

14 Block, P. and J. Ochsendorf. "Thrust network analysis: A new methodology for three-dimensional equilibrium." *Journal of the International Association for Shell and Spatial Structures,* 2007. 48(155): 167–173.

15 Tibert, et al. "Review of Form-Finding Methods for Tensegrity Structures." *International Journal of Space Structures,* 2003. 18(4): 209–223.

16 Eigensatz, M., et al. "Paneling architectural freeform surfaces." *ACM Transactions on Graphics,* 2010. 29(4).

17 Pottmann, H., et al. "Freeform surfaces from single curved panels." *ACM Transactions on Graphics,* 2008. 27(3).

18 Glymph, J., et al. "A parametric strategy for free-form glass structures using quadrilateral planar facets." *Automation in Construction,* 2004. 13(2): 187–202.

19 See note 9 above.

20 See note 16 above.

21 See note 5 above.

22 Bachmann, H. *Vibration problems in structures: practical guidelines.* 1995: Birkhäuser.

2.4

"Discover processes where the parameters are insufficient"

Interview between Mark Pauly (Computer Graphics and Geometry Laboratory, LLG, EPFL, Lausanne) and Yves Weinand, February 10, 2015

Yves Weinand: Something I'd like to discuss with you is the prototype presented in this chapter, Modular Timber Structures, which has never been published before. There is also Dr. Ivo Stotz's thesis about fractals and iterative details, developed by architects and computer scientists at IBOIS. We created this pavilion with civil engineering students, where we simply changed geometric parameters. That's what we're doing all the time: we try, fail, and retry. It's an iterative manner of working with civil engineers. This is something we could try to highlight in our discussion. The first question is: What did we do here? We developed this structure as architects, and then we controlled the contact zones by mechanical means, as we have in the models. Then we improved the geometry by slicing pieces deeper in relation to one another to add additional contact points, thereby creating another form of global geometry in order to stiffen the structure. I can do this when I work with students, but they are not able to reach this point very often. Sina Nabaei, who was working on this project at the beginning, was aware of this aspect, because he understood the geometry. He made it very clear at the outset and we haven't been able to advance further yet. I would like to know: How would you approach this in a different way? This is just one solution; there is no general framework about this geometry and it's more from the point of view of a civil engineer, who was trying to stiffen the structure a bit further. There's already something here; it's a case study. Do you see it as defining a general framework to describe this type of geometry and to optimize it?

Mark Pauly: First, it is an iterative process—we use our intuition and we try. I think that's something that will remain and that's probably a good thing, because when you design this kind of thing, you're not only looking at solving a problem, you're also looking at defining a problem. You often don't know exactly what you want, and this is makes it harder. This is the process of searching for the right constraints and the right parameters. I think these types of problems are inherently changing, because they involve justifications. The problem is well understood when you have continuous space and continuous energy and you want to optimize them. But when you have connections and you have many potential ways of connecting them together, I don't think there is a general solution, where you can simply play with the system to find an optimal solution. The individual piece that you have may be locally optimized, the local shape is interacting with the global geometry and that is absolutely fundamental to look at. At this point I don't see a solution for this. We're still working on very specialized solutions: the justifications can't be determined in advance, nor can they be fixed with regard to the shape, the geometry, the size of the elements, the manner of changing the connections, or the interaction between the elements. I don't think there is a system that allows you to dynamically modify these things and obtain any kind of meaningful results, because you have a kind of fragility in the sense that if you change things very slightly in terms of the connectivity, it'll have a huge impact on the structural performance.

Y. W.: We probably didn't really manage to model the contact zones. It was more a global geometry change. We added more contact points because we modified the global geometry and this already helped structurally. As a computer scientist, would you take this as a case study? I would actually like to take this project further. It's there and nothing has been done, but I think this kind of structure—just plates—could be very interesting in architectural applications, for example in lighting. As directors, how could we define something like a research plan that would also be suitable for one of your doctoral candidates in computer science? Would you encourage this type of topic, or how would you define the approach? I would like to take a civil engineer and show him this project. He would probably repeat the research until he gained rigidity. I think we're a bit lost, as we don't have a general framework, as you said, we have not really addressed topological questions yet.

M. P.: The first thing I see—and this is perhaps my perception as a scientist—is the possibility of abstracting this structure into general performance. Basically, what I see here is a framework that is constrained. You have many small performances that interact with each other and these are local interactions, but they create a global dependency. Let's say you have hundreds of plates or hundreds of elements, and then you can encode the interrelations towards a topology of the system. You can define their local interactions, even experimentally, whatever contact forces you have. Your aim would be to have a system that allows you to interactively change the shape of individual elements, the shape of the overall structure, and all the connectivity, plus get feedback on structural performance and assembly constraints.
I would try to have a general description of this kind of element; this means a specific way of joining two pieces together that have certain properties, geometrically and structurally. However, you could think of many other ways of connecting pieces together. My first approach would be to render this in more abstract terms. On the one hand, this type of construction has a very general structure beneath; on the other hand, it's also important to make progress to study a specific instance. You don't want to investigate all possible ways you could connect things and try to solve all these problems together. I think at this stage, as we don't have a tested solution for these topological assemblies, it would be good to start with a concrete example and maybe limit the variations. By maybe resizing the plates, you can redesign and change them, and so on. You can parameterize the connections in certain ways, so that you have a fixed space in which you could operate. The parameters are fixed, but the numbers and types of parameters are not fixed. This is a design task. At least start with a fixed set of variables and build a system that could work with them and then discover processes where the parameters you decided to work with are insufficient. You need to build such a system, a sort of bottom-up approach, starting with something simple and adding complexity as you go. But at the same time, keeping the opposite perspective in mind, the sort of abstract representation that you could use for this, because you don't want to solve one specific problem, you want to keep the general problems that are more useful for the application. I would use a dual approach: you look in very general terms at elements that are connected in a very abstract language that could be easily mathematically represented and, at the same time, come from a bottom-up approach and use wooden plates or shapes that are connected in an abstract way, and bring these two approaches together. You can learn both from the abstract point of view and the general point of view, but also from the interaction system, from the concrete properties of your design, because you won't necessarily be able to think of relevant factors up front. You may understand once you start to play with these things, when you break down some of them, or with material tests and so on. Then you will realize that there are parameters that you haven't thought about and that you need to include in your system. To me, this is an approach I would like to take. It's often hard to impose this on students. My experience is that students either have a bottom-up kind of mind-set or a top-down kind of mind-set, and they like to progress in a linear fashion and find it difficult to look at things in different ways. I think as a tutor you can give them suggestions.

Y. W.: When you speak about parameters, are they geometric in nature or are they other sorts of parameters?

M. P.: Of course, many of them are geometric, such as the size, the thickness of these plates, and the angles. But you can also put in material properties: the strength of certain materials, the friction parameters between materials, and so forth.

Y. W.: Do these parameters equal constraints in this case? When you speak about mechanical constraints, for example?

M. P.: By parameters, I mean aspects you can vary; for instance, the thickness of the plates is a variable. In my case, the geometric constraints would be that the angle between two plates cannot exceed 45° because of the limits of the fabric technology. These are fixed parameters, for example, in the case of simple constraints. With more complicated constraints, if you want to assemble these things, then there is only a certain path whereby you can move the pieces and perhaps there is a global dependency. In terms of materials, you may say the maximum stress that the piece can support would be a mechanical constraint. To me, these are the parameters that define a space. This is a trivial example (he draws something): You have a two-dimensional space; let's say you have an x variable and a y variable. Then you specify constraints as a function, so things are particularly free, and then you say something trivial, for example, the sum of the two would be equal to one. And it gives you this two-dimensional space. You start with two degrees of freedom and you have one constraint. Of course it's not always the case that the degrees of freedom are independent or that the constraints are independent.

It's a challenging modeling task. First of all, you have to define your degree of freedom and your constraints, and then model them in a mathematical way in order to test an optimization. The challenge for design is that neither the degrees of freedom nor the constraints are fully understood. You start with something that you think is reasonable, you build your solution, you explore things and you find while designing that maybe you are actually missing a constraint, maybe you have too many constraints, or perhaps you have to introduce new degrees of freedom. For example, in this case, you decide at the beginning that the thickness of the plate is fixed, but you then you realize that with this fixed thickness, you cannot obtain what you want and you have to introduce a parameter, that is, the thickness of the plate, so that you have more freedom in design. As you mentioned at the beginning, this whole process needs to be iterative. You won't know the degrees of freedom at the beginning; you won't know the constraints. If these two elements are not modeled correctly, either your model will be over-constrained and you'll have no more freedom to design, or it will be under-constrained and every solution will be possible, but then you won't know which one is good anymore. You need additional thickness criteria in order to make a choice. If your constraints don't restrict the design space, then you don't gain anything through an optimization process, because it doesn't tell you anything. When you design something that looks nice, but doesn't work mechanically, how would you know what you have to change in order to make it work? This is exactly the challenge here, because if you have a multi-dimensional space you can have thousands of degrees of freedom. Many of the designs will be poor. These are totally different things. How could you build a system that lets you understand the different ways to achieve a meaningful design? Should you make the plates thicker, should you lower the height, should you change the degrees, or something else? This remains unsolved. Because of this iterative process, what I would like to do in the future is to better understand the consequences of introducing new constraints or removing constraints. Sometimes you have constraints that are conflicting and there is no visible solution. How do you understand which of these constraints you need to lose and which to keep? If you have hundreds of constraints, you cannot simply modify each of them individually. It's very difficult from an optimization point of view. We have to come up with another solution. It's challenging and that's why I think you have to build it up slowly. The pavilion is a good example to begin with, because it's not overly complex. It's not combining many different things together. There is certain regularity to the problem and the scale is appropriate for this type of problem, because I believe you would have to start small.

Y. W.: Of course, I'm interested in defining mechanical constraints and to see when you compare those two systems: This one is far better in vertical loading, or this one is not good at all in horizontal loading. When we think in terms of optimization, it's quite hard to see because every load case is already very different structurally. Different types of structures react differently to different types of load cases. That would be more like a geometrical study, which gives its structural performance. My feeling is that the geometrical problem is easier to tackle or to define than the mechanical one.

M. P.: Yes, because the topology is fixed. For example, let's say you want to optimize something like this for a vertical load, you have this prototype, and you see that it performs badly in terms of sustainable load. What are you supposed to do? This is exactly the kind of thing you would say, regarding your constraints and how much vertical load you can tolerate. The connections are not strong enough, so you have to tie them so that they can support more load. That's a very complex constraint, because it doesn't necessarily affect every piece. Let's say the function to evaluate your design and its effect gives you the vertical load tolerance, the maximum vertical load you can sustain. This is a very complicated function, but you can evaluate it, you can run your design through a simulation and determine the result. In order to design, you have to work on that function, you have to figure out how to modify x to increase the load. This would be an example of a complicated constraint, as opposed to a simple calculation that can be easily evaluated. Figuring out how you would need to change an axis in order to increase that load optimally is highly complicated. Currently you would probably do it intuitively to make things different, but coming up with the system does something like this automatically. I think it's very difficult, unless you clearly fix the parameters. If you only allow for parameters, let's say some geometric parameters like the sizing of the plates, maybe the thickness, then that's a tool, as you have a sort of continuous space. But as soon as you allow for topological changes, I have no idea how to project this in an optimal manner.

Y. W.: In order to be prudent we should begin by trying to define certain parameters. In terms of coming up with a research plan, do you think that would be possible?

M. P.: The first question is: What are your design objectives? In this case, maybe you want structural performance. You have some boundaries and, for example, this is the amount of load we want to tolerate. You can vary certain things, depending on what you decide to address at the beginning. You have a list of parameters that you could change. Is this something that is difficult to specify at the onset if I want to find the optimal angle to get the optimal performance?

Y. W.: But you can't say it in advance, it's really strange.

M. P.: So you don't have the physical model to simulate this?

Y. W.: No. I know this parameter is the most important one, but in the end, I think it's probably not solved.

M. P.: But you did run simulations, right?

Y. W.: Yes, from them you have the results and then, for example, you get deflection. Then we can register the structural basis, the stiffer one for example. But that's all. Even the interactions between the parameters are unclear.

M. P.: What you have now is a system in which, if you change parameters, you can run it and it gives you an answer?

Y. W.: Yes.

M. P.: So, you have an oracle basically. What I mean is, you choose a point in your parameter space and it gives you a quantity indicating how good it is. In theory you could sample all the possible combinations of parameters and then look for the one that is the most compliant. In theory, though not in practice.

Y. W.: Yes.

M. P.: That's already significant, because you can evaluate this function, but you can't invert it, you can't say which parameter is the best. What you really want in optimization in mathematics is to find the argument for maximizing or minimizing that function. Right now, all you could do is to evaluate f. If x has many dimensions, it becomes computationally attractive. If you only have one parameter, you can simply sample twenty simulations. But already, you have this function, but it is often not given or you don't know the function. In such a case, I think there are different strategies. One has to make the function evaluation efficient enough to show that your project is valuable, so you can do a search; or you can try to simplify this function, for example, by looking at geometric properties that you can evaluate more easily. I think this last approach is an interesting alternative. The question is: Looking at the simulations, can you gain some geometric intuition? Let's say your energy function looks overly complicated (he draws), but you can find a simpler function and then you can decide to go back to your original function. You can return to a complex function, but you can start with a simpler geometric formulation you can optimize, one that you can search for globally. You'll never be guaranteed to find this pick. You may find some other picks on this function, which are not as good, but at least if you do it carefully, you can usually prove that you are going to stay within a certain optimal boundary. There are many optimizations, which are exponentials, so you cannot hope to find the right solution, but you'll have a certain approximation that will guarantee that you to stay within a certain boundary. That means, though, you're guaranteed to be within five percent of the optimum result. You wouldn't know if you achieved the best possible result, but at least you'd know it's not the worst. It's a challenging and intellectually difficult task to simplify such a model; simplifying a model is not something you can easily automate.

Y. W.: When you speak about form-finding and optimization in a mechanical way, the Block research group, for instance, defines an ideal function for specific load cases. But if you change the load case, you would need another form. So we would rather choose a parabola, or something similar, if the main goal is efficiency for taking vertical loading. The form optimization—having more load points—would, however, increase the flexion rigidity for horizontal forces. Thus geometric modifications of a given form can vary depending on whether the modifications tend to be able to carry vertical or horizontal forces. The parameters that need to be considered can be of a different nature regarding different load cases. Even if you defend that, you will have a global stability and a local approach (not rigidity). I still don't see exactly how I could have a global approach here in terms of structural engineering. There is another thesis at the ETH in Zurich written by Thomas Kohlhammer about reciprocal frames. Here the optimization process consists of taking out bending moments.

M. P.: I think a good approach would be to abstract certain physical properties and map them into a geometric space. You generalize the model to optimize it, and it's beautiful, but you have to challenge this first abstraction step. You then find the optimal surface for a perfect vertical load, and I guess in practice that also means it's a terrible surface for a slightly vertical load, but you lose any solutions that might work. This has to be checked.
I think the only advantage is that it gives you initial points for further investigation. It doesn't solve the whole problem, but the whole problem has to be attractive enough for certain cases. It's interesting in itself to look at attractiveness. You can solve problems for constraints graphically with twenty or fifty increments, but it's impossible with a thousand. There are many solutions that seem to work well on twenty examples, but they don't scale. It's not just that the computer is faster and you can do all the optimizations that are

possible—no, you have to increase by a factor of a million times before anything works. This is a challenge. It's interesting to study the problem of model scale; this is what I mean when I'm also keeping this global perspective. You can solve this at a small scale and you can learn quite a bit, but if you don't understand the global aspect of it, and for example, see that your solution has an exponential complexity, then you will never be able to use it on a real-scale model. You can argue that it's still informative to look at it, but the solution is not precise. That's why
it's good to go back to the global function; whatever you want to apply is a real example. Space parameters start from two degrees of freedom to five hundred, raising the question: Is this approach still valuable? Many forms are beautiful in two dimensions, but they don't work when scaled up by ten, or fifty, or two hundred. This is a challenge, but it may just be important to reflect complexity for certain tasks. For others you have to accept simplification and sometimes you have to hope for the best. You have to hope that whatever comes up—what's actually tested—is still good enough. I don't think there is a way to simulate such a shape in terms of formal, mechanical, and structural properties in a way that you could invert the simulation for form-finding. It will always be an approximation, anyway. The question is: Are your approximations relevant? You have to verify your experiment.
I think at some point you have to go back to physical tests. Of course, it would be nice if you could optimize this link between your physical experiments and virtual designs to make them more aligned. If your virtual production and your physical tests have no relationship, then you have no meaning in this model.

Y. W.: You see, there is a subjective motivation. Its design is a little bit different and we would like—with regard to the present publication—to objectivize our research. We would like to increase our objectivity. Whether this is actually possible is something that is still unclear to me.

M. P.: We don't think it's at all possible. What do you mean by "objectivity"? You could say that we have a shape that has a certain performance, let's say structurally. We could show that this performance is better than any other alternative. Maybe in certain cases, you can even prove it's the optimal performance. Even so, it may still need to be designed. I would say the structural performance is just one criterion. At some point, you may want to sacrifice structural performance for aesthetics, or cost, or other aspects, but I'm viewing it from the perspective of the engineer. At the very least, you want to be able to know if this is a well-performing structure or not. You don't want to just guess performance by how it looks, though I don't know if there is a better solution or not. I would like to fully understand this aspect.
I think this is important, because if you knew these things, then you could make meaningful decisions. You could say, "OK, I accept these degrees of structural performance, because I know I gain something for it, and I don't break down the structural performance for nothing, because I have the wrong design tool or whatever." This would be something you have already understood through your design: What you sacrifice for what gains. I think for that you will need to have a more complete picture of the design basis and its performance.

Y. W.: We often have a strong correlation between form and structure. At least, we say this because we feel it. But we cannot explain it in a deeper sense.

M. P.: Is that really true? For example, we can say that an arc would structurally have a more robust performance than any other form.

Y. W.: I think there is something here, which is convincing for the correlation between structure and form, but this is just one take on an enormous number of topological possibilities and it's not so clear. I would like to have more clarity here.

M. P.: I guess what you're saying is that there are some basic principles that can lead to better performance, but these are only general principles.

Y. W.: Yes, and we have the cohabitation of different structural systems in one, etcetera, but we select a sort of mixture satisfying different criteria, which are for performance motivation and design, and we end up having entry in forms without being able to explain this in a more comprehensive manner. Are you more attracted by this or that, or is this just the same research question for you when you consider the different scientific aspects?

M. P.: That looks different to me, in the sense that the structure is fixed here. I guess the question for this type of thing is, do you just want to optimize the form, or would you also like to optimize the topology?

Y. W.: I would also like to have more information about the ideal size of these pieces.

M. P.: Again, it's a matter of abstraction. But, as you go through the abstraction to more concrete sizes, there could be some additional topics because the connections, the topology, or the parameters are different.

Y. W.: Years ago, we were fascinated by all that we could do in terms of introducing new visual/ geometrical factors such as nurbs and meshes. But then there was a rebound effect; an overload of images, random images, images without meaning. We didn't want to continue down that route. Now it's clear that we have a sort of general tool and then very subjectively, we select a different modification, which we then stick to. But this is a very subjective process. Using this tool, for example, we can add an architectural or structural solution. We're far less enthusiastic about the wide range of different types of curves we can really produce. There is this motivation to travel down a very general route for this pavilion, but the selection of certain images gives a new interpretation to an existing vault structure. You have an atypical view of something you know or something that is a bit familiar to you. I think that's the reason why this image works so well in communication. When it's more free-style, like the table, for example, it doesn't have the same impact, even if it's made using the same tool. As an introduction, I would like to have something documented about this.

M. P.: If you have desired constraints, the offset surfaces come back to the notes "beam torsion free" mathematically. Then it introduces a very clear mesh. Mathematical theories will help you to design. As soon as you move away from these types of surfaces, you're going to run into this problem. If you don't know that, then there is no way to fix this with just a cap-modeling tool. You really want an optimization that brings you back to this space to avoid these problems. This is true for many examples. We have local considerations that lead to global dependencies. You know that you want to change the position of one of these plates or its orientation, because you don't like it. It's very difficult to modify anything, because somehow, when you modify things you usually do it from a local viewpoint. You like the form and you want to just change things, but you can't, because when you change something here, it will affect the entire thing. If you do design in such a space without a global optimization in the background, then your results will almost be left up to chance, because maybe it works out for some strange reason, or you accept that things can change in reality, but you don't truly understand it. You don't know what you could do differently to avoid this effect. What you are saying here is probably also a question of perception.
To be honest, this problem requires a sort of interdisciplinary effort, which is always a challenge in itself. To really make something like this meaningful, we can completely ignore all of the physical aspects, the structural aspects, the material constraints, and we can purely abstract the question to its geometry, which is fine if you keep to a toy scale. Thin wood models support the structure anyway, so there is no issue. Of course it's a limited approach, which cannot be generalized to large-scale constructions. I think you need somebody to look into the structural aspects; you need somebody to look at global geometric aspects, and somebody else to

look at material aspects, and hopefully somebody will solve this optimization problem in some way. It would be very challenging to put together the right team, find the right people, and make sure that PhD students can make a thesis, and that all the constraints are of a university standard. I think at some level, if you want to have results that can be applied on a larger scale, you cannot avoid going down this route. The approach based on a toy world, where we can abstract physics into geometry, simply doesn't scale. On one hand it's convenient, because we abstract a lot of problems and you can do mathematics or some computer science models, but on the other hand, it's frustrating, because you're stuck in this toy world. My ideal approach would be to pull together the right people and somehow suffer through this additional complexity of managing a bigger project, of having highly skilled people and different communities and all the things that ensure the likelihood of a project's success. That's the big challenge here. I don't think there is a good model. I can't point to many examples where we could say here is a structural engineer, an architect, a computer scientist, and students at the university level, working together and making the work successful.
Perhaps there are more examples in industry, I don't know, but I also doubt there are many. Possibly that's the challenge we have to face. We have to educate people to be able do this research.

Y. W.: While speaking with you and with Helmut Pottmann I had the impression that you would prefer to start your investigations from a specific situation with its own specific constraints and probably physical constraints.

M. P.: We did move a bit into this direction to add more physics, but it's still toy physics in many ways.
It's still insufficient to bring in simple elasticity models, gravity forces, from an engineering point of view.

Y. W.: This is precisely what Sina Nabaei did.

M. P.: We're still trying to stay in this world of real-time feedback. We want to do optimization. If you have an insufficient system that take hours to give a single answer, then if you do optimization, you need an answer every ten milliseconds and we need to invert that function. We're coming from a starting point of purely geometric design, to make it interactive and make it work faster. Now we're adding physics, but without moving too far away from offline computation. We still want to do something inspired by physics or some plausible physical behavior added to these interactive processes.
You can say here's the real physics that is too slow, we need to make it faster or simplify it in order to make some interactive process. However, we started from some interactive process and we added as much as we could without losing interactivity. The rationale is that you are still in the early phases of concept design, where you would simply like to find a shape or a form that is a starting point towards the final structure.
We try to give you as much feedback in terms of physics as we can afford without claiming that it is already meaningful for the structural properties.
I think it's an interesting question of what you can add, in what way, to inform your design and your form-finding with as much structural physics as possible.
Once again it's a question of how you design and what you want. Right now, we're probably mostly looking at this initial form-finding, exploring shapes rather than finding them with structural properties in mind. But this to me is the ultimate goal. In the end, you want to get all of the analysis tools you have at your disposal, including structural performance, maybe energy performance, and heat gain—all the things that you can't predict in a final given design. You'd like to bring them into the earlier design stage in some form or another; but this is a lot of work.

Y. W.: From speaking with Matthias Kohler, I know that there are so many interesting processes. The students try to learn something from the process, but we need a sort of final goal or a synthesis of one kind or another. We're looking more to the synthesis or to the end product. In many cases, process observations concern additive processes. Here we aim for a synthesis, where the global form is considered.

M. P.: That's true, but I think that the approach as a process is interesting in itself. The process you design or that you want could be beautiful, but is simply not realizable. That's an interesting, but also a very complicated process; you design a logic of connections that abstracts the problem in some sense. When you say, "Here's my connection logic. From this logic I can explicitly specify what needs to be satisfied for a structure to be constructible." If you construct a Lego model, you know exactly how to connect the blocks. If you change the design a bit, then suddenly the structure might become unstable; again, it's very difficult to explore this.

Y. W.: Well, we probably will end up having both considered: the process but also the synthesis. I'm looking forward to exploring this with all of you, taking advantage of our NCCR research collaborations.

2.5

Modular pavilion: a structure for the Paléo Festival

Marielle Savoyat

Design and execution Project design and execution:
IBOIS, Laboratory for Timber Constructions, EPFL, Swiss Federal Institute of Technology, Lausanne, Switzerland
Bastien Thorel, student, Prof. Yves Weinand and Sina Nabaei, researchers

Execution 2008–2009

Location Paléo Festival, Nyon, Switzerland

This structure was initially designed by the student Bastien Thorel in the Weinand studio at IBOIS, the Laboratory for Timber Constructions, EPFL (Swiss Federal Institute of Technology, Lausanne) during the academic year 2008/09. The goal of the exercise was to design a structure inspired by textile mesh for an architectural program, which would then be proposed as a structure for the Paléo festival in Nyon, a town on the shores of Lake Geneva in Switzerland. Thorel proposed a structure based on a discrete geometry composed of interlocking wooden panels. Only two types of panels were used to create this structure, which also acts as an envelope and an enclosure. A basic module in the shape of a folded V was connected to another basic module, also V-shaped and mirroring the first shape. Each module included four U-shaped cutouts, commonly called semi-wood, which slide into each other to create an arc.

The result of this spatial structure—light, lofty, aesthetic, and easy to install—made it particularly suitable for a temporary function in the context of a summer festival, like Paléo.

A prototype of this structure was first created at the EPFL, to test both the structural feasibility of the concept and the resulting architectural language. The project necessitated the development of a digital cut with five saws to create the U shapes, as well as the development of the connection details and the overall structure. The basic module was composed of 21-mm-thick, 3-ply panels.

A three-dimensional digital model was useful for studying the structural behavior of the modular arc as undertaken by IBOIS researcher, Sina Nabaei. The structure is a good example of a complex global final form, built using simple wood-wood connection techniques and folded and interlaced modular panels. Its ingenuity lies most notably in the fact that the modules support themselves reciprocally. The global form is determined by the local geometry of the connection.

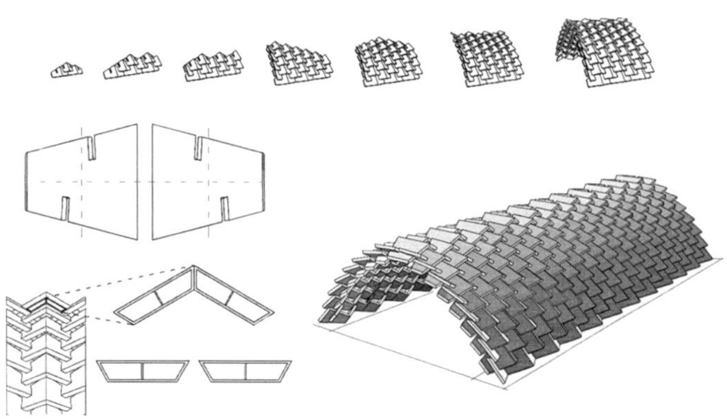

Fig. 1

Fig. 1 Axonometry of the complete arch

Fig. 2 EPFL Campus 2009: Shell prototype

Fig. 3 Section of a half arch

ent 5G
ts AAsco

Fig. 2

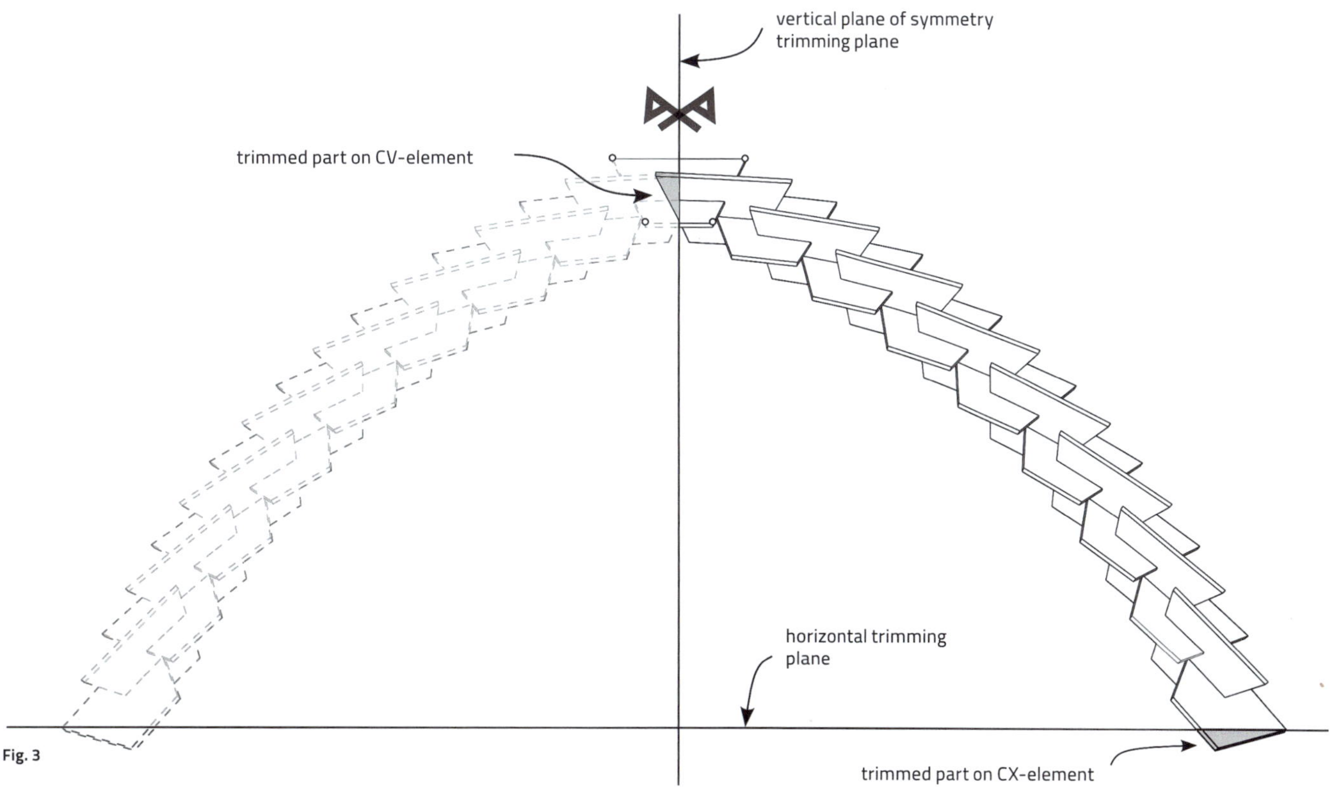
vertical plane of symmetry
trimming plane
trimmed part on CV-element
horizontal trimming
plane
trimmed part on CX-element

Fig. 3

3 Active bending

3.1

Actively bent and woven structures

Yves Weinand

The tests carried out on IBOIS investigations can be broadly divided into four types of structures:

Structures created from flat panels:
- Origami-like folded structures
- Free-folded panels, e.g. developed from iterative functions

Actively bent structures:
- Structures resulting from linear supporting elements, such as rib shells
- Structures formed from planar elements, such as braided structures

If the aim is to create a curved or double-curved surface, the above-mentioned structures can be utilized. When progressing with a structural design, it is also necessary to take the possible connection techniques into consideration. Particularly in actively bent structures, new types of connections are possible that would be unfeasible in the structural types previously discussed. It is possible to change, or even to do away with, the connection node altogether. In the parabolic fence posts depicted in fig. 1, the rotation of the linear element results in a double-curved surface. The consequence of this is that the connections vary in nature.

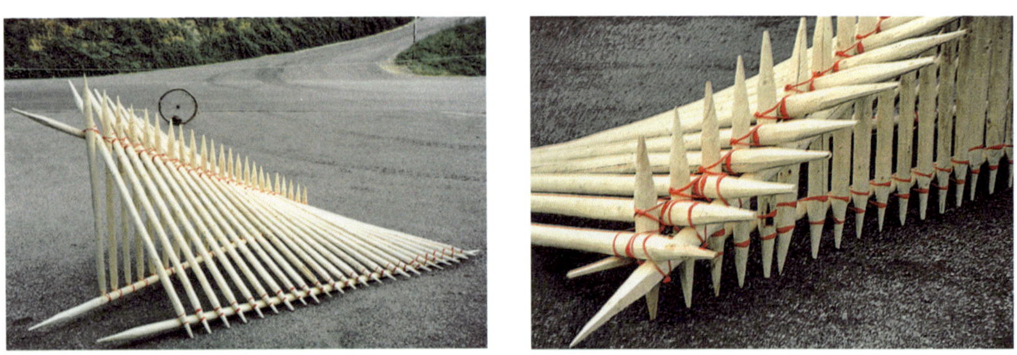

Fig. 1 a and b Parabolic fence posts

Fig. 2 Glass cones generated from triangles

Fig. 3 3-D chart of tower and axial system of profiles: the cross-section of the triangular frame is shown in green; the cross-section of the inner tube is shown in red. The axonometric shows linear connecting elements—which are not incorporated into the node—the triangular, self-contained carrying frame, and the glued frame structure of the glass panes.

Fig. 4 Elevation of the completed frame system

Fig. 2

Fig. 3

Fig. 4

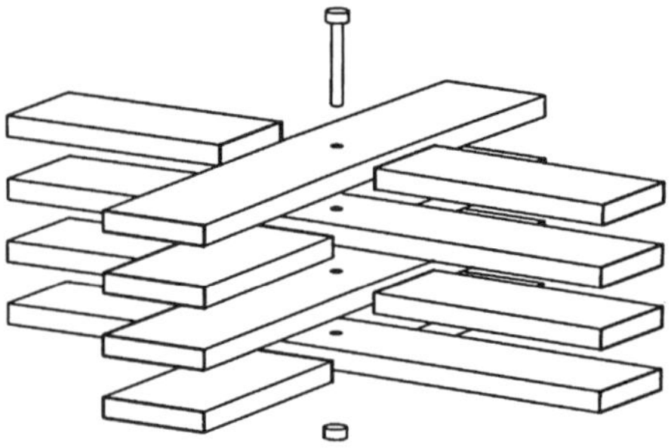

Fig. 5 a

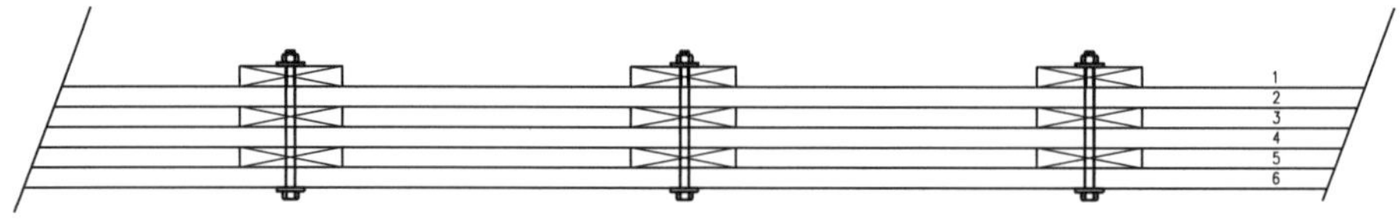

Fig. 5 b

Fig. 5 a and b Diagram of the nodes connecting a ribbed shell

Fig. 6 The rib is composed of several layers. With an active connection of the rib, the individual layers have a tendency to slide against each other. The connections should help to avoid, or at least reduce, this displacement.

Fig. 7 The stiffness of the connection must be calibrated. The shifting of layers in relation to one another generates transverse forces in the connection, which are tested here.

Fig. 6

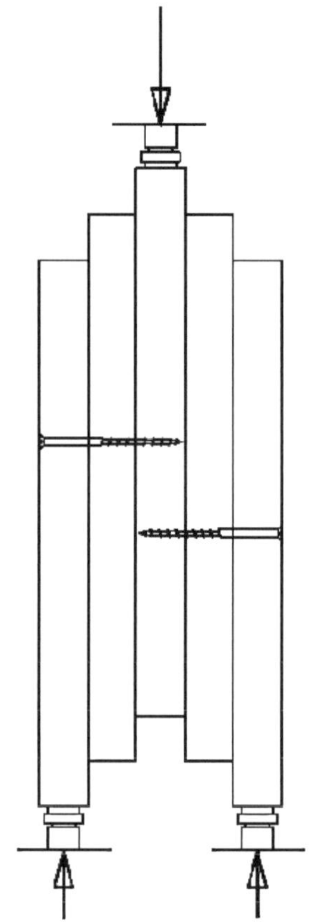

Fig. 7

In the example shown here, the connection is made from rubber, which is able to adapt to the specific, local form due to its elasticity.

Another example of the development of connections is demonstrated on a glass cone in the Belgian town of Lommel. Here, the node connections at the corner junctions of the triangles are removed and replaced by cross connections along the sides of the triangle. The flow of forces thus no longer passes through a connection node, but is rather directed around the node. The effect of the forces within the surfaces and their transmission along the edges of these surfaces is preferred to the transmission of the forces through single nodal points.

The system described here highlights the relationship between construction principle and connection detail. Depending on the choice of the overall system, the connection type varies. To begin with, structural systems made from linear elements were differentiated from structures composed of planar surfaces. The two types of structures resulted in different types of connections; this fact is essential for the developed structures explained below.

The most widespread type of structure in the category of actively bent systems is the ribbed dome. Well-known examples of ribbed dome buildings are the halls of the Federal Garden Show in Mannheim (Mutschler/Otto), the Polydôme at the University of Lausanne (Natterer), and Savill Garden in London (Happold). The main load-bearing elements or ribs are actively bent elements that are not interrupted at their connection points but continue, uninterrupted, with components of up to 50 meters in length. Their cross-sections thus fully retain their strength and are not weakened. Ribbed domes consist of continuous supporting elements, as opposed to disconnected support systems—such as reciprocal systems—where the structural parts are interrupted and the nodes have to be connected to one another. A possible variation of a ribbed dome connection is shown in fig. 5 a and b.

The series of examples shown here is based on standardized surface areas. We are interested in learning how a certain size of surface area (or footprint) proposed by an architect could be built over, using the ribbed dome system. This work is described below by Claudio Pirazzi. The following fig. 10 a and b illustrate the construction of a prototype.

The ribbed shell constructions presented here reference, on the one hand, felt fabric and, on the other hand, woven items, such as knitting patterns, braiding, and weaving.

The geometric view of this pattern shows that a stabilizing "system effect" strengthens these structures. The close proximity of the elements to one another results in a high friction effect before the system can be deformed. This property was of particular interest to us. It raised the question of whether this material-specific assessment might be used from an engineering perspective to stabilize and reinforce structures.

If you start to "weave" wooden materials, you have to deal with various systems. Fig. 13 illustrates a braided structure that develops in the plane. The spatial interpretation of this woven structure is illustrated in fig. 14. A spatial knitted pattern is created by the braiding pattern along the bottom or along the third axis. However, the extrapolation of this local principle in a curved structure diverges

Fig. 8

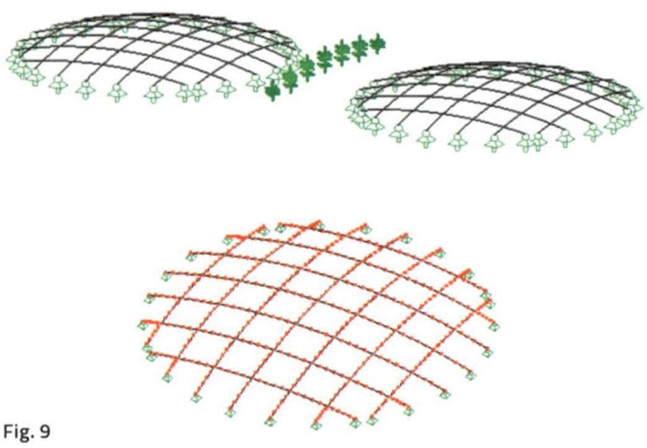

Fig. 9

Fig. 8 The IBOIS team paid special attention to the continuous formation of the ribs when they built their experimental ribbed shell.

Fig. 9 Modeling of geometry as finite element models, while taking into account the connection stiffness between the layers.

from the original concept. The arched truss shown in fig. 15 is unable to replicate the rigidity of the basic braiding pattern at this scale.

The example shown here fails because exclusively linear components are used. For this reason, actively bent and twisted components are also used, as illustrated in fig. 16 a and b. Leonardo da Vinci's system of reciprocal self-stabilization is inserted at the top of the image. As shown at the bottom of the image, the components are forced into both actively bent and planar forms. Here, the structure is subdivided into a primary weave and a cross weave direction, represented by a linear element.

The targeted structural optimization of the geometry remains a focus. The weave presented here shows that deeply woven strands not only are interrupted but are also shifted. This shift happens spontaneously, and results in the shifting of the network strands in relation to one another.

In the fall of 2008, the architecture studio at the EPFL focused on the theme of woven structures. Some of the results have already been published in *Wood: Material or Form: Transformation of Structural Logic* (2014).

Markus Hudert started to manipulate forms of textile modules, then to multiply them and shift them. One of his first successful attempts is depicted in fig. 19 a. Through the active bending of two planes and their rotation into a single, unified system, a compact, seemingly simple form is generated. The wood panel surfaces were bent and twisted, thereby producing bending and torsion moments, building up residual stress.

It is fascinating to examine the standardization of this support system. If the panel edges lie in a parallel and horizontal

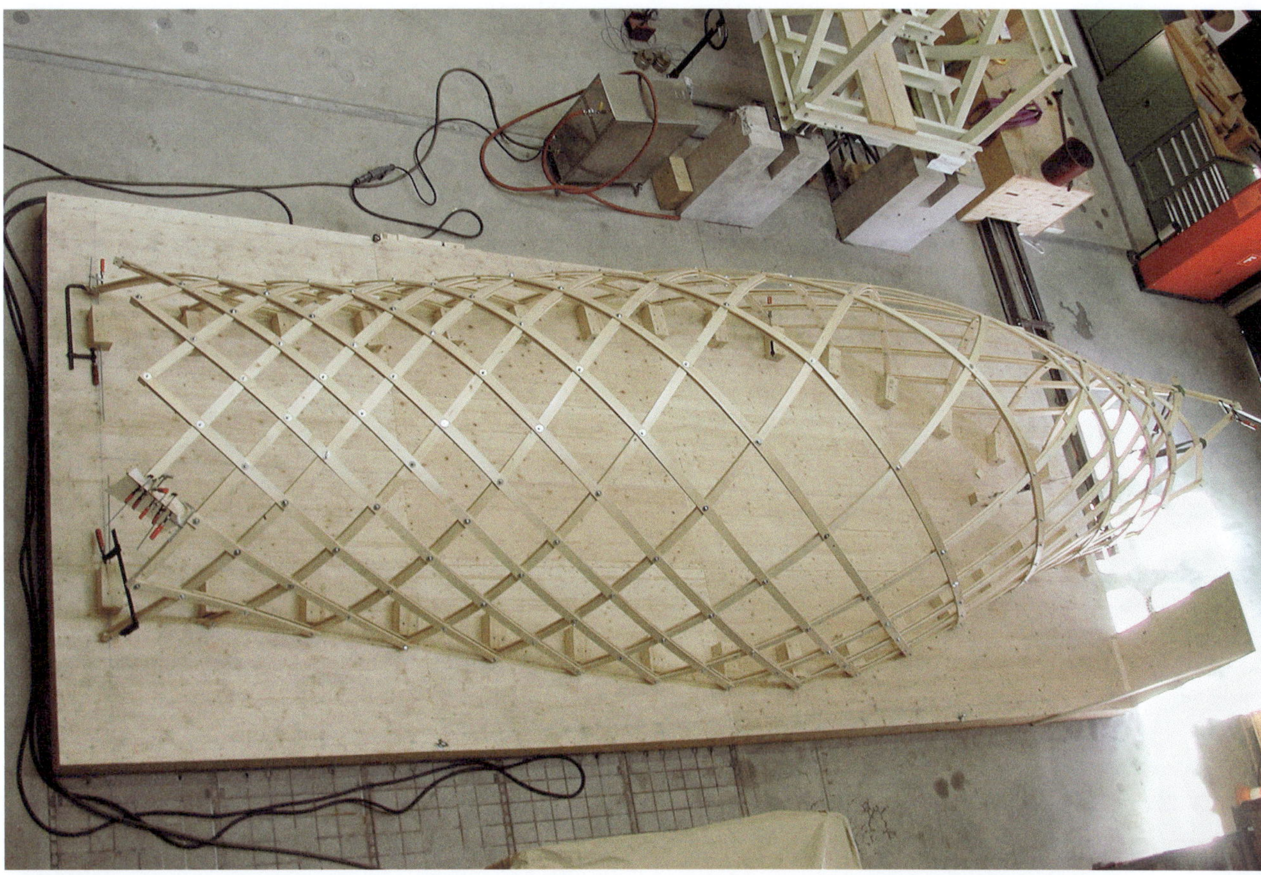

Fig. 10 a and b Construction of a free-form ribbed dome at IBOIS, EPFL, based on the use of geodesic lines on a free-form surface.

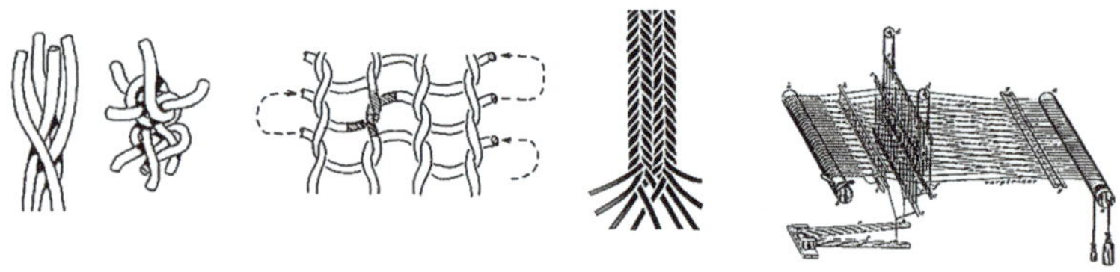

Fig. 11

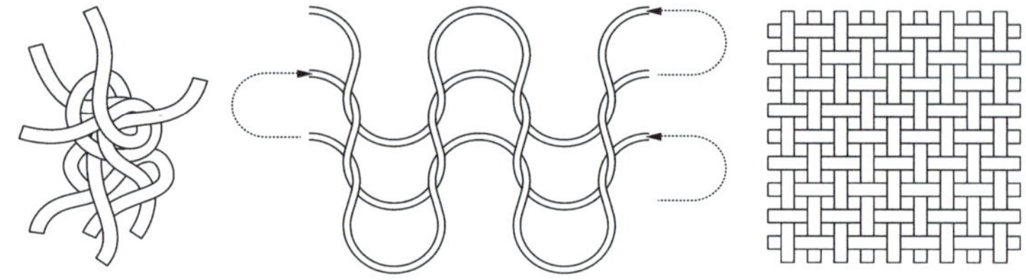

Fig. 12

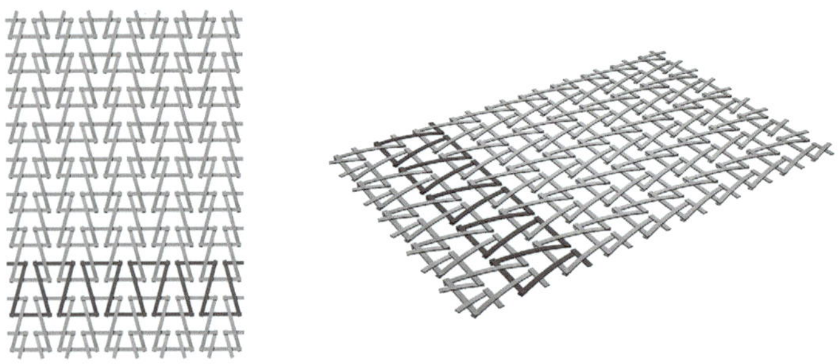

Fig. 13

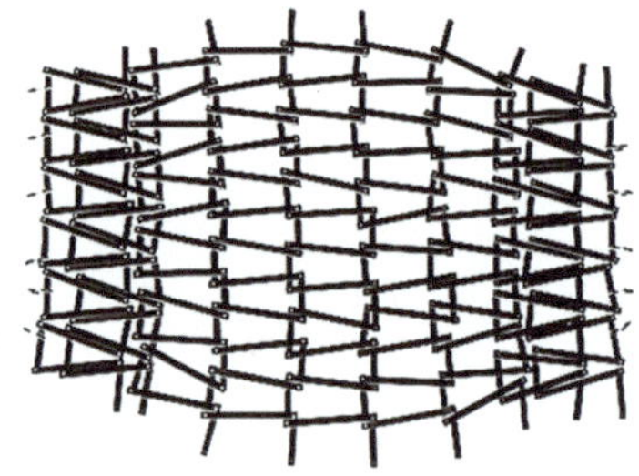

Fig. 14

Fig. 15

Fig. 11 Summary of the braiding and weaving patterns

Fig. 12 The three considerations are portrayed diagrammatically.

Fig. 13 Representations of basic planar braiding patterns

Fig. 14 Spatial representation of basic braiding patterns

Fig. 15 Representation of a stave mill, drawn from the basket weave.

Fig. 16 a and b Reciprocal system: actively bent weave

Fig. 17 Fish Market in Tokyo: woven roof structure Kerto plates; span 21 m; Atelier Weinand /EPFL 2009. Student: Sophie Carpienteri

Fig. 18 Student project from the workshop "woven structures," EPFL, 2008

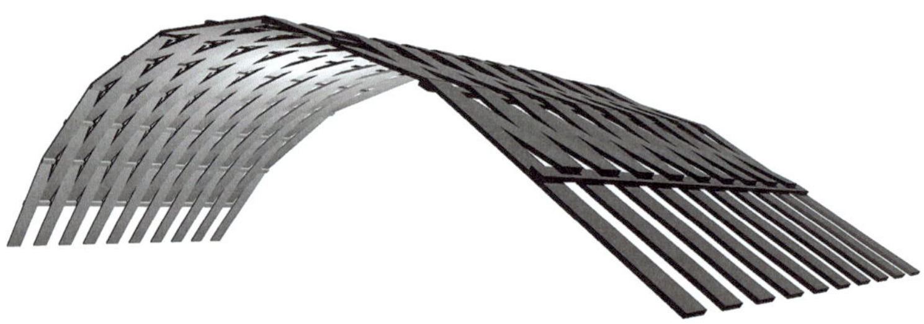

Fig. 16 a

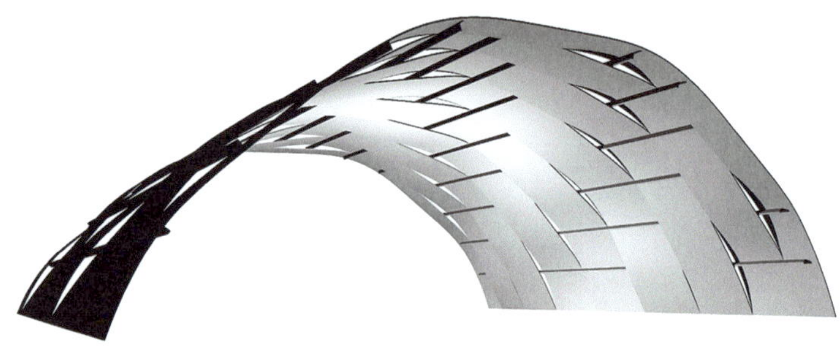

Fig. 16 b

Fig. 17

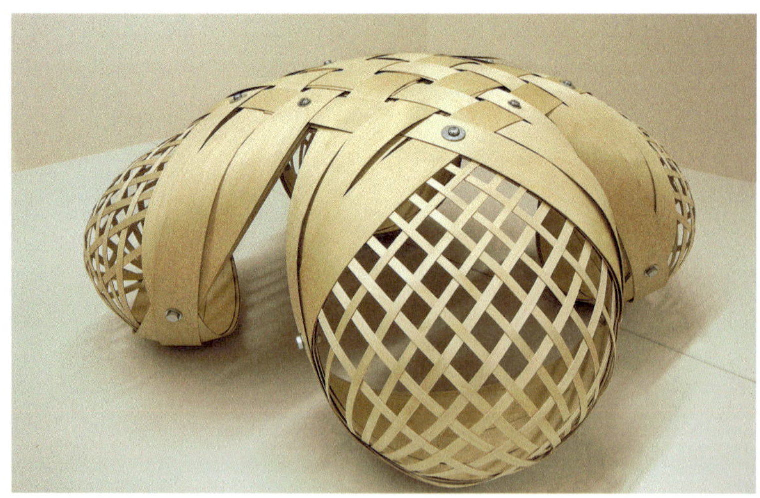

Fig. 18

Fig. 19 a

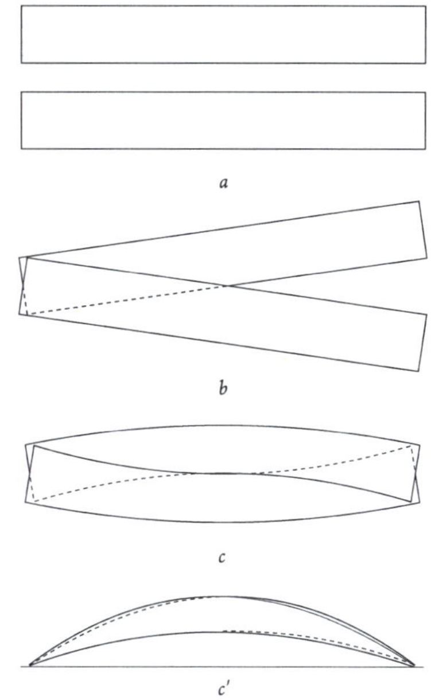

Fig. 19 b

Fig. 20 a

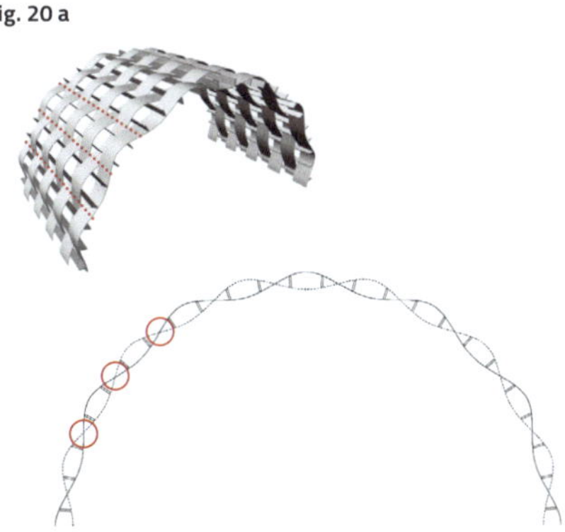

Fig. 20 b

Fig. 20 c

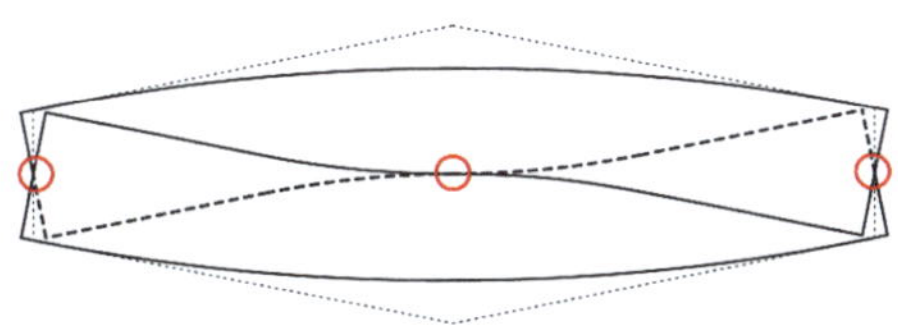

Fig. 21 a

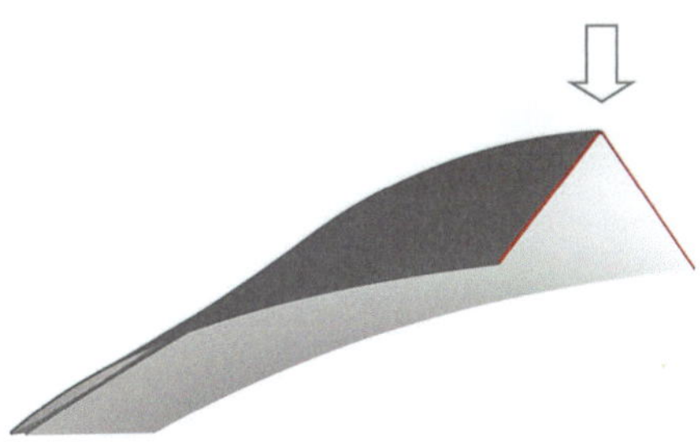

Fig. 21 b

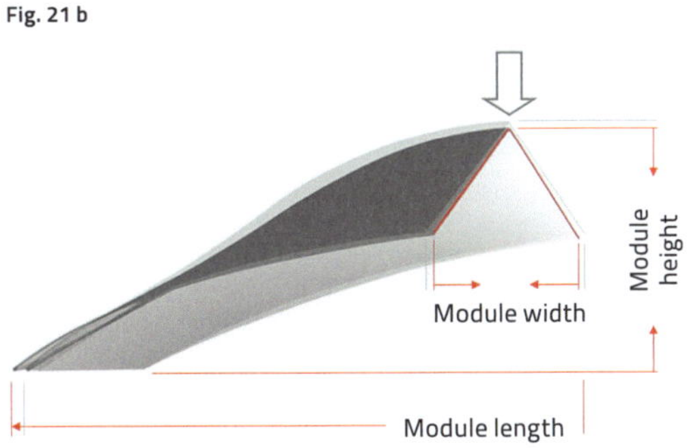

Fig. 21 c

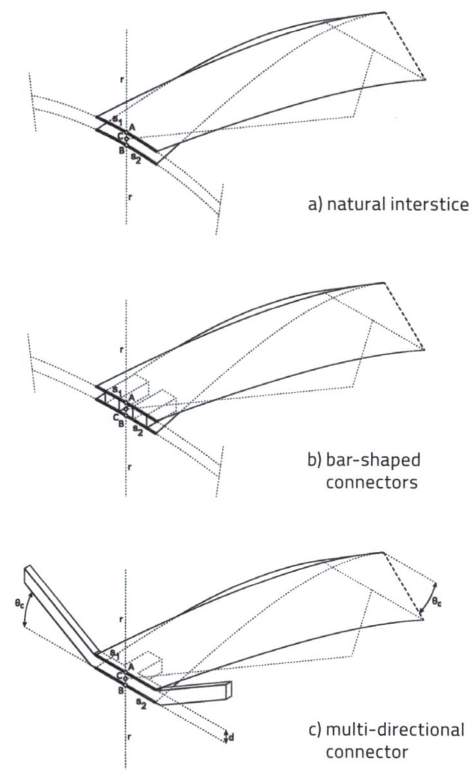

Fig. 22

Fig. 19 a and b Textile module, EPFL, 2008

Fig. 20 a, b, and c Superimposition of nodal points and offset of network elements in relation to one another

Fig. 21 a, b, and c Illustration of static behavior: fixed or contact points, applied force increases structural height

Fig. 22 Geometry of fabric module

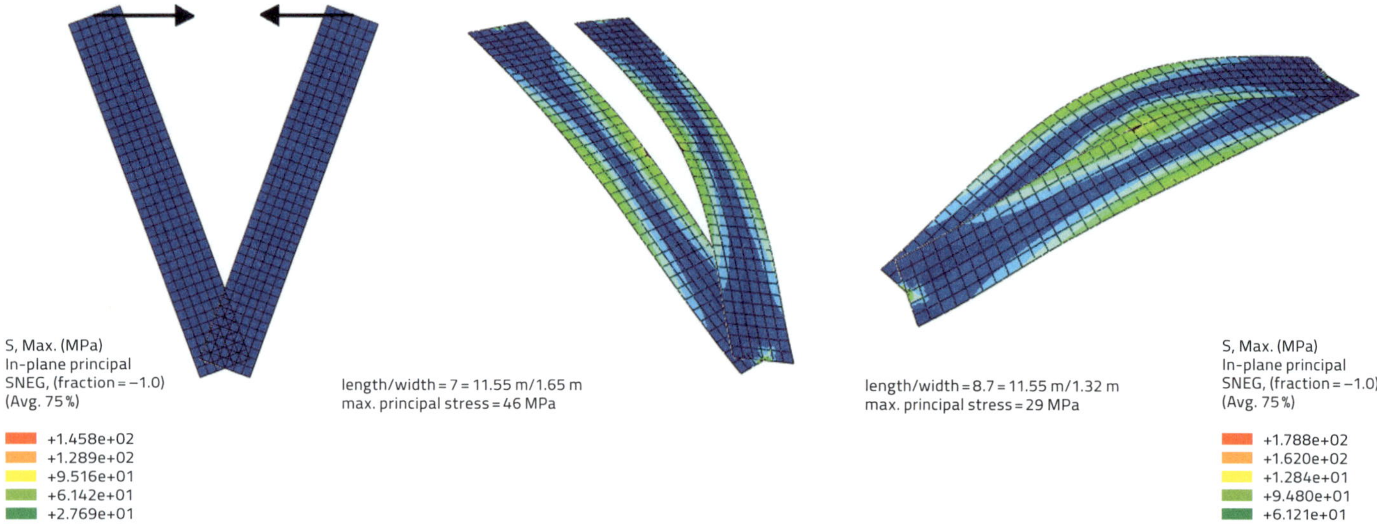

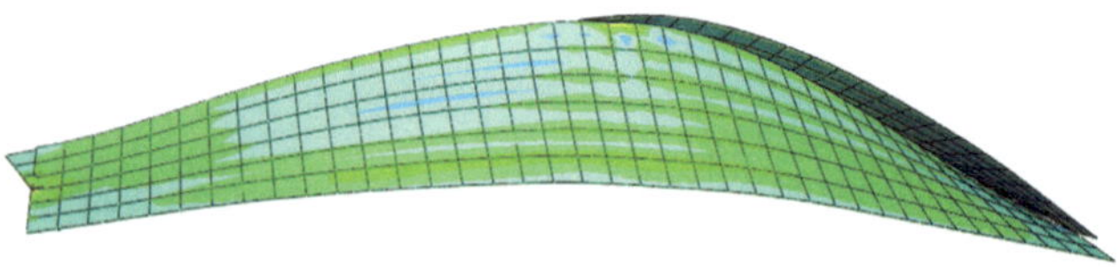

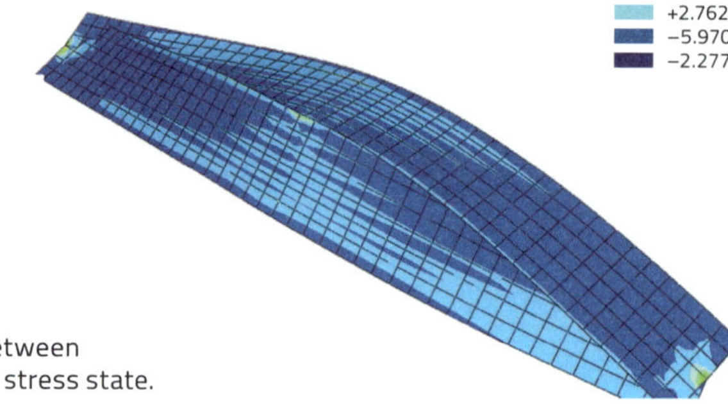

Fig. 23 Finite element model of the active connection of a fabric module. Comparison between two column widths: 1.65 m and 1.32 m. The narrower the plate, the lower the residual stress state.

Fig. 24 The calculated values on a physical model can be verified with the aid of a test form.

position, then the panels are forced into an inclined position by mutual bending. The panel reaches a certain height and forms a kind of hat or triangle in cross-section. Fig. 21 a shows the contact points in plan view. The cross-section described is superimposed in red in fig. 21 b and c. If an external force is added to this bent active module, then it increases in length and contracts in cross-section.
Thus the cross-section gains height, and therefore structural height. From an engineering perspective, it is a kind of active system that increases in rigidity when the external force is increased.

The geometric position of the nodal points and the panels in relation to one another directly determines the final shape of the fabric module and its stress state. The contact conditions shown in fig. 1 transform into peripheral conditions if a layering of textile modules is envisaged. This observation concerns the primary weave direction and also leads to the emergence of the cross weave direction.

Fig. 2 illustrates this cross weave, which is created with linear slats that serve at least two functions:

- Rigidity of the woven structure in depth
- Peripheral conditions of a module

3.2

Geodesic lines on free-form surfaces—optimized grids for timber rib shells

Claudio Pirazzi and Yves Weinand

In order to optimize grids of timber rib shells with regard to the bending stress of the boards due to initial curvature, GEOS software was developed at the Swiss Federal Institute of Technology in Lausanne (EPFL) between 2002 and 2004. The construction of a prototype in the summer of 2005 proved the reliability of the assumptions upon which the program is based.

The following article describes the steps that were taken, starting from the generation of the form and the design of the optimized grid to the final construction of the prototype. Loading tests were carried out to evaluate the structural calculation model. Finally, the comparison between calculated and measured deformations will be discussed briefly.

Keywords ***rib shells, active bended structures, geodesic lines, multi-layered beam structure***

1 Introduction

Although concrete and steel are the most common materials for the construction of lightweight spatial structures in modern architecture, timber has recently had a well-deserved renaissance. In addition to glue-laminated timber, screw-laminated timber has increasingly been applied to rib shells. During the past two decades several spatial structures of this type were constructed.[1,2] The transparency of their bearing behavior and their aesthetic architecture fascinate experts and laypeople alike.

The ribs are made from laminated timber boards, which are joined together with the aid of pin-like fasteners, like screws or nails. In contrast to other rib shell structures, which were made from square timber sections,[3,4] relatively thin boards were used with a thickness of between 16 mm and 35 mm. These laths are inexpensive and construction can be executed in a relatively simple way without sophisticated techniques. The extra cost due to manual labor for the assembly can be compensated for by rationalization methods during the planning and manufacturing process. For geometrically demanding structures this method of construction is a viable alternative to glue-laminated timber rib shells.

In order to reduce the stress due to initial curvature, the ribs on the surface are arranged according to geodesic lines. Thus, bending of the boards across the strong axis, which causes unfavorable stress, can be avoided. Ideally the boards are only subjected to bending and torsion across their weak axis. In addition, this approach allows the use of straight boards. A geodesic line on a surface is defined as a curve, where the normal vector of both curve and surface are parallel or non-parallel at each point. The shortest distance between two points on a surface is always a geodesic line. In plan, a geodesic line represents a straight line. The term "geodesic" is derived from the Greek ge (earth) and daiesthai (to divide).

Geodesic lines on simple, regular-shaped surfaces can be determined by analytical means. Geodesic lines correspond to helices on cylindrical surfaces and to great circles on spheres. On free-form surfaces—currently enjoying great popularity in contemporary architecture—

the determination of geodesic lines is far more complex. In order to satisfy this demand and to improve automation of the production process, GEOS software was developed in close collaboration with the Laboratory of Timber Construction (EPFL/IBOIS) and the chair of Geometry (EPFL/GEOM). This software calculates grids of geodesic lines on free-form surfaces and provides all geometric data necessary for computer-controlled sawing. The project was financed by the Swiss National Science Foundation (SNF).

In order to examine the reliability of this program and its precision concerning the assumptions made, a free-form shaped timber rib shell prototype has been designed with GEOS and constructed at the IBOIS. The realization of this prototype was financed by the "holz 21" fund, a research program at the Swiss Federal Environmental Agency (BAFU).

2 Form design and calculation of geodesic lines

The form is defined by manipulating control points of cubic Bézier-polynomials within a certain user-defined number of parallel cutting planes along the surface. A Bézier surface is calculated based on these polynomials. Starting points and end points of each geodesic line and the connectivity of the lines have to be defined for the calculation of the grid **(Fig 1)**. The iterative calculation of the geodesic lines is based on a L-BFGS-B algorithm. The tangential vector is used as a control variable. This quasi-Newtonian targeting turned out to be the most efficient method for calculating geodesic lines on mono-patched, free-form surfaces.[5]

The user has no direct influence on the topology of the resulting grid, but can manipulate the starting points and end points and define the connections of the geodesic lines. Thus, the result is not always compatible with structural and architectural demands like the regularity of the grid or the minimum radius of curvature of a single board. It has to be adapted by an iterative process. In general it can be stated that the spacing of the meshes tends to widen in significant convex areas, whereas it tends to narrow in concave hollows. However, the existence of geodesic grids with homogeneously spaced meshes on complex free-form surfaces is not always evident.

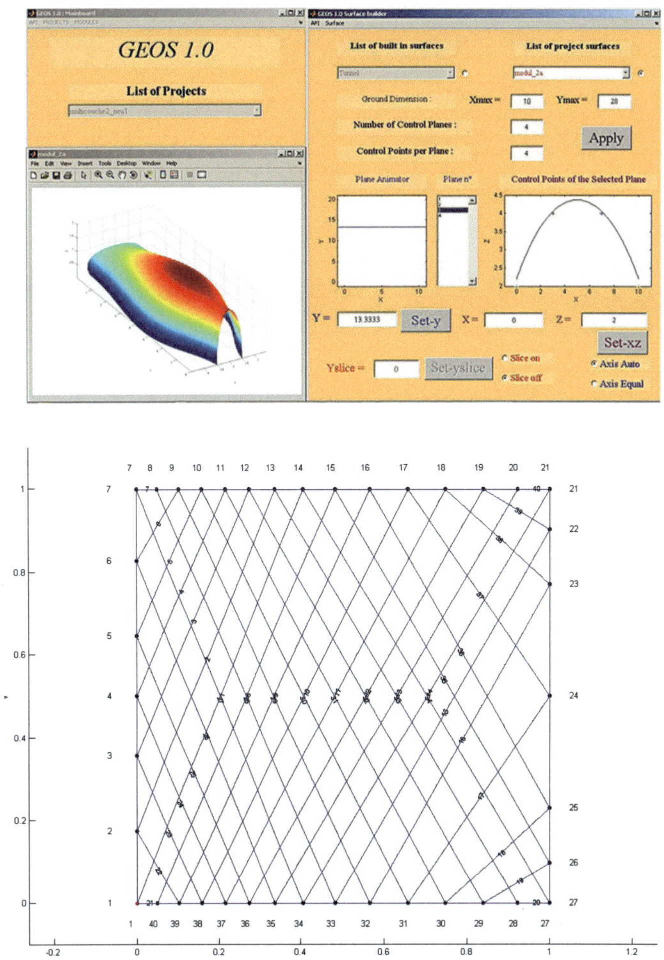

Fig. 1

Fig. 1 Software GEOS; design of the free-form surface by means of control polynomials (above); definition of starting and end points of the geodesic lines and their connectivity in a u,v-parameter range (below)

Fig. 2 Design of the timber rib shell prototype; generated mono-patched free-form surface (left); geodesic line model (center); multi-layered model (right)

Based on the calculated grid of geodesic lines, a multi-layered model is developed according to the number and the dimension of the laths **(Fig 2)**. This is achieved by extrapolating the layers normal to the surface inwards and outwards. The midlines of the extrapolated layers no

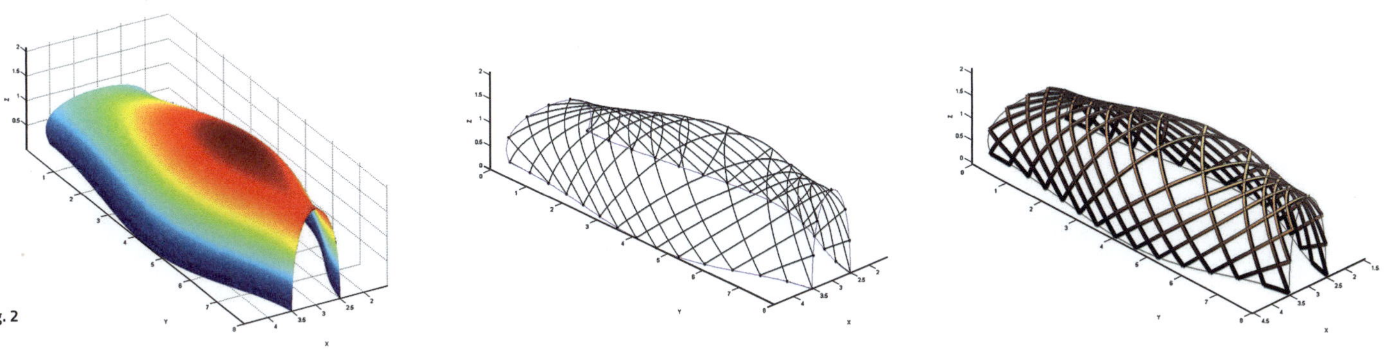

Fig. 2

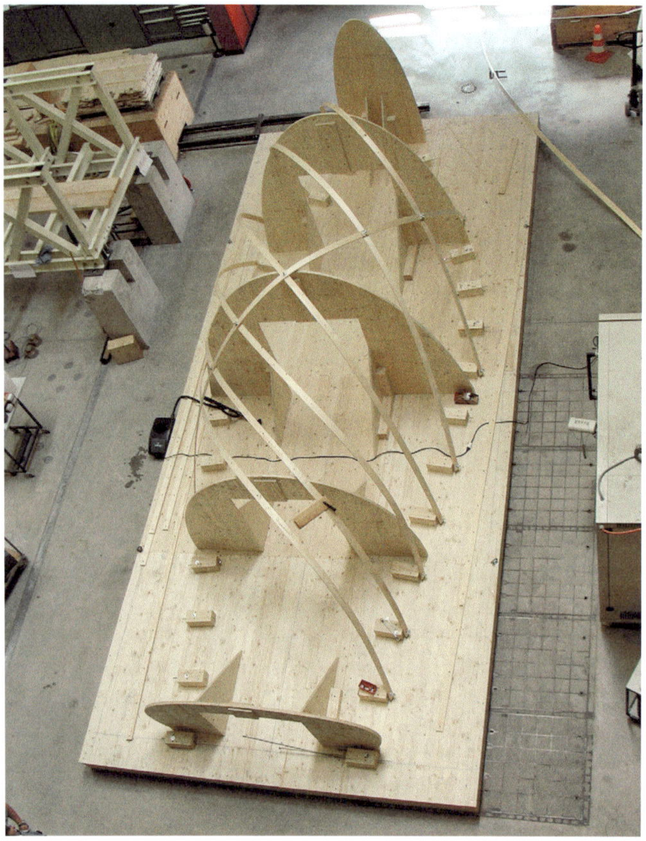

Fig. 3

Fig. 4

longer correspond to geodesic lines. They are neither part of the initial surface, nor do they represent geodesic lines for the enveloping surface to which they belong. However, the error resulting from this inaccuracy seems to be negligible with regard to the relatively large ratio of the effective radius of curvature R_0 and the thickness d of one single board, usually applied in practice. The smaller this ratio is, the more important the inaccuracy becomes, with the risk that the boards of a rib will no longer be parallel and that the predrilled holes at the intersections will not fit precisely. Therefore, the front section of the prototype has been designed with a relatively small radius of curvature, with a minimum ratio R_0/d of 100.

3 Construction of the prototype

Thin timber laths made from Swiss spruce wood (Picea abies), with a rectangular section of 12 × 60 mm were used for the construction. Due to the tight curvature of the front part of the shell, high quality wood for the laths is imperative. Careful visual grading guaranteed an excellent strength class with only a few knotholes. The modulus of elasticity was determined for five random samples. The average value is 14,800 MPa. The ribs consist of four laths: two of them are continuous and the remaining two are considered as intermediate in-fill layers. Continuous layers of diagonally crossing ribs intersect them. Both directions are offset by the thickness of one lath. No finger joints were required. The longest lath is 6,740 mm and the shortest is approximately 160 mm. In total, 792 pieces were mechanically sawn with a five-axis, computer-controlled Créno saw. The data was exported from GEOS in dxf. format.

Beginning with the inner layer, the laths are fixed at their starting points before they are curved across scaffolds and connected to the corresponding end points. The five transverse scaffolds have been planed in order to ensure a maximum number of fixed points in space. In general, these supports are unnecessary because the expected form is automatically obtained by successively connecting the laths at their intersection points by means of bolts (10 mm diameter). The continuous inner layers in both directions could thus be realized in a relatively short amount of time, followed by the installation of the two continuous outer layers **(Fig 3)**. In general, it can be stated that even after assembly of the fourth layer, the predrilled holes at the intersections fitted precisely about 85 percent of the time. This confirms the reliability of the calculation and shows that the required precision is kept even for significant curvatures.

Fig 3 Construction of the prototype

Fig 4 Intersection points before (above) and after (below) assembly of the intermediate in-fill layers

Fig. 5 Load tests; uniaxial displacement transducer (left); loading with two, single loads (center); load measuring (right)

The intermediate in-fill layers were mounted after placing all continuous layers in both directions. Screws with a diameter of 4 mm were used to connect the layers. The effective distance of the doubled-rowed fasteners is about 50 mm. Fig 4 shows a detail before and after the assembly of the intermediate in-fill layers. The in-fill layers are approximately 2 mm shorter than the distance between the continuous laths at both ends.

4 Data

Lath, rectangular section:	12 × 60 mm (602 running meters of high quality Swiss spruce)
Number of pieces:	792 (including intermediate in-fill layers)
Timber volume:	0.434 m³
Dimension of the shell:	8000 × 3000 × 2060 mm
Developed surface:	≈ 36.0 m²
Base:	≈ 18.6 m²
Number of bolts:	202 M10
Number of screws:	≈ 3000 ∅ 4.0 mm

5 Load tests

In order to get a better understanding of the load-bearing behavior of the structure, load tests were carried out before and after the assembly of the intermediate in-fill layers (construction phases C1 and C2). For both phases, three symmetric load cases with single forces on different intersection points were examined. The forces, introduced by a cable winch, were applied normal to the surface. Uniaxial deformation of a total of seven intersections around the charged points was measured (Fig. 5). The displacement transducers were adjusted so that they measured deformation normal to the shell's surface and parallel to the introduced load in the case of the charged intersections. The test results show that, for relatively low charges, the structure undergoes hardly any plastic deformation at all and reacts absolutely symmetrically. A distinct increase in stiffness could be shown after mounting the intermediate in-fill layers. This is not only true for load cases, where the structure shows flexional behavior, but also for load cases that are essentially transmitted by standard forces. Certainly, the flexional stiffness of the ribs increases considerably if the intermediate in-fill layers are mounted and screwed down. Thanks to the double-curved surface, the effect of the intermediate layers on the stiffness of the structure was estimated to be far less significant. In most cases examined, the deformations were halved by assembling the intermediate in-fill layers.

6 Structural analysis

The comparison of the measured deformations wmes with the calculated deformations wcal enabled the evaluation of the calculation model based on a framed load-bearing system. The structural analysis was carried out with SAP2000 V9.16 NL software from CSI-Berkeley. Based on the geometric data exported as a dxf. file from GEOS, the structure was programmed in an external text file and then imported. The generation of data by means of external text files allows one to carry out modifications in a relatively straightforward way and ensures a more general overview of the data.

The load-bearing behavior of the structure is fairly complex. The engineer has to take various effects into consideration. The extent of the effect on the structure is basically unknown. This makes modeling fairly difficult. Apart from the influence of the rotational stiffness of the ribs' intersections (distortion of the diamond-shapes in their plane), the initial slip modulus of the screwed connection is largely unknown. Due to different displacement behavior of the connection, the characteristic values of this modulus—empirically determined for standard connections and given in the technical specifications—cannot be applied without modification.[6] Preliminary tests were carried out to determine the initial slip modulus in this specific connection. Furthermore, the initial bending of the laths has a pre-stressing effect on the structure's stiffness. This effect—ideally favorable—is disregarded.

As the intersections of two ribs are not entirely rigid, both directions are modeled separately in the form of two substructures. The degrees of freedom are subsequently coupled by means of kinematic linking ("local-constraint" option). This allows the release of a single rotation about the local axis normal to the surface at each intersection and thus the simulation of the distortion of the diamond-shapes in their plane. According to the method of shear analogy[7–10] two subsystems—A and B—are introduced. This method is applied to the modeling of the construction phase C2 in order to take the flexible compound cross-section of the ribs into consideration. Subsystem C, introduced for stability problem by Scholz,[11] is neglected due to minor normal forces. Therefore, the model consists of four substructures, two for each subsystem (Fig. 6).

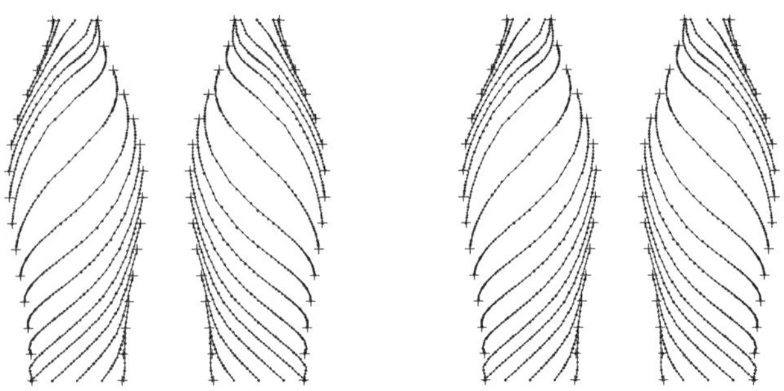

Fig. 6

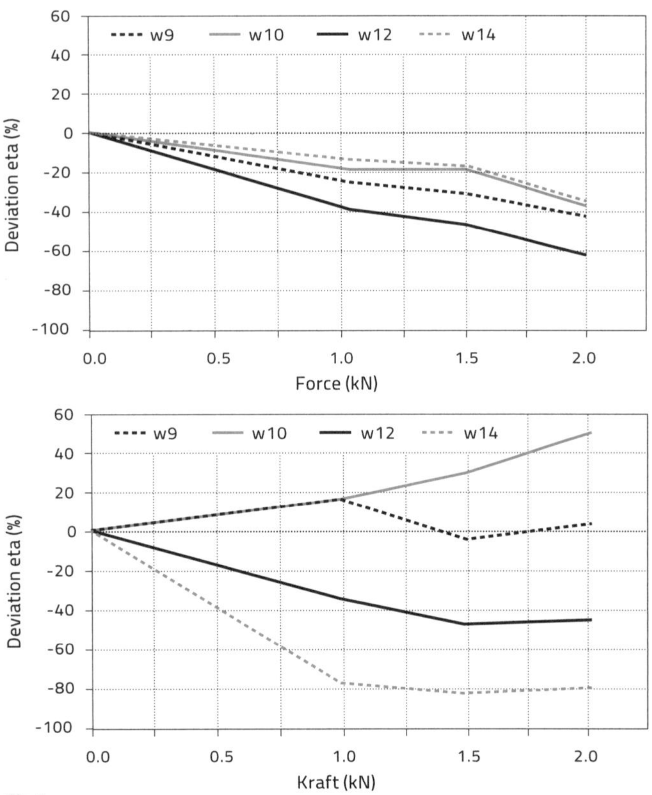

Fig. 7

Fig. 6 Four substructures used for modeling of the construction phase C2 in the floor plan. Linking was by means of kinematic constraints (not in view).

Fig. 7 Comparison between measured and calculated deformations of two load cases of construction phase C1 (above) and C2 (below). Directly charged joints are shown in dotted lines.

The following assumptions were made:

- Isotrope and ideal elastic (Hook) behavior of timber
- Geometrical non-linearity (P-delta effect)
- Linear slip behavior of the mechanical fasteners with an initial slip modulus Kser,1 of 500 N/mm, according to preliminary load-tests carried out at the IBOIS [12]
- Rotation about the frame 1-axis is considered to be constant between two intersections
- Initial state of bending stress is not taken into account
- No fixed restraints of the substructure B (slip between the laths is not hindered at the restraints)
- Torsion stiffness due to the build-up compound section is not activated
- Weakened sections at the intersections are not taken into account

For two demonstrated load cases of construction phases C1 and C2, the relative deviation between measured and calculated deformation is shown in fig. 7 (before and after fitting the intermediate layers). Observing the former, one can see that with an increasing load level, the deviation increases noticeably. This is due to nonlinear effects, which were not taken into account in the structural analysis. However, for small charges, a strong accord between calculation and reality (with relative derivations mostly below 30%) was observed. Even the comparison of the final construction phase C2—far more complex in its structural behavior and with additional compound stiffness—shows satisfactory accordance.

7 Conclusions

The GEOS software, developed to calculate grids of geodesic lines on free-form surfaces, was tested by the construction of a timber rib shell prototype. The precision of the calculated geodesic lines of the computer-controlled prefabrication of the laths and its assembly was found to be highly satisfactory and augurs well for an application of this type of structure at a larger scale.

Load tests were carried out in order to evaluate the structural computer model. For the observed load cases, the tests show that the structure is approximately twice as stiff after assembly of the intermediate infill layers. Good agreement between the measurement and the calculation, at least for relevant loads, has been observed. This is even more pertinent with regard to the complexity of the treated structure. Several parameters, such as the initial slip modulus of this specific connection or the level of rotation at the intersections (no in-plane shear stiffness) have not yet been completely clarified. Further reflection on the influence of these parameters on the structure's bearing behavior is necessary.

GEOS provides an important tool for the design and the realization of timber rib shells. It contributes to clarifying current uncertainties in the design planning process and will greatly improve the confidence of engineers and architects in conceiving and realizing this type of challenging lightweight spatial structure.

Literature

1 Natterer, J., N. Burger, and A. Müller. "The roof structure 'Expodach' at the World Exhibition Hanover." *Proceedings of the Fifth International Conference on Space Structures.* University of Surrey, United Kingdom, 2002. 185–193.

2 Natterer, J., N. Burger, and A. Müller. "Holzrippendächer in Brettstapelbauweise: Raumerlebnis durch filigrane Tragwerke" ("Screw-laminated Timber Rib Shell Roofs: The Experience of Space through filigreed Structures"). *Bautechnik,* 2000, Vol. 77. No. 11. 783–792.

3 Happold, E. "Timber Lattice Roof for the Mannheim Bundesgartenschau". *The Structural Engineer,* 1975, Vol. 53, No. 3. 99–135.

4 Kelly, O.J., R.J.L. Harris et al. "Construction of the Downland Gridshell." *The Structural Engineer,* 2001, Vol. 79, No. 17. 25–33.

5 Rozsnyo, R. "Optimal Control of Geodesics in Riemannian Manifolds." Doctoral thesis, EPF-Lausanne. Switzerland, 2006.

6 Pirazzi, C. "Zur Berechnung von Holzschalen in Brettrippenbauweise mit elastischem Verbundquerschnitt" ("On the Calculation of Screw-laminated Timber Rib Shells"). Doctoral thesis, EPF-Lausanne, Switzerland, 2005.

7 Ibid.

8 Kreuzinger, H. "Verbundbauteile aus nachgiebig miteinander verbundenen Querschnittsteilen" ("Structural Members made out of semi-rigidly connected built-up Sections"). *Ingenieurholzbau,* Karlsruher Tage 2000, Forschung für die Praxis Vol. 28, 2000. 42–55.

9 Kreuzinger, H. *Mechanically Jointed Beams – Possibility of Analysis and Some Special Problems.* International Council for Research and Innovation in Building and Construction, Working Commission W18. Venice, Italy, 2001.

10 Scholz, A. "Ein Beitrag zur Berechnung von Flächentragwerken aus Holz" ("A Contribution to the Calculation of Spatial Timber Structures"). Doctoral thesis, TU Munich, Germany, 2004.

11 Ibid.

3.3

Apparent simplicity and modular complexity in timberfabric structures

Markus Hudert

The research work "Timberfabric: Applying Textile Assembly Principles for Wood Construction in Architecture" explores the interplay between assembly procedures and the properties of the involved components as well as the potential of this interplay as a generative factor in architectural and structural design. More specifically, it examines how principles of textile assembly techniques, as applied to elastically deformable timber elements, can be employed for the development of an innovative unit based timber construction system. The starting point in this is the so-called Timberfabric Module which consists of two interlaced strip-shaped elements and in which various textile properties are combined. Through the combination of several such modules, different support structure configurations of varying complexity can be generated.

Keywords ***experimental timber construction, assembly and material driven design, material computation***

1 Introduction

This paper is based on the doctoral research "Timberfabric: Applying Textile Assembly Principles for Wood Construction in Architecture,"[1] which the author undertook at IBOIS between 2007 and 2013. First, this research explored the interplay between assembly processes and the properties of the components involved, and second, the potential of this interplay as a generative factor in architectural and structural design. More specifically, principles of textile assembly techniques were examined—in combination with elastically deformable timber elements as components—to try and find out how they could be employed for the development of an innovative unit-based timber construction system.

With the above-mentioned goal, this research adopts an empirical approach that is subdivided into three parts. The first part determines the timberfabric module as a basic unit for the modular structural system; this is the developmental goal. The second part examines the properties of this module and the implications of its large-scale production. The final section, which forms the basis of this contribution, systematically explores different possibilities of combining multiple modules into more complex structures, and examines how the connectivity between their components can be established and optimized.

The production of physical models and prototypes plays a crucial role throughout the process. Physical modeling allows direct measurement of the material's elastic deformability and the impact of the proportions of the basic elements. Furthermore, it contributes to a general understanding of the geometric foundations of the developed structures, as well as of their mechanical properties. The models and demonstrators resulting from this process are examined with regard to a series of select evaluation parameters. The overall impact of transferring textile assembly principles is one of the issues discussed. Analogies and differences between the developed and actual textile structures are identified. The insights derived provide the basis for a modular construction system for self-supporting building envelopes. It is shown that the interplay between material properties and assembly techniques not only generates a specific architectural form of expression, but also contributes to the spatial and structural qualities of buildings.

2 The basic unit of timberfabric structures

Timberfabric structures are modular, wooden constructions that can be configured with varying levels of complexity. Their basic unit is known as the timberfabric module. This module was developed based on the unit cells of woven fabrics, as well as on the principle of arranging elements in a helical order, as is applied in yarn twisting and some forms of braiding. The unit cells of woven fabrics are comprised of two interlaced sections of yarn and are of interest because of their modular qualities and the reciprocal relationship of the components involved. The timberfabric module is likewise composed of two interlaced lamellar elements that support each other in a reciprocal manner. Taking inspiration from the techniques of yarn twisting and braiding, these two components are assembled to achieve a helical disposition. Along with the components' properties, this manner of assembly generates the overall shape of the module.

3 Multi-module configurations

Based on the geometrical and mechanical properties of the timberfabric module, different strategies of how to combine several modules into more complex structures were developed. By systematically applying these strategies, various kinds of one-, two-, and three-directional configurations were generated. Although the first two configuration types could be implemented as structures in their own right, the third variety has the greatest potential as applications for load-bearing building envelopes: the use of multiple layers not only provides an improved structural performance, it also permits greater control of the structure's degree of transparency and thus the quantity and quality of incoming daylight. As mentioned above, the timberfabric module is the basic unit in all cases. Hence, the factors that influence the geometry of an individual module are likewise relevant for the overall geometry of multi-module assemblies.

4 One-directional timberfabric

Different methods can be employed to combine several modules into one-directional assemblies. The resulting assemblies are three-dimensional and have a width that is relatively narrow in relation to the length of the structure. With regard to textile assemblies, one-directional timberfabric can be compared to twisted fibers in yarn, or certain types of braided textiles, where the latter "[...] are considerably greater in length than in width or in diameter."[2] However, by definition,[3] braided textiles involve at least three yarn elements, while the timberfabric module, which is the basic unit for one-directional timberfabric, consists only of two elements. Another analogy can be found in the twisted fibers in yarn. The development of the one-directional timberfabric is an intermediate step towards two- and three-directional structures, where the one-directional structures act as higher-order units.

4.1 Geometry and structural performance

In an individual timberfabric module, the greatest cross-sectional height is located at the module's center, whereas the lowest height is located at its extremities. In assemblies with two or more modules, lower cross-sections are not only located at the support points, but become part of the arch-shaped structure. As a result, the structural height of an arch-shaped assembly such as this varies between zones of higher and lower strength. In one-directional configurations, additional elements are required to connect of modules in the longitudinal direction. In addition to their function as connectors, these elements also have the potential to reinforce the zones of lower strength at the modules' extremities.

5 Two-directional timberfabric (2DTF)

Two-directional timberfabric structures are produced by aligning several one-directional units perpendicular to the span. Two types of one-directional units can be used that are inherently either able to link or unable to link in the lateral direction. Regarding their connectivity and coherence, more requirements arise in two-directional configurations.

5.1 Additional connectors in the lateral direction

In interconnected configurations, coherence in the lateral direction is an intrinsic property. In two-directional timberfabric based on self-contained assemblies, supplementary connectors are necessary in order to establish a coherent structure. These can be designed as independent connectors in addition to those that establish connectivity in the longitudinal axis. Alternatively, according to a more inclusive approach, these connectors can be employed to fulfill both tasks simultaneously.

5.2 Alignment of laterally adjacent modules

The manner in which laterally adjacent modules are aligned is relevant for the ways they can be connected. Two options relating to the properties of the timberfabric module are considered, where one stronger and two weaker zones were identified. Due to their geometry[4], these two weaker zones are also the most suitable positions for connector elements in the lateral direction to the span. Resulting from these considerations, two possible constellations of laterally adjacent modules can be envisaged: either fully aligned or rotated in relation to one another by the length of half a module.

5.3 Minimum distance in a lateral direction to the span

In a lateral direction to the span, the minimum distance between the central axes of two modules corresponds to the measurement of their width, which is determined by the dimensions of the basic panels. In this case, the two structures touch each other at their lateral edges. To avoid this, the minimum distance is set to twice the width of the components used in the modules.

In interconnected structures, this distance is defined by the length and angle of the lengthening elements, whereas the angle is in turn determined by the geometry of the timberfabric module. The geometry depends on the connection detail and the proportions of the panels. The global geometry of each of the variants of two-directional timberfabric presented here can be approximately described as the linear extrusion of a cross-section profile. However, one can also imagine two-directional structures that are extruded along a curved line. This could be achieved if the individual arch-shaped units were based on tapering components. As a secondary effect, the global cross-section of such structures would have a non-continuous curvature.

One example of a two-directional configuration is shown in the following figures. This configuration is comprised of mutually rotated, self-contained, higher-order units and is implemented with V-shaped connector elements.

6 Observations on one- and two-directional timberfabric

A number of observations can be made regarding one- and two-directional timberfabric, some of which are listed below.

Fig. 1

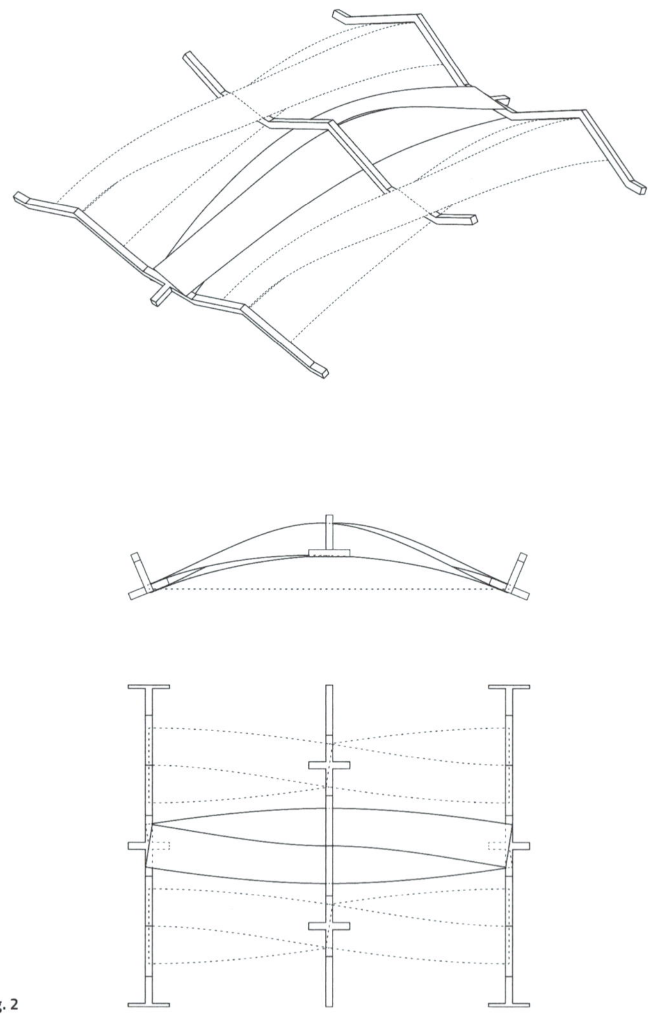

Fig. 2

6.1 Self-contained and interconnected configurations

If a two-directional structure is based on repeat units that are exclusively composed of modules, additional connector elements are required to establish coherence in the lateral direction. In structures with laterally connectable units, this is not the case. However, connector elements in a lateral direction can be employed as reinforcing elements.

6.2 Number of connector types

Connectors in the lateral direction can be designed as a second type of connector in addition to those that establish connectivity in the direction of the span. Alternatively, a single type of connector that is able to fulfill both functions is conceivable. Each approach has implications for the assembly procedure and sequence.

6.3 Textile aspects and analogies

One-directional structures can be considered as repeats. Similar to woven textiles, a warp and a weft direction can be identified in two-directional structures. Expandability is perpendicular to the span direction. This can also be considered as analogous to active and passive "thread" systems. Although one-directional build-ups can be used as freestanding structures or as structures in themselves, they are more often considered as intermediate structures that define or describe the character of two-directional structures. In that sense, they can be interpreted as a physical implementation of a repeat. Repeats are graphic representations or descriptions of the pattern of a woven fabric.

Fig. 1 Prototype of a two-directional configuration with multi-directional connector elements

Fig. 2 Connection between modules in longitudinal direction

Fig. 3 Connection between modules in lateral direction

Fig. 4 In this study, the principle of multi-directional connectors is transferred to a three-directional configuration.

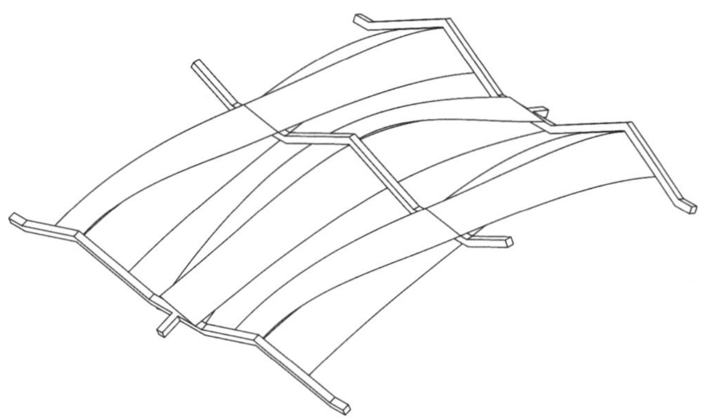

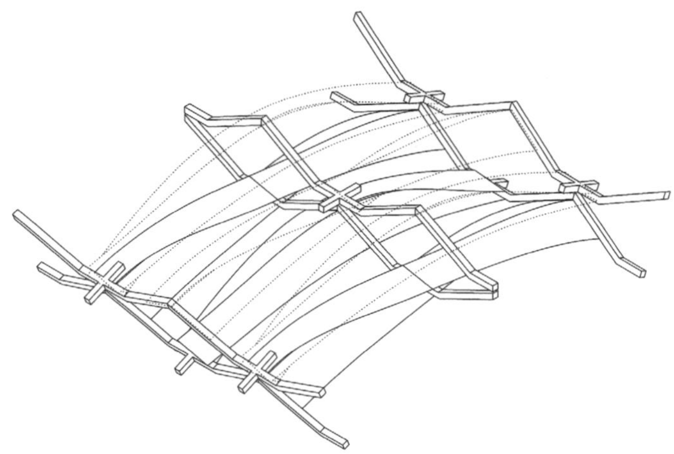

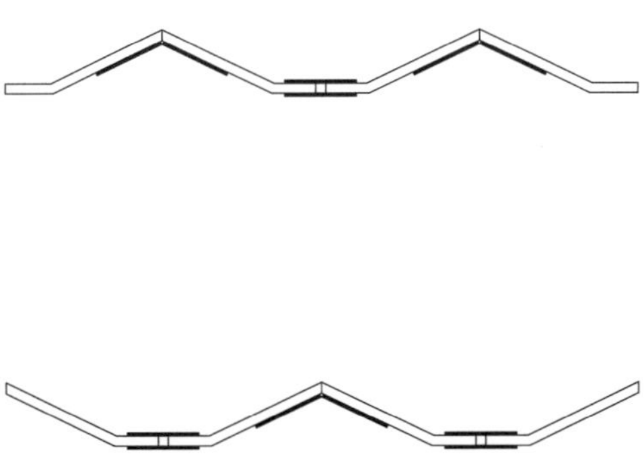

Fig. 3

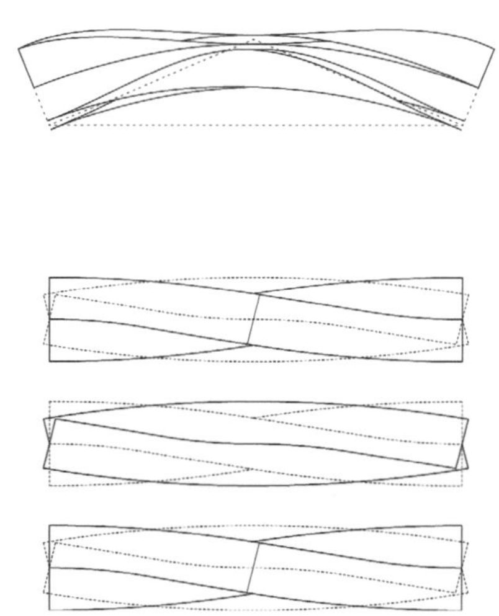

Fig. 4

Fig. 5

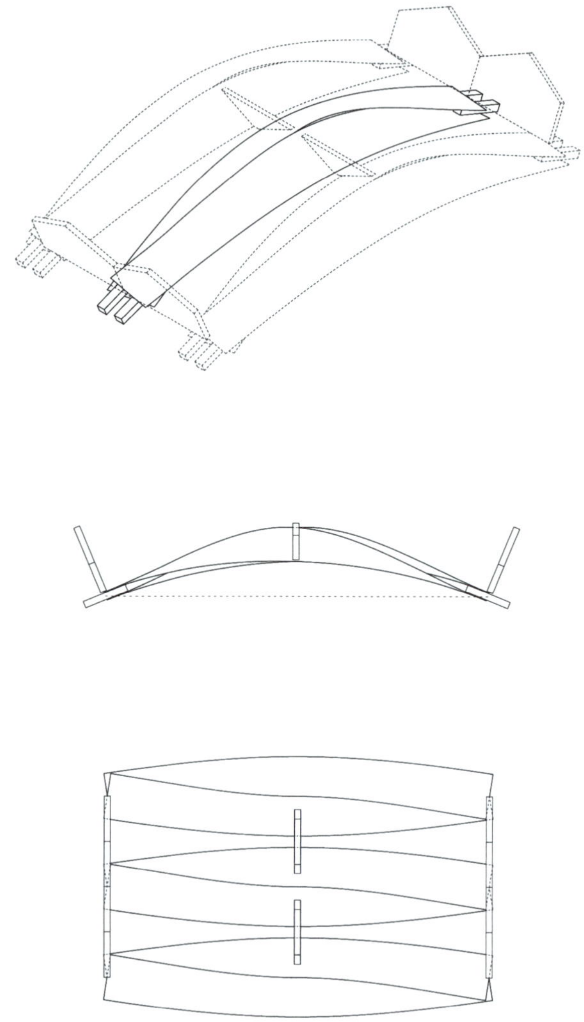

Fig. 6

Fig. 5 Intermediate-scale prototype of three-directional timberfabric, configuration study one, photographed near completion

Fig. 6 Connection of modules in longitudinal direction

Fig. 7 Connection of modules in lateral direction

Fig. 8 Connection of modules in radial direction

7 Three-directional timberfabric

The previous section showed how multiple modules and their basic components can be combined into one- and two-directional timberfabric assemblies. For two-directional assemblies, two modes of aligning laterally adjacent modules have been identified with regard to potential additional connector elements in a lateral direction. In three-directional timberfabrics, these principles are taken to the next level by superposing and connecting two layers of two-directional timberfabric of identical or different construction. To make this possible, the exterior layer needs to be based on modules with a larger radius or apothem. Several other factors that need to be taken into account are discussed below.

7.1 Offset between layers

In all variants except one, the superposed layers circumscribe a cylindrical segment and have a common central axis (that runs in the x-direction) while the exterior layer is perpendicular to the interior layer. The offset between the two layers is achieved by increasing the size of the modules (the elements' width is neglected here) that are designated to be on the exterior layer.

7.2 Minimum offset

Technically, the value of the offset can be freely chosen. However, several aspects have to be considered in the process. If the offset is too small, the units cannot be rotated about each other, as this would cause a collision between the low-lying extremities of the upper module and the prominent center of the lower module. As a self-imposed rule of design, the offset has to be large enough to allow for two modules to be superposed while being mutually rotated against each other by half a module.

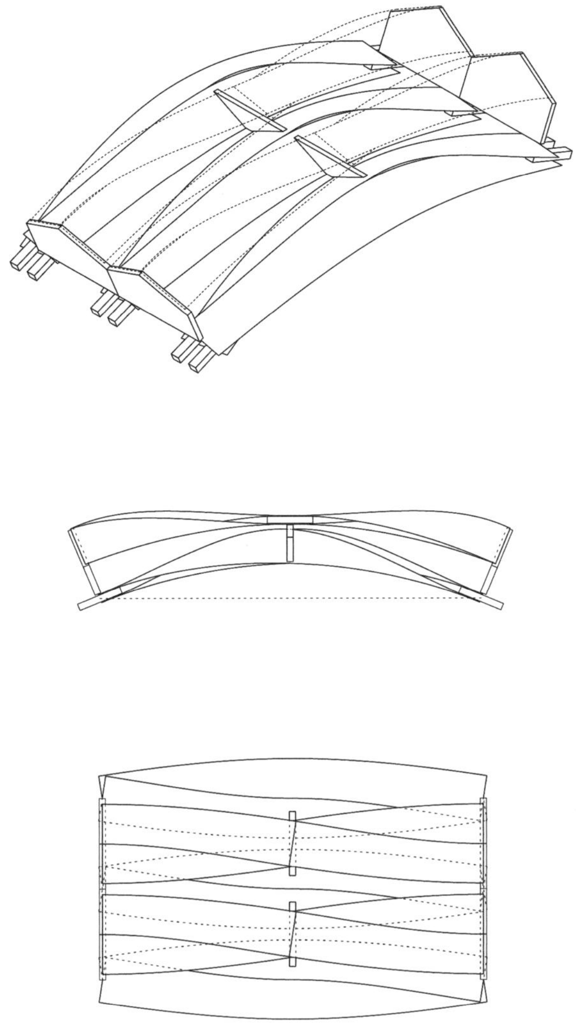

Fig. 7

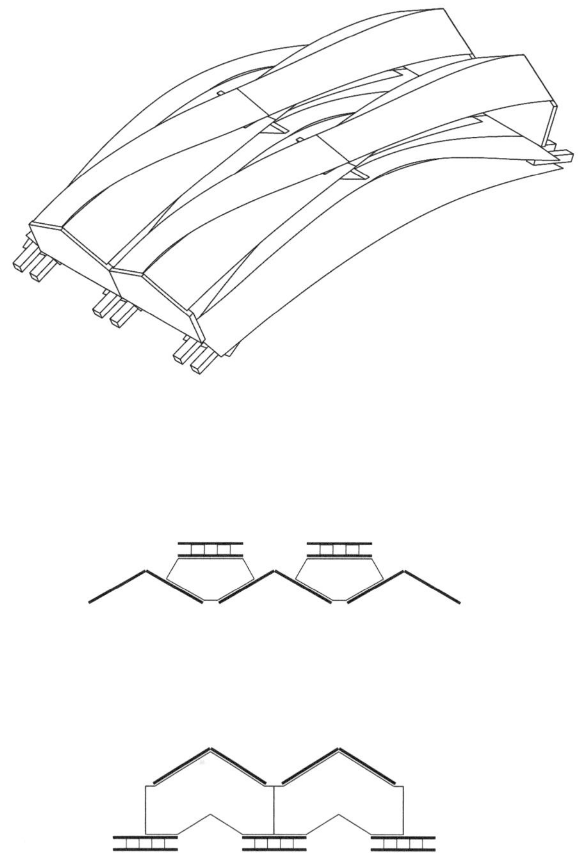

Fig. 8

7.3 Maximum offset

In principle, there is no upper limit for the offset distance between two layers. One could even argue that a greater distance is desirable as it results in a higher structure overall. On the other hand, the combined height of the two layers is only effective if they work as an entity, which in turn depends on the connections between them. Among other considerations, these connections have to be moment-resisting, in order to prevent relative displacement of the layers in the span direction in the case of horizontal loads, as such a displacement lowers the overall rigidity of the structural system. Greater distance leads to greater leverage and hence also increases the demands on the connections' resistance. Furthermore, increasing the distance also has a visual impact: it decreases the visual coherence of the two layers, which to a certain degree can be understood as analogous to the structural logic.

7.4 Configuration principle of vertically adjacent modules

The character of the layers and their alignment relative to one another is defined with regard to their constituent modules. Five basic principles of alignment have been established and form the basis for the timberfabric configurations subsequently developed. The alignment principles are conceived in order to provide for connections between the low- and high-strength zones of the modules.

8 Timberfabric demonstrators

As part of the research, various configuration studies of three-directional timberfabric have been carried out, two of which are discussed below.

In this configuration, the layers are mutually rotated around O and shifted in the x-direction. The interior layer is built up of a series of identical arch-shaped units of higher order that are placed next to each other. The center-to-center distance of these units in the x-direction corresponds to twice the width of the modules' constituent elements. Each of these arch-shaped units consists of three entire modules. The exterior layer is composed of a series of units similar to those of the interior layer. The units of the exterior layer have an overall length of three modules. Unlike the units of the interior layer, they consist of two plain and two halved modules, where the plain are located in the center of the unit and the halved are located towards the support points.[5] The module geometry and the distance between the one-directional assemblies together determine the shape of the connectors' cross-sections. The interstice between the two layers and the constellation of their vertically adjacent modules make a dual connector approach more appropriate than an integrated one. Two different types of connectors are used for longitudinal and lateral connections. The overall configuration can be varied without being constrained by the connectors.

The approach taken in this study of twisted superposition is to use the same pattern on both layers. The overall construction is relatively simple. Each layer consists of a sequence of independent "interbraided" arches. The arches of the interior layer equate to a length of three timberfabric modules and sit perpendicular to the edges of the structure. The exterior layer is also composed of a sequence of "braided" arches, again with an arch-length of three textile modules. However, this layer's arches are built up quite differently from the interior modules. Each arch starts with a half timberfabric module, followed by two complete ones, and ends with a half module again. This approach results in a shift-

Fig. 9 Large-scale implementation of three-directional timberfabric, configuration study three, view from above

Fig. 10 In the longitudinal direction, the modules are connected by disc-shaped elements.

Fig. 11 In the lateral direction, the modules are connected by two variants of custom-made connectors: their shape results from the interstice between the interior and the exterior layer.

Fig. 12 Due to their customized geometry, these elements can likewise establish the connection between the two layers.

Fig. 9

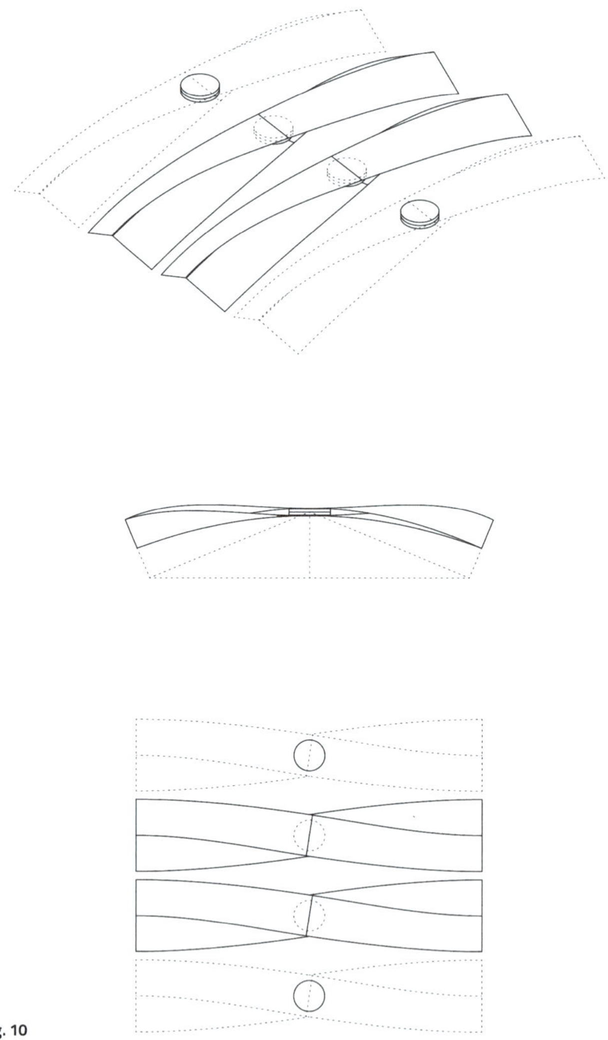

Fig. 10

ed pattern on the exterior layer and allows the connection of weaker and stronger zones between layers rather than merely within each layer, thus increasing overall structural performance. The major difference in comparison to the interior layer is that the exterior layer's arches are not perpendicular to the edge. The solution proposed by this study is similar to the structure that was developed in the first double-layered prototype. This approach succeeds in improving the connectivity of the separate arches by the simple means of rotating the span direction of the exterior layer relative to the interior layer. In the original, double-layered prototype, one arch of one layer was connected to two arches of the other layer, while in the solution employing rotation, one arch of the exterior layer could be connected to five arches of the interior layer. In this way, an improvement in connectivity and continuity is achieved. As in the other examples, with the exception of the interlaced-layers solution, the sequence of the layers is adaptable. This configuration can be summarized as follows:

- *Configuration principle*
  Mutual rotation of superposed units around O and around a central vertical axis.

- *Construction of interior layer*
  The interior layer consists of a series of self-contained higher-order units of one type, which units in turn consist of three entire modules.

- *Construction of exterior layer*
  The exterior layer is made up of a series of self-contained higher-order units of one type. Here, the unit consists of two entire and two halved modules.

- *Relation of the two layers and their constituent modules to one another*
  The two layers and their constituting units are rotated around each other around a central, vertical axis.

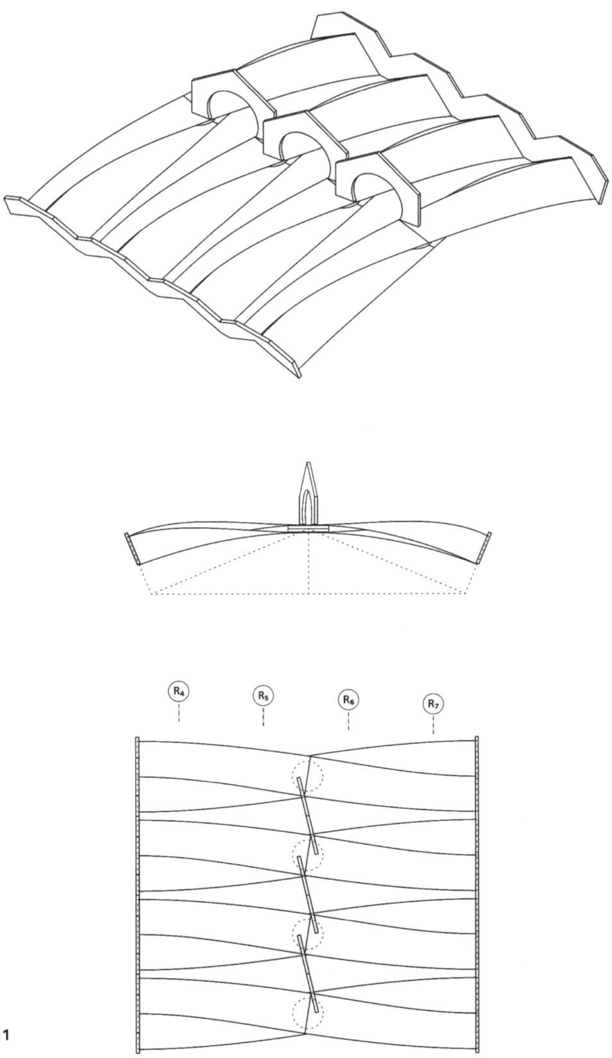

Fig. 11

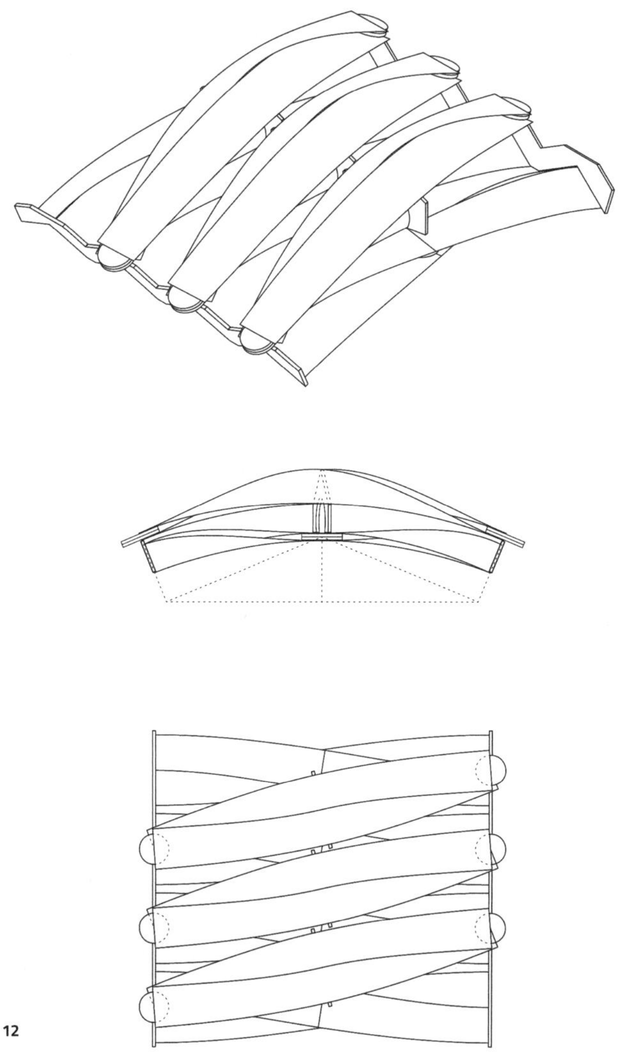

Fig. 12

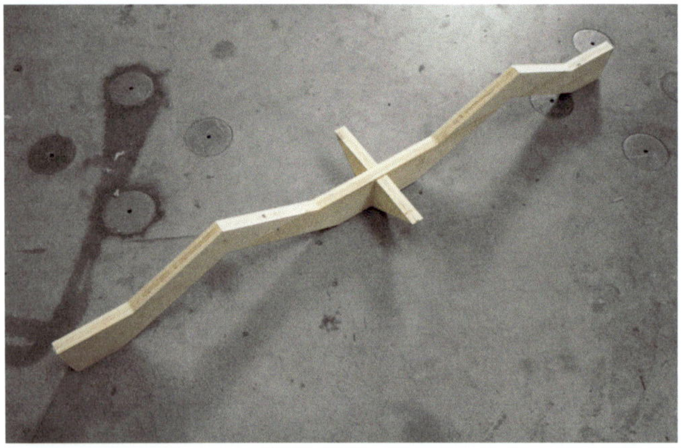

Fig. 13 Different types of connector elements for longitudinal, lateral, and multi-directional connections

8.1 Impact of layer configuration on connector elements

Due to the mutual rotation of the interior and the exterior layer, the cross-sectional profile of their interstice limits the use of multi-directional connectors at certain locations. Here, multiple one-directional connectors are employed.

9 Timberfabric structures with alternative global geometries

The timberfabric configurations developed thus far share a fairly simple global geometry. In part, this is due to the construction of the one-directional configurations, for which until now—with one exception—only identical modules have been used. As a consequence, the configurations developed all have an arch-shaped profile with constant curvature. The two- and three-directional configuration studies are essentially linear arrays of such one-directional units. Logically, their global cross-sectional profiles display identical geometrical properties.

In order to obtain differently shaped timberfabric structures, several strategies can be imagined. For example, cross-sectional profiles with non-constant curvature could be achieved by using modules of different sizes. Another approach could be applied in multi-directional configurations that are based on independent, one-directional units. By arranging these units along a curved instead of a straight trajectory, double-curved structures could be created. Furthermore, one could gradually increase the size of the modules in the respective units, which in turn would increase their span and height. This would result in an overall geometry that can be described as part of a cone. A further approach towards alternative global shapes could be the use of modules that are made of specifically formed, custom-made panels.

References

1 Hudert, M. "Timberfabric: Applying Textile Assembly Principles for Wood Construction in Architecture." Thesis no. 5553, EPFL, Lausanne, 2013.

2 Seiler-Baldinger, A. *Textiles: A Classification of Techniques.* Bathurst: Crawford House Press, 1994.

3 Beech, S.R., C.A. Farnfield, P. Whorton, and J.A. Wilkins, eds. *Textile Terms and Definitions.* Manchester: The Textile Institute, 1986.

4 In this context, the term "geometry" refers to the shape of the cross-section.

5 In order to maintain the geometry of a bisected module, a triangular spacer piece is placed at the intersecting line to keep the module's deformed elements in position.

3.4

"Shaping by bending is very simple"

Interview between Jan Knippers (Institute of Building Structures and Structural Design, ITKE, Stuttgart) and Yves Weinand
October 12, 2015

Yves Weinand: Right from the outset at IBOIS, we have dealt with fractals, as well as with folded systems. The latter, however, could not really be implemented. During the 1970s, there were attempts to develop folded structures. I truly believe in the potential of folded systems.
Our goal as engineers is to implement these types of structures. Architects are rarely able to, as they have to create 3-D drawings for their architectural competitions that only superficially represent structural optimization or topological considerations.
In the research represented in this publication, structures, for example derived from fractal geometries, will be analyzed. Here, we draw on subjectively selected iteration forms with systems that directly imitate existing configurations found in nature. We adopt a similar approach when developing woven systems. Although woven systems can be found in nature, we avoid copying them directly and rather try to interpret these geometries or systems. Directly copying or interpreting nature are issues that you analyze at your institute. You referred to the structural principle of sea urchins that you have adopted. What do you mean by this?

Jan Knippers: Of course, we aren't trying to directly translate a biological principle, like the sea urchin, into a technical process, though we do see the benefit in testing our approach to design in relation to what we see in nature. We can only design what we are able to simulate, as we always invent something that does not yet exist. We have to predict the structural capacity, stability, and integrity. And we are only able to do this if we are also able to calculate and simulate things. This is our "design space," so to speak. Basically, it works exactly as it already did back in the nineteenth century, only now it is far more complex and much larger.
And that is why I believe that it is interesting to work with biological forms, as the criterion of predictability is totally irrelevant. It all develops in a process of self-organization and identification with a few critical elements. This is a completely different approach from ours, as we work with the principle of predictability based on simulated calculability. I think it's interesting to analyze natural structures in order to gain awareness of one's own approach to design and construction rather than to make a direct reference. It isn't our goal to build a large sea urchin; rather, we strive to expand the variety of possibilities and to validate our approach.

Y. W.: In publications from the Stuttgart scene, it struck me that bionics is used as an argument for the generation of form. It wasn't previously clear to me whether nature should be directly copied. You've just explained how you prefer to distance yourself from what you see in nature and to use what you observe as an educational tool, thereby creating more space for reflection in which to define what might be constructible.

J. K.: To create my own design approach, to construct, and to reflect, I look at structures that are completely different from my own, as a biological tool, so to speak. So I challenge my approach to the principle of design as an engineer on a meta-level. It's also interesting to find a connection to woven structures. The theme of shaping by bending is actually very simple. It is the most basic approach to the creation of nonplanar geometries. It was nevertheless never utilized, as one couldn't calculate it. Only now do we have the methods of calculation and the simulation tools to calculate large deformations. These tools have only recently become accessible to the average engineering practice, so this approach has only now become part of my design process, because we are able to calculate it.

Y. W.: We proceeded from basics. We bent two panels, but not in any relation to nature. Our approach is intuitive; it is an inductive method. I am particularly interested in discussing applications of this method—that is, how it might actually be possible, in addition to bending these panels, to create more efficient structures that are also far more flexible than structures usually proposed by engineers. What advantages do you see in these multilayered woven or twisted systems that you first presented in 2010 at the first ICD/ITKE pavilion in Stuttgart? What advantages do you see in terms of structure?

J. K.: We didn't approach this theme from a structural standpoint, as you might have expected, but rather from two different directions. One aspect was related to textiles, and to the benefits of accessing the potential energy in the bias of the textiles. Almost ten years ago, we constructed a simple structure made of fiberglass slats and we stretched fabric over it. The other aspect was to investigate the use of different geometric configurations in mobile structures. The bending allows one to shift between different geometric scenarios without altering the system. This aspect of relativity, the adaptability of the geometry, and the connection with the membranes—these were basically our first forays into the area of bending. From this, we developed the idea of applying bending, thereby creating geometrically complex configurations, such as the ICD/ITKE pavilion from 2010.

Y. W.: From a strictly engineering standpoint, what are the structural benefits of these actively bent systems? If one does not consider the process but rather the result, I can well imagine that one could obtain an extremely flexible structure that can bend strongly. Thus you can deal with large deformations. Light structures such as these, for example in earthquake zones, would not even absorb the horizontal energy, as a reinforced concrete structure would. Aren't there also formal benefits here too?

J. K.: In the ICD/ITKE pavilion, we found that by introducing bending stress, the structural capacity increases and, with it, the tensile strength and system rigidity. We utilized this effect in other projects, such as the synthetic facade for the Expo pavilion in Yeosu, South Korea, in 2012. It consists of four strips that are bent upward, thus creating the climatic structure. Due to the upward bending of a strip, a bending stress is produced. Since the strip is only 8 mm thick, it stiffens and becomes more stable, even against wind loads. Here we used the principle of bending to respond to different geometric configurations. Additionally, the system becomes more rigid due to the bending.

Y. W.: That's a good application.

J. K.: Unfortunately, though, none are made from wood. The most interesting aspect of active elements, however, is that different geometric configurations are accounted for. Members are attached to the upper side and act like a compound spring and span the membranes. The elements can be bent and can thus track the sun. The elastic energy is used to pre-stress the textile strips.

Y. W.: Can you actually develop support systems from these support structures, in the traditional sense of load transfer? Do you have an idea of the kind of deformation in the upcoming 2016 version of the ICD/ITKE pavilion?

J. K.: I don't know that yet. We use bending to create the geometry of the modules, and we use the adjusted rigidity—that is, where the bend is most prominent and the element at its thinnest—to obtain the least resistance. The forces are dissipated in the shell, which consists of several layers woven together. We basically use many individual boards made up of composite modules. In principle, our current system is fairly similar to the previous one, only we no longer assemble each panel from a planar element that is bent. So we use bending as a way of modeling.

Y. W.: Woven structures can also be superimposed, that is, double-layered and offset spatially. That is how to obtain the greatest rigidity in the center of the element, where the structural height is at its maximum, and you can place this zone above the weak points of the next layer. You overlay the weak points with strong points in several layers. It is fascinating when a third weave direction is required. That means that we first weave twice in a principal direction, similar to knitting. Then you continue along the third weave direction. This third weave direction is what defines both the constraints and the support conditions of the individual panels. It would be fascinating to be able to weave on a large scale and to be able to connect the entire structure using robotics.

J. K.: I have always wondered how structures deal with local pressure at the contact points. Are there any problems with pressure across the grain?

Y. W.: As a rule, we haven't had any problems—except when we were working with larger Kerto panels that already broke at 18 N/mm² bending. They really shouldn't have failed, but they did, due to the fact that there were coincidentally too many random knots concentrated in one place. We developed a small parametric system in order to be able to select different geometric parameters, so that changes could be made to the overall geometry as required. But this has remained a prototype.

J. K.: We have noticed this even in the simplest office building, where we frequently propose timber structures. But then at some point in the planning process the use of timber is rejected, for either structural reasons, thermal mass, or budget. Although we work with ambitious architects and contractors and start by specifying a straightforward timber structure, at some point a decision is made against the use of timber. It is very difficult to find a project where the conditions enable an innovative step. I try to incorporate research from the university into our practice, for example the composite structures used in the pavilion. But it is difficult to find the right constellation of architect, client, and design brief.

Y. W.: Basically, I agree with you. It has also been my experience that many timber projects are not realized, even though the chances of implementation initially looked promising. But perhaps something is currently happening. I have the impression that many contractors are strongly influenced by their clients, who would like to work with wood. That is, they would like their projects to be visibly environmentally friendly. It is now common knowledge that steel and reinforced concrete may not be as sustainable as timber, even though they are great materials. Only recently a contractor approached me with a request to build a hall using a new system.

J. K.: After the timber shells of the ICD/ITKE Research Pavilion 2011, in 2014 we were able to build the exhibition hall of the National Horticultural Show in Schwabisch Gmund by Achim Menges. This development allowed us to put our ideas into practice. There are now a series of projects on the horizon where this technique will probably be used. But in other cases this step has proven extremely difficult. For example, our office has tried a few times to realize timber bridges, but this has proved futile.

Y. W.: I'd like to return to the material itself. You've just shown that you yourselves laminate here. But for the exhibition hall in Schwabisch Gmund you decided to use Hess panels. Today there are also Pollmeier panels, which are very good. The company, Pollmeier, has built an impressive new facility. At last there is an alternative to Kerto. Kerto was first introduced to Germany in 1995, and we built a small library with Kerto panels at the RWTH in Aachen. As a civil engineer, what requirements would you give the manufacturers of the panels today?

J. K.: We've built a whole series of projects using cross-laminated timber panels. At the IBA Softhouse in Hamburg, we tried to work with wooden ceilings in a bid to become more independent from the manufacturers. The argument we used there is that laminated timber construction can be calculated using normal design standards, that is to say, independent of specific authorizations, and thus is feasible for smaller companies. That's why we try to promote this construction method despite its structural limitations.

Y. W.: Shouldn't composite construction methods be encouraged in timber construction? For example, floors built from a combination of concrete and timber that provide an acoustically better result at a lower price? The timber industry often promotes a purely wood-based doctrine—in reaction to the other industries. But don't we need a stronger relationship to the existing construction industry? Perhaps something is happening with general contractors who increasingly integrate timber into their schemes.

J. K.: There is a strong lobby for timber construction in Britain, as well as here in Baden-Württemberg, in the Black Forest, which has been replanted after large sections of pine forest were destroyed by a number of storms. There is a political lobby that strongly supports timber, but it has not been able to translate this support into concrete regulations, which would, in turn, result in the re-evaluation of construction methods. We have had discussions in Freiburg about where a new football stadium ought to be built. There is a vague desire to use timber for the stadium, but when it comes to maintenance, investment, and construction, the use of timber is immediately called into question by the contractors and stadium operators. Thus, there is a discrepancy between the general attitude on a meta-level and the legal requirements that a building project has to meet. What is missing is an intermediary layer, which would provide the contractors with the appropriate support.

Y. W.: Timber is a good material and I hope that we'll soon be ready to propose larger wooden structures, such as stadiums. The logistics exist for the use of timber in housing, but there are no logistics for contractors that would allow for innovative or hybrid structures made from wood. Classical structures with laminated beams are proposed time and again. Perhaps not enough has yet happened to allow new structures to develop.

3.5

Geodesic lines for shell forms—a playground installation at the Vallée de la Jeunesse (Lausanne, Switzerland)

Marielle Savoyat

Design	IBOIS, Laboratory for Timber Constructions/ EPFL, Swiss Federal Institute of Technology, Lausanne, Switzerland; Prof. Yves Weinand and Dr. Claudio Pirazzi, researcher GEOM, Laboratory of Geometry/EPFL, Swiss Federal Institute of Technology, Lausanne, Switzerland; Prof. Peter Buser and Roland Rozsnyo, researcher
Project execution	Charpentes VIAL SA
Completion	2006
Location	Vallée de la Jeunesse, Lausanne, Switzerland

This architectural structure in curved wood is based on the geometry of geodesic lines. The idea was developed in conjunction with Claudio Pirazzi's doctoral dissertation,[1] and undertaken at IBOIS, the Laboratory for Timber Constructions, along with Roland Rozsnyo,[2] at GEOM, the Laboratory of Geometry, both at EPFL, the Swiss Federal Institute of Technology in Lausanne. A geodesic line represents the shortest line between two points on a curved surface, or lines made up of simple curves. The curvature achieves the weak axis of inertia of each rectangular section and not the strong axis of inertia. However, the geodesic line follows a helicoidal line, where some torsion is also experienced.

In the context of this research project and under the auspices of the two laboratories, the researchers developed a program capable of determining the geodesic networks on free-form shapes These networks, in turn, enabled the definition of geometric data necessary for a digital cut. In this way, new options were found in terms of the use of timber for the creation of curved structures.

Fig. 1

Fig. 1 Modelization of the pavilion

Fig. 2 View of the pavilion in its Kindergarden situation

Fig. 2

A large-scale prototype of a ribbed shell made of laminated, screwed boards was created to test the potential of this program. The chords of the shell followed geodesic lines, avoiding all double curvatures and reducing the important constraints initially posed by curvatures, such as bending at the fortified axis. The slats are therefore solely utilized for a unique moment of flexion, based on their weakest axis. The meshing of optimal wooden ribbed structures can be achieved, thereby reducing the risk of buckling.

It is interesting to note that different dispositions of geodesic lines on the same free-form surfaces are possible, and that a subjective element can appear in the architectural design, notably regarding questions of aesthetics, construction, or spatiality. This paves the way to a wealth of new potential in the creation of contemporary architecture.

The resulting prototype structure from this research, called "Géoline," was installed in the garden of a day-care center in Lausanne, where it now serves as a playground installation and playhouse for children at the Vallée de la Jeunesse.

References

1 Pirazzi, C. "Zur Berechnung von Holzschalen in Brettrippenbauweise mit elastischem Verbundquerschnitt." Thesis no. 3229, EPFL, Lausanne, 2005.

2 Rozsnyo, R. "Optimal Control of Geodesics in Riemannian Manifolds." Thesis no. 3481, EPFL, Lausanne, 2007.

4 Form-finding and mechanical investigations of active bended systems

4.1

Optimization of double-curved shell structures

Yves Weinand

The scope of research presented here concerns the interface between architecture and structural engineering. The simultaneous openness to both inductive and deductive methods is an important aspect of the manner in which research is conducted at IBOIS at the EPFL. The functional diagram above shows that both empirical and rational tools are used for brainstorming; physical experiments, digital models, and mathematical and analytical descriptions are all used simultaneously to solve problems. Initial ideas are described in principle by one or another of these methods, and jumps in scale are deliberately staged, where both structural and architectural considerations come into play. The synthesis is achieved in the context of a structural optimization process.

Why do hybrid structures no longer appear in civil engineering textbooks? Why is there no theory supporting the superimposition of different structural systems in a single structure? One of the aims of this publication is to stimulate the interest of civil engineers in the question of new global forms. Architects' tools are highly representational; therefore, the presentation of space is greatly simplified. Engineers' tools have a high level of mechanical precision, but their spatial representation is less user-friendly. In other words: even if tools are increasingly better able to represent space, the process of mechanical verification remains incremental.

In this context, the work of Sina Nabaei presented here aims to develop a simulation model for shells and support structures, enabling the creation of simulations during the structural design process that are based on physical characteristics.

Nabaei also presents a robust and stable simulation method for calculating woven timber structures. While the work presented by Markus Hudert has been empirically developed "by hand," the tool developed by Nabaei uses mechanical constraints to control the form-giving process.

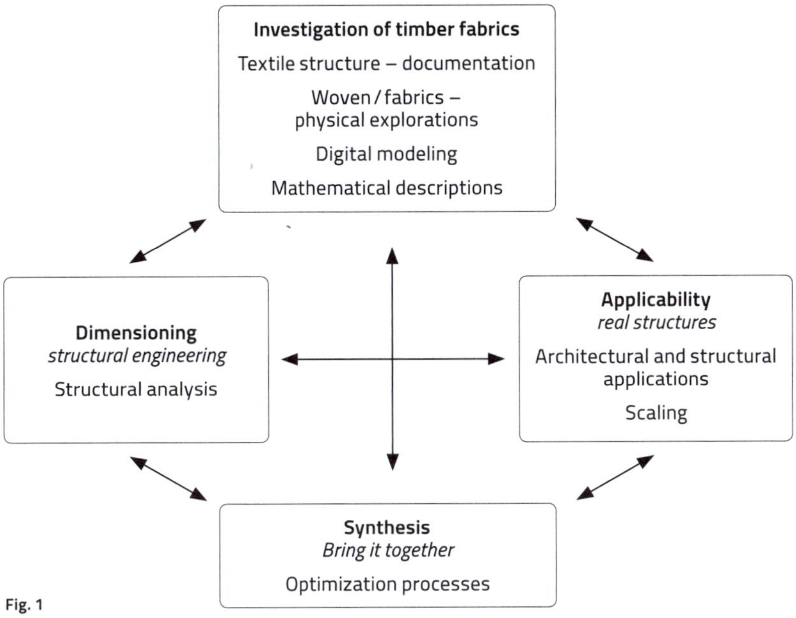

Fig. 1

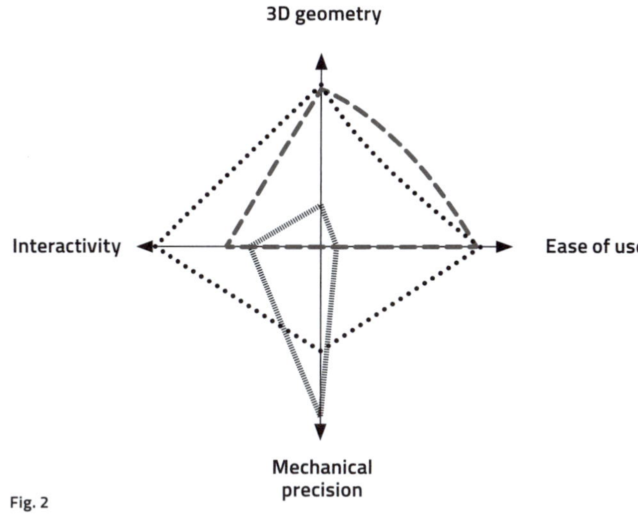

Fig. 2

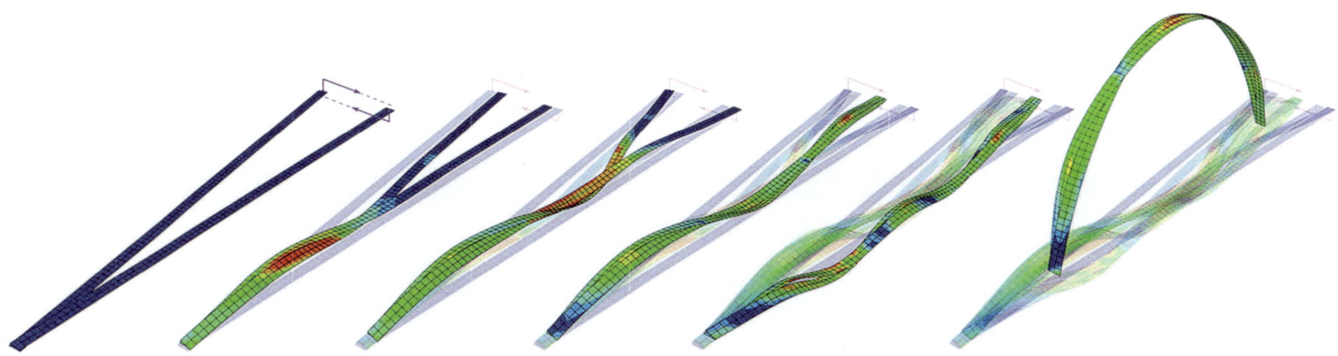

Fig. 3

Fig. 1 Functional diagram, IBOIS Research Cluster

Fig. 2 Diagram illustrating the interaction of existing tools

Fig. 3 Textile module formed by interweaving

Fig. 4 a and b Individual strands are formed mechanically

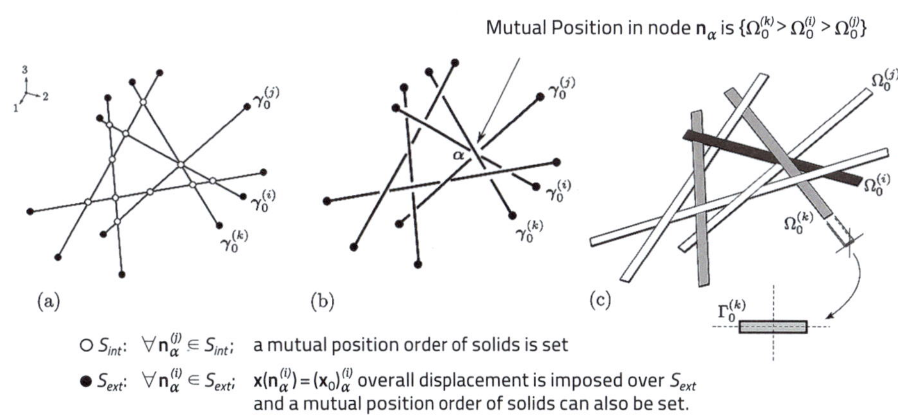

Fig. 4 a

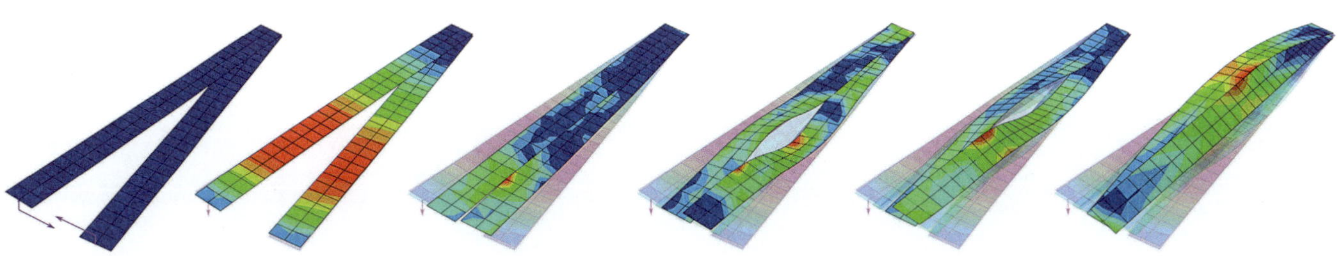

Fig. 4 b

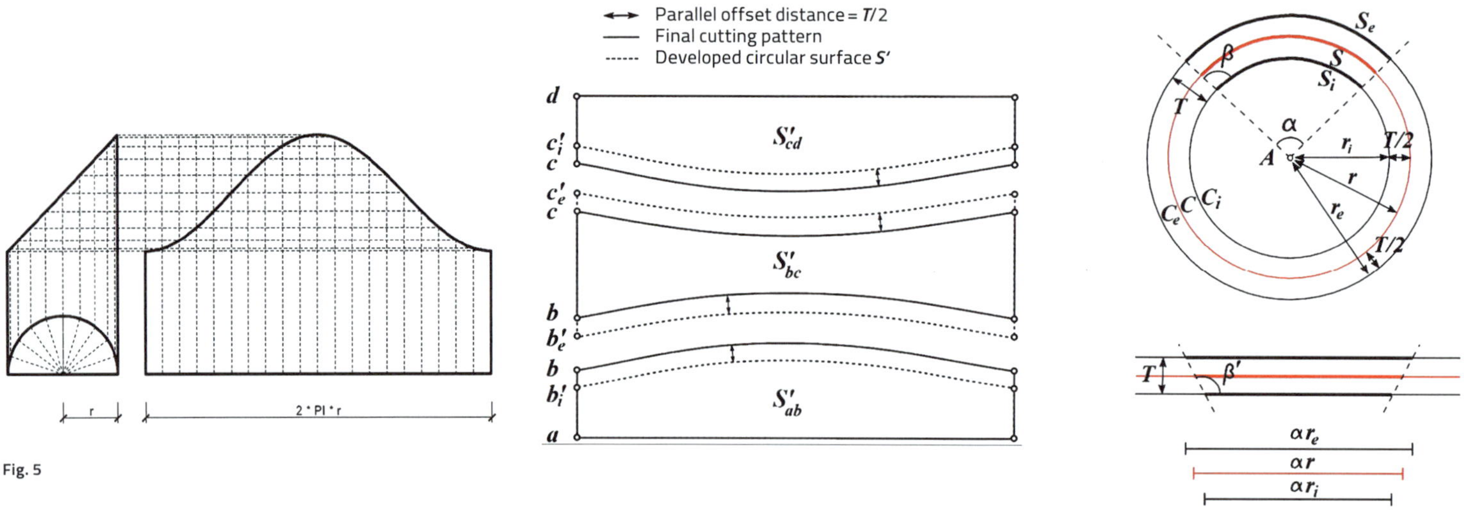

Fig. 5

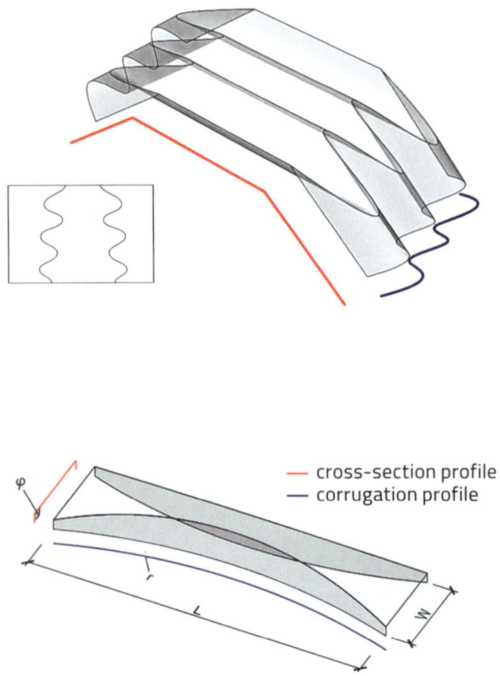

Fig. 6 a

Fig. 6 b

Fig. 6 c

Fig. 7

Fig. 5 Geometric structure and detail: laying the edge geometry flat, multi-layered structure, determination of the central plane

Fig. 6 a) A folded form made from actively curved panels, b) a box girder made from actively curved panels, c) Detail of box girder in b)

Fig. 7 Examples of actively curved shell structures

As the previous examples show, it is clear that form and structure can exist in a close relationship to one another. Two other examples presented here also illustrate the potential relationship between form and structure in order to optimize structural behavior. The origami design tool presented in Chapter 1 makes it possible to generate folding processes. Originally, the flat panels created the necessary preconditions. During further experiments, it was found that transverse and longitudinal profiles are also able to generate structures with curved surfaces. A folded structure formed with curved panels and an active box girder made from composite curved surfaces are shown here.

The ability to construct with panels that curve in opposite directions resulted in a new pavilion project, which was commissioned by the Accademia di Architettura in Mendrisio in 2013.

Initially, the principle is studied with regard to convex and concave panels that are connected in two opposing directions. Civil engineering master's students at the EPFL were asked to critically observe the given geometry and to optimize the structure using only geometric adjustments. For this purpose, parameter studies were undertaken where not only span widths but also, for example, the depths of the indentations were varied. The educational aim was to direct engineers toward a consideration of the overall form of structures.

Even with a constant panel thickness, this construction demonstrates a tremendous bending stiffness in the corners of the frame. In analyzing the causes of this phenomenon, it was ascertained that bending stiffness is ensured by the existing structural strength. Because the surfaces are rotated in relation to one another, they cannot rotate around an axis. The bending moments that are generally absorbed in the corners of the structural frames are not activated, as the tension from the axis of rotation is deflected by the curved surfaces.

A 1:5-scale prototype was built with multilayered panels. For this purpose, a form was built that corresponded to the chosen curvature. In order to achieve this, several layers of panels are forced into the curved shape and glued in place. Thus, it is not about actively bent panels, but about timber panels that are produced in curved forms.

An important aspect of the project development is the formation of timber–timber connections along the edges. The inclination of the panels in relation to one another affects the nature of the timber connection. Within the framework of this prototype, two different interlocking geometries were selected, parameterized, and CNC-milled.

Fig. 8 Potential applications of the main structural system

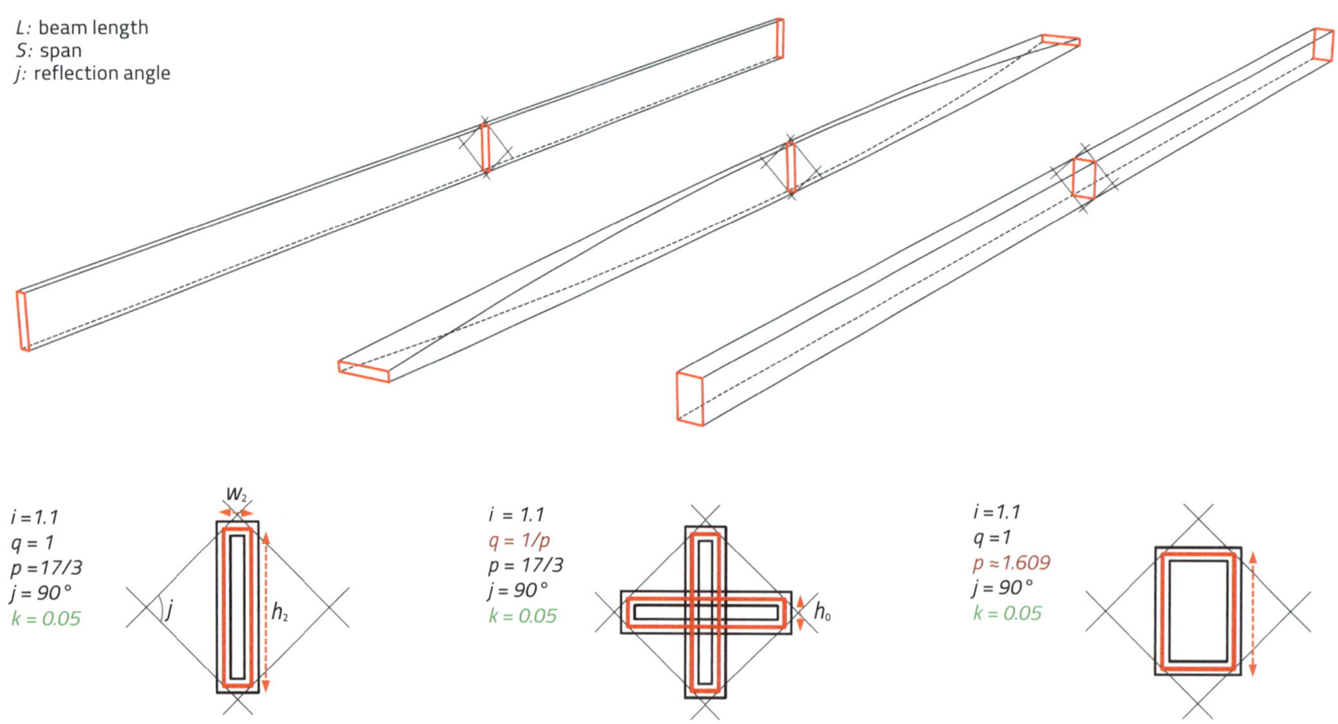

Fig. 9 Study of the support parameters

Fig. 10

Fig. 11

Fig. 10 Test equipment for measuring deformations

Fig. 11 Test girders made at IBOIS

Fig. 12 Test results of the three volumes

Fig. 13 Geometric parameters of the pavilion (plywood model)

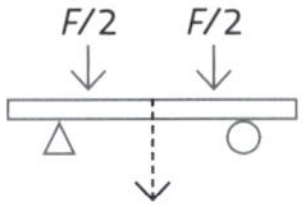

Prototype	4m deflection* (mm)	4m max load (kN)	8 m deflection** (mm)	8 m max load (kN)
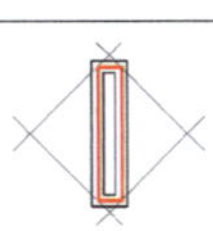	16.3	9.6	14.3	
	15.9	9.1		
	16.7	9.0		
	17.7	8.2	15.0	>37.2
	16.5	10.4	15.4	>35.9
	17.1	9.5	15.6	>35.6
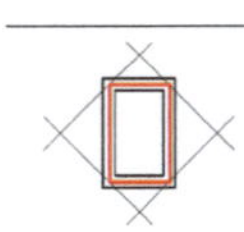	23.9	7.9	20.3	
	23.6	8.1		
	20.8	8.1		

Fig. 12

* $F = 2$ kN ** $F = 4$ kN

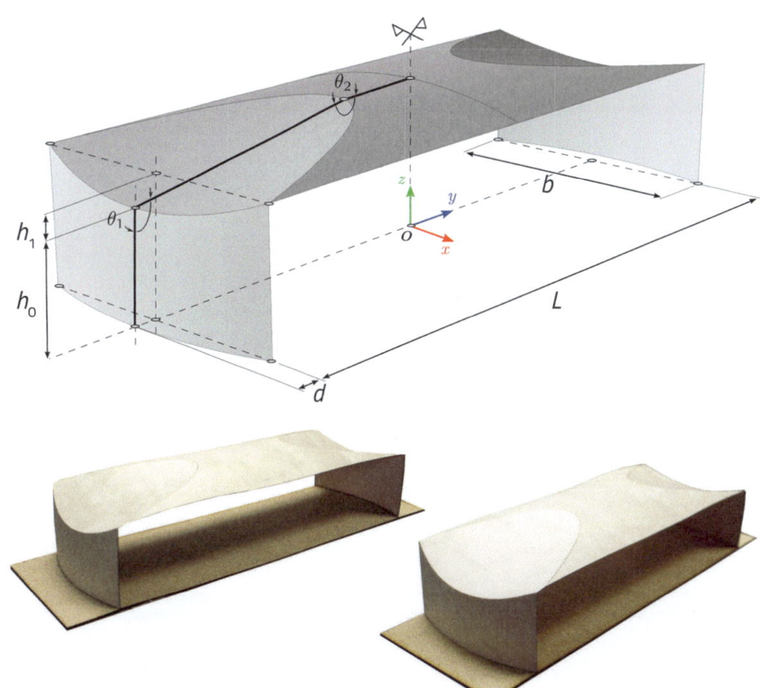

Fig. 13

Fig. 14 a

Fig. 14 b

Fig. 14 c

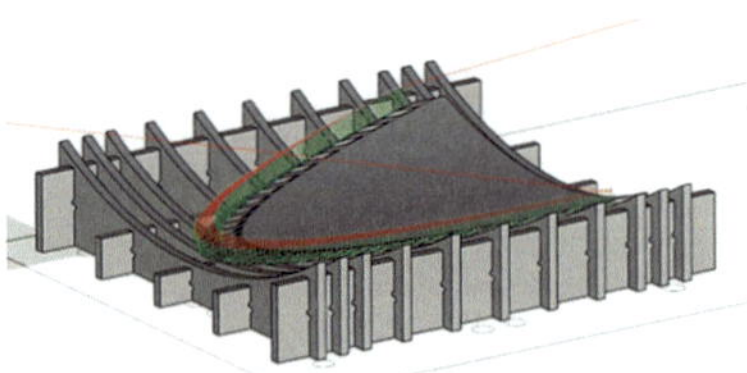

Fig. 14 d

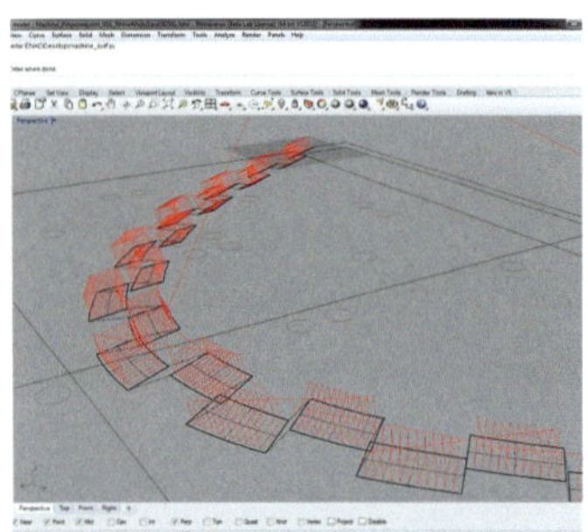

Fig. 14 e

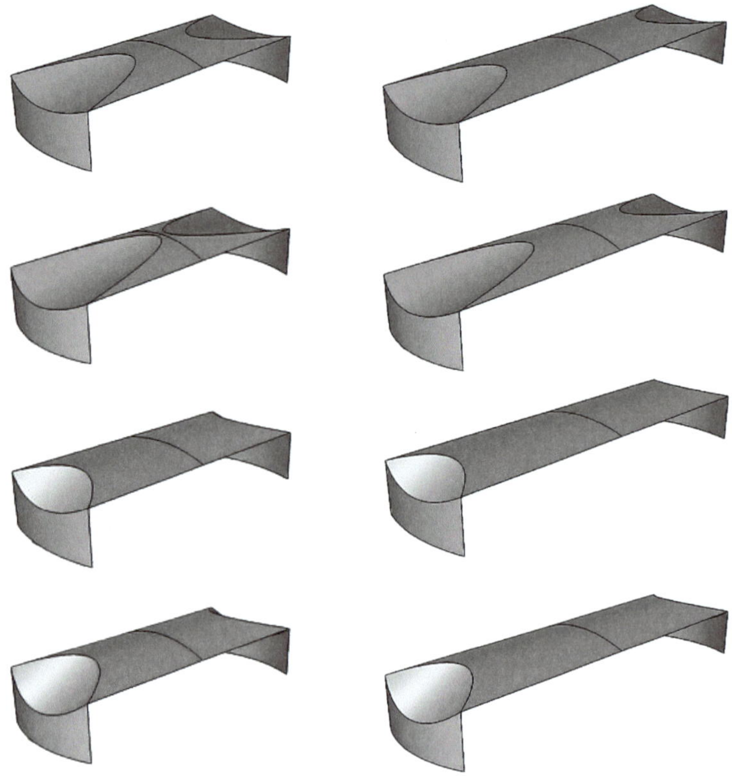

Fig. 15

Fig. 14 a–f Production of curved timber panels: parametric geometry of the timber–timber connection

Fig. 15 Parameter studies on a singular basic geometry

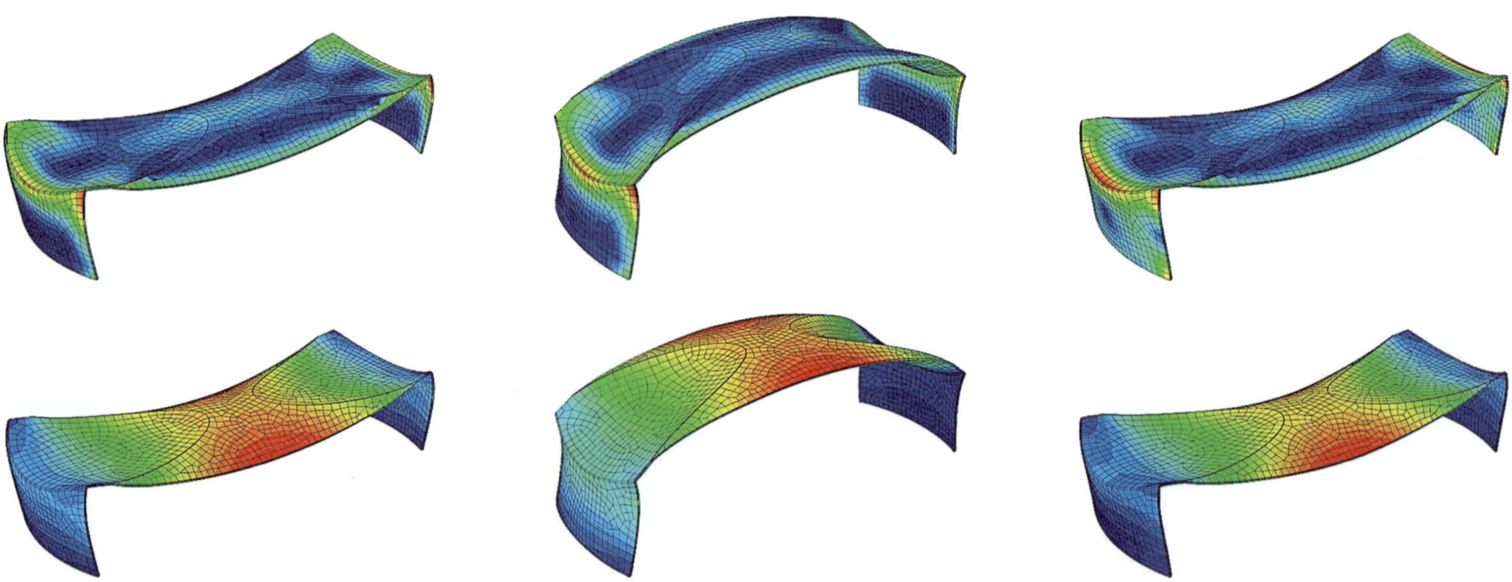

Fig. 14 f

On this basis, a pavilion with a span of about 13 meters was developed. Convex and concave panels exhibit the same curvature. This measure simplifies the manufacture of panels, as there is only a single form. The finite element model shows the panel thickness and the degree of deformation of the structure. The timber–timber connection could be fully modeled as the panels are laminated. The 77-mm-thick panels are produced under a pneumatic pressure pad. Subsequently, the panels are cut along their curved contour. The geometry of the timber–timber connection can also be milled in the same machine path. The resulting time savings are relevant in the development of this typology of structure.

Fig. 16 Assembled pavilion

4.2

Experimental and numerical study of the structural behavior of a single timber textile module

Masoud Sistaninia, Markus Hudert, Laurent Humbert, and Yves Weinand

This essay investigates an advanced class of timber structure termed timberfabric, which has potential applications in roofing, facade, and bridge construction. The development of timberfabric structures originates from the approach of harnessing the structural, modular, and quality of textiles in timber construction.[9] Timberfabric structures comprise repetitive arrangements of one or more structural unit cells, called textile modules. When properly designed, one obtains a modular and lightweight structure with interesting and unusual geometric and structural qualities.

This paper focuses on the single timber textile module. Based on the finite element (FE) method, a reliable procedure is proposed for modeling the overall assembly process of the textile module. In order to compare, textile module prototypes are constructed at both large and intermediate scales with different assembly conditions. The proposed geometrical, nonlinear FE model allows the evaluation of the stresses that are induced during the construction process and which may affect the structural integrity of the module. In particular, the risk of failure during assembly is identified using the anisotropic Tsai-Hill criterion. The structural behavior of the timber textile module is then investigated through bending tests using the constructed prototypes. During the loading procedure, the vertical deflections are measured at different locations on the prototype surface by means of external displacement transducers. Using the FE model, the corresponding deformed shapes are simulated by applying the bending loads on the pre-stressed textile module. Experimental displacements and FE predictions are compared to see if they correspond favorably.

Keywords ***timber textile module, construction stresses, experimental method, finite element modeling.***

1 Introduction

Timber is a versatile construction material that is abundant in many regions of the world. Moreover, it is a renewable resource that can be processed and assembled in energy-efficient ways. Recent studies[1,2] indicate that the use of timber as construction material results in buildings with a better environmental performance than conventional materials. With regard to present-day concerns over globally increasing energy consumption and simultaneously decreasing resources, wood holds a distinct advantage over other construction materials such as concrete or steel. This, in turn, should increase the interest of the research community in expanding the range of applications of timber structures.

Examples of modern but well-established timber architectural forms include folded plate structures,[3,4] lattice structures (e.g. timber lattice roof for the Mannheim Federal Garden Show),[5,6] and multi-reciprocal frame structures.[7,8] Forms such as these present clear advantages over more traditional, flat-surfaced roofing structures, in that the efficiency and load-carrying capacity of the structure is increased and its weight reduced.

A new type of timber structure, called timberfabric, has recently been developed at IBOIS,[9,10] the particular structural properties of which emanate from the principle of weaving techniques. The development of timberfabric has been driven by the concept of incorporating qualities specific to textiles, such as modularity and the mutual support of constituent elements of textile fabrics into timber construction. Timberfabric structures have a broad potential for architectural applications thanks to their versatility, adaptability, and the qualities directly linked to their structural make-up. They are based on the repetition of a structural unit cell, the textile module **(Fig. 1a)** which results from applying textile assembly principles to timber components. Fig. 1b shows the double-layered timberfabric structure, representing a single example of the many possible configurations of textile modules.

The single textile module presented in fig. 1 a provides a structural shape of particular interest for

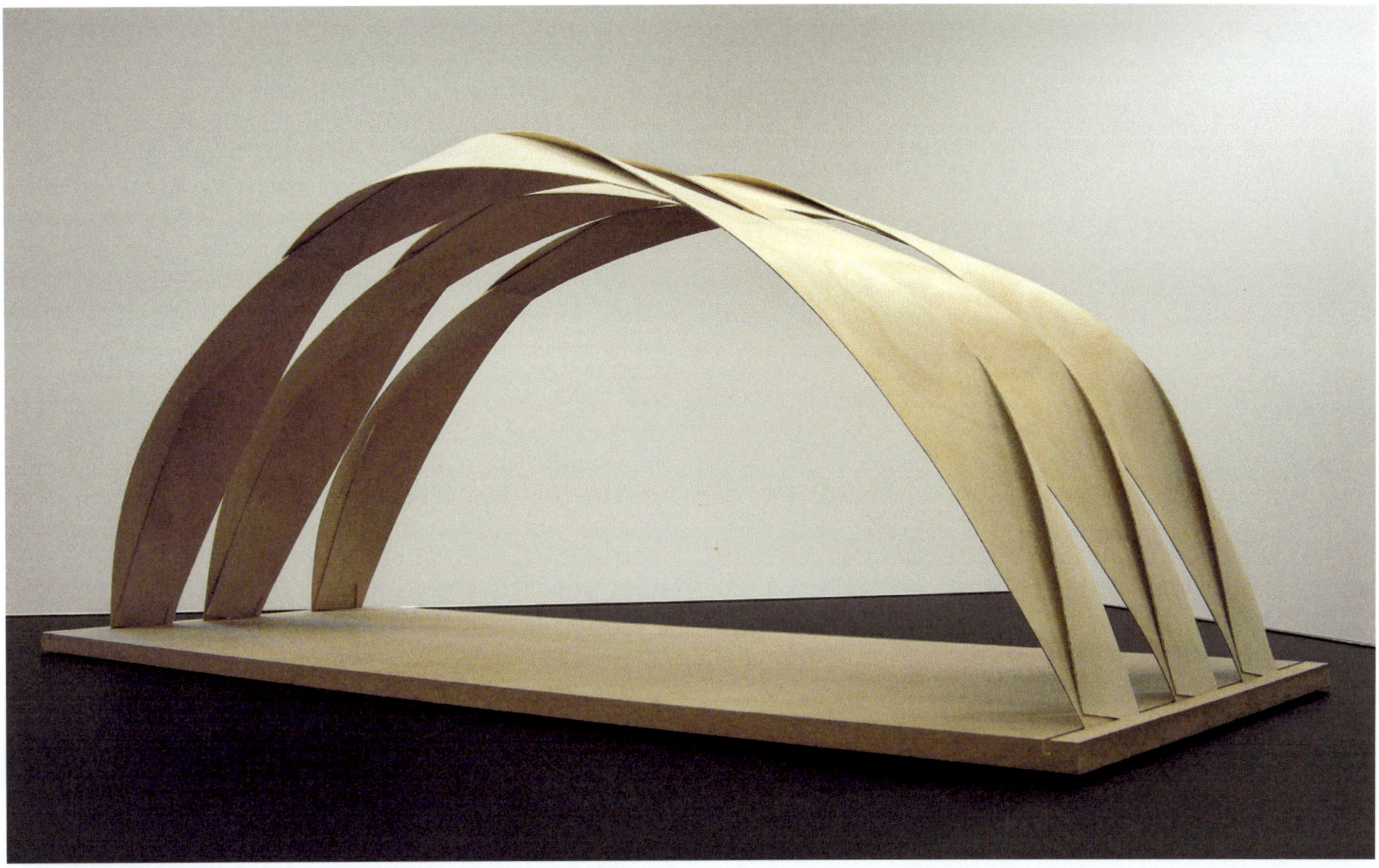

Fig. 1 Model of the structure

this study. Briefly, it consists of two mutually supporting thin panels that become curved during the assembly process, as illustrated in fig. 2. Consequently, construction (or residual) stresses are generated during the fabrication of the module. Their amplitude typically depends on the constitutive material, the size of panels, and on the assembly conditions. The use of poor quality material or inappropriately dimensioned panels, or a combination of both, may even cause premature failure of the textile module during the assembly process. The construction stresses can be evaluated by means of a finite element (FE) model that takes the different fabrication stages into account **(Fig. 2)**.

This paper focuses on the fabrication process and structural behavior of a representative single textile module in a bending-load configuration. The proposed approach is both experimental and numerical. It involves the fabrication of two prototypes at two different scales (intermediate- and large-scale) under different assembly conditions, as discussed in the first part of section two. The second part of section two is devoted to the bending test setup and the measurement equipment required.

Despite the numerous finite element (FE) models already available for braided textile composites,[11–15] the numerical study of the timber textile module requires special attention, as its analysis is complicated by the

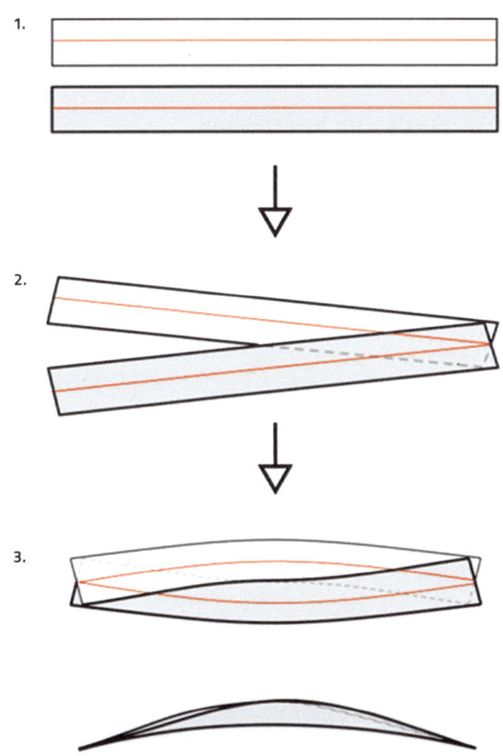

Fig. 2 Design principle for the textile module

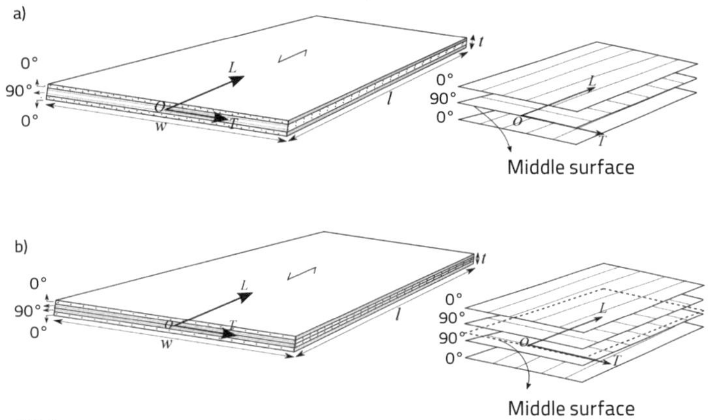

Fig. 3

particular geometry and assembly conditions encountered. Because of the large deflections and rotations experienced by the module during the fabrication stages (Fig. 2), a geometrically non linear FE model that aims to accurately reproduce the geometrical shape of the textile module is developed (section 3). It is anticipated that this model should permit a representative evaluation of the construction stresses involved in fabrication. In section four, vertical displacements measured at several locations of the prototype surface during the bending tests are compared to the finite element predictions. Finally, the structural behavior of the textile module is discussed.

2 Experimental investigations

2.1 Materials and specimens

Two textile modules have been constructed for this study: one is formed from two GFP laminated wood panels with a length of 12.320 m and a width of 0.770 m, while the other is formed from two TeboPly™ Okoumé plywood panels with a length of 2.34 m and a width of 0.24 m. The GFP and Okoumé panels used to fabricate the textile modules are respectively supplied by the companies Schilliger Holz AG (Switzerland) and Thebault (France). As schematically depicted in fig. 3, the panels consist of symmetric orthotropic laminates, with a three-ply [0/90/0] GFP and four-ply $[0/90]_s$ Okoumé with a thickness of 33.0 mm and 6.3 mm respectively. As is customary, the plywood is produced from rotary cut veneers, bonded with an adhesive (synthetic) resin under high-pressure conditions. In each case, the outer layers of the veneers are of equal thickness, with the same grain direction along the longitudinal axis (L-axis) of the laminate. As depicted in fig. 3 a, for the three-ply configuration, the axis of symmetry passes through the center of the 13-mm-thick core ply, and its grain is directed along the transverse axis (T-axis) of the laminate. For the four-ply model, the two core plies of equal thickness (i.e. 2 mm) are glued together with their grain direction running perpendicular to the longitudinal axes of the face veneers (along the T-axis). This glued interface corresponds to the plane of symmetry (central surface) of the plywood. This even-layered arrangement has been proven to increase efficiency in veneer manufacturing and grading (less variability), although it does increase the cost of production.

Both the GFP and Okoumé panels are composed of orthotropic layers (veneers) with their principal material axes coinciding with the longitudinal (L) and transverse (T) geometric directions of the panels (Fig. 3). On a macroscopic scale, they can be treated as homogeneous orthotropic materials, with the L and T axes as the principal axes of the equivalent material.

Homogenized elastic material properties, provided by the manufacturers under standard conditions (20 °C temperature and 65 % relative humidity), are reported in Table 1, where the L and T subscripts refer to the longitudinal and transverse directions respectively. Table 1 also provides the longitudinal and transverse moduli of the Okoumé panels, measured in the environmental conditions of the laboratory. Tensile tests were carried out in a standard crosshead testing machine (in displacement control with a crosshead speed of 2 mm/min) approximately eight months after receiving the

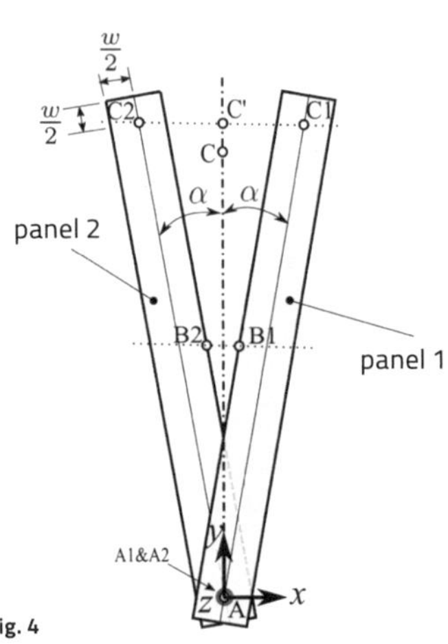

Fig. 4

	Characteristic values of strength at fifth percentile (MPa)								Modulus of elasticity (median values)		Shear modulus (median values)
Plywood	Bending		Compression		Tension		In plane shear		Tension and Compression (MPa)		In plane shear (MPa)
	$f_{m,L}$	$f_{m,T}$	$f_{c,L}$	$f_{c,T}$	$f_{t,L}$	$f_{t,T}$	$f_{v,L}$	$f_{v,T}$	E_L	E_T	G_{LT}
GFP (3 layers)	15.8	2.6	7.3	4.7	4.8	3.2	1.5	1.5	6667	4333	720
Okoumé (4 layers)	35	32.4	10	28.5	6.1	17.4	7	7	2398 4592[a]	6852 7288[a]	552

[a] *Averaged values obtained in the laboratory.*

Table 1 Mechanical properties of GFP and Okoumé laminated wood panels from the manufacturer. The average values of the elastic moduli reported in Table 1 are calculated on the basis of measurements obtained using four samples with the direction of the grain of the face layers parallel to the loading direction, and two samples with the direction of the grain of the face layers perpendicular to the loading direction. These values will subsequently be used in the finite element simulations.

The GFP textile module will hereafter be referred to as the large-scale (TM1) specimen and the Okoumé textile module as the intermediate-scale (TM2) specimen. Besides their differences in size and material properties, the modules are also assembled differently. For prototype TM1, the ends of the panels are mechanically linked by means of a pinned connection, while hardwood wedge connectors that restrain all degree of freedom are employed to build the second prototype, TM2. The construction process of the two prototypes is described in detail below.

2.2 Construction of the textile modules

2.2.1 *Pin connection*

A pinned connection is used for the assembly of textile module TM1. Practically, the two panels are connected to each other at their ends using a system of steel-threaded rods with nuts and washers. The starting point for the assembly procedure is illustrated in fig. 4. For convenience, the global reference coordinate system (x,y,z) is introduced, such that the vertical z-axis crosses panel 1 and panel 2 at points A1 and A2 respectively. These two points have the same geometric position, taken as the origin A of the global system. Strictly speaking, point A1 belongs to the lower face of panel one, while point A2 is on the upper face of panel 2, with the same in-plane coordinates $L=w/2$, $T=0$ in the local frame (L,T) of the panels (Fig. 3). The middle surfaces of the panels are positioned parallel to the plane (x,y) and symmetrically rotated about the z-axis by an angle $\alpha=10°$. Then the panels are linked to each other at points A1 and A2 using the steel-threaded rod (not represented in fig. 4), thus restraining their relative displacement at A while keeping the rotation free along the threaded-rod axis. Note that the threaded-rod axis initially coincides with the z-axis, but will not retain this direction after assembly.

Now the opposite points C1 and C2 (Fig. 4) will be considered, with local in-plane coordinates $(L=l-w/2, T=0)$, such that $AC1=AC2=l-w=11.55$ m. The line (C1C2) lies perpendicular to the y-axis, bisecting it at point C'. In other words, one can trace an isosceles triangle (AC1C2), the base of which is (C1C2). Practically speaking, the panels are conjoined at points C1 and C2 by means of cables to a third common point C on the y-axis. The distance AC is typically less than AC'. The panels that become curved are pin-connected again at C by means of a second threaded rod. During this operation, the midpoints B1 and B2 of two panel edges come into contact at point B,

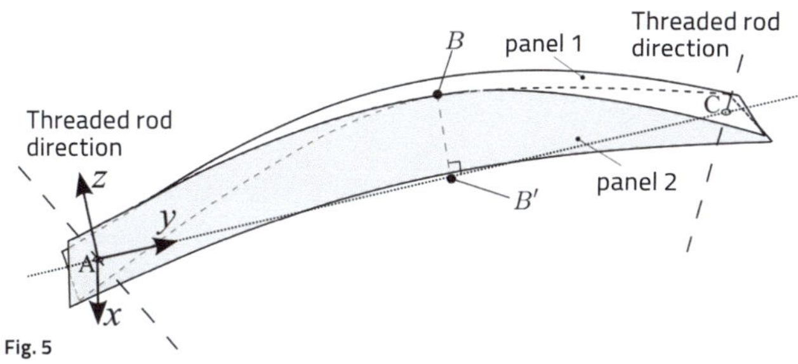

Fig. 5

Fig. 6 a

Fig. 6 b

Fig. 3 a) GFP and b) Okoumé plywood layups and local coordinate system (L,T).

Fig. 4 Assembly procedure and associated coordinate system.

Fig. 5 Textile module geometry with curved panels

Fig. 6 a) Prototype TM1 and steel connector used to fix it on the floor b) Failure during the assembly process which corresponds to the maximum height of the resulting structure (Fig. 5). The maximum height (i.e. distance BB') and span (i.e. distance AC) for TM1 are 1.50 m and 10.05 m respectively.

It is worth noting that the two threaded-rod axes and point B belong to the vertical plane (*y*, *z*). By denoting B', the vertical projection of B onto the *y*-axis, one has an isosceles triangle ABC of an equal height to the line segment BB' and base AC = 2AB'. Moreover, the two threaded-rod axes are symmetrically oriented with regard to the axis BB'.

The two ends of the textile module TM1 are firmly fixed to the ground by means of a special steel connector (Fig. 6a), thus restraining all displacements and rotations along the corresponding panel edges. Various assembly conditions can readily be imposed on the textile module panels, accordingly modifying its final shape and structural behavior. The selection of the type of connectors to be used depends on the assembly conditions under consideration.

Finally, it should be noted that apparently minor alterations to the assembly conditions may have a significant influence on the strength of the structure. An early prototype (TM0) was composed of panels with characteristics similar to those of TM1 except that the width *w* was equal to 1.65 m. This prototype was broken in two after the nuts were tightened in the final stage. Both panels failed in two parts in the middle (Fig. 6b) due to the stresses induced during assembly.

2.2.2 *Wedge connection*

A second type of connection, referred to simply as a "wedge connection," is employed for the construction of prototype TM2. As indicated in fig. 7, two identical sets of three wedge elements are used to connect the panels to one another and to simultaneously affix the entire structure to the ground. The different elements are manufactured by cutting a 20-cm-long block of hardwood with a cross-sectional area $w/2 \times w/4 = 12 \times 6\,cm^2$ along two oblique cutting planes with different orientations (Fig. 7). The resulting uppermost (*Wedge 3*) and lowermost (*Wedge 1*) elements thus have one cross-section at right angles and the other obliquely oriented in relation to the lateral faces. Before cutting, three 10-mm-diameter holes were drilled along the length of the wooden block, as indicated in fig. 7, for the insertion of a tightening system composed of threaded rods and nuts.

Positioning the structure on the ground is achieved by means of the supporting element *Wedge 1*. The intermediate element (*Wedge 2*) permits the relative position of the panels to be accurately controlled on the constrained areas (assembly conditions). The third wedge element (*Wedge 3*) is used to connect the panels together by simply tightening the nuts, thus aligning the three wedges vertically. Moreover, this locking device allows the module to be fully clamped on the ground. The positions of the two oblique (cross-sectional) planes can clearly be described by means of six assembly parameters ($\theta_{x1}, \theta_{x2}, \theta_{y1}, \theta_{y2}$, *d*, *d'*) as presented in fig. 8. The origin O of the reference coordinate system (*x,y,z*) is placed at the center of the bottom section of *Wedge 1*. Fig. 8 illustrates the final shape of the module, where the vertical *z*-axis intersects panel 1 and panel 2 at points A1 and A2, the respective coordinates of which are (0, 0, d + d') and (0, 0, d'). Unlike the prototype TM1, these two points do not occupy an identical position, because of the introduction of *Wedge 2*. It should also be noted that the threaded-rod axes keep a fixed orientation corresponding to the vertical *z*-axis during the assembly. Moreover, A1 and A2 are centroidal points for the contact areas *p1* and *p2*, whose orientations—with respect to the *x*-axis and *y*-axis—are given by the two angles θ_{x1}, θ_{y1} and θ_{x2}, θ_{y2} respectively.

Similarly, the opposite points C1 and C2 are centroidal points for the contact areas *p1'* and *p2'* as defined in fig. 8. Point C1 is now located under point C2 and their respective coordinates are given by (0, *s*, *d'*) and (0, *s*, *d* + *d'*). Denoting O', the intersection of the vertical axis passing through these points and the *y*-axis, one obtains s = OO'. In this case, the angles $-\theta_{x1}$ and $-\theta_{y1}$ (i.e. the angles $-\theta_{x2}$ and $-\theta_{y2}$) orientate the plane *p2'* (i.e. the plane *p1'*) with respect to the *x*-axis and *y*-axis.

As was the case for prototype TM1, an isosceles triangle OBO' can be constructed where vertex B is defined as the common position of the contacting points B1 and B2. Again, B' denotes the vertical projection of the vertex B on the opposite side OO'.

Fig. 9 shows a TM2 module that is affixed to a thick wooden base using the previously described wedge connections. The maximum height BB' corresponds to 0.36 m, the span s=2.056 m and the assembly parameters are chosen as $\theta_{x1} = 17°$, $\theta_{x2} = 11.5°$, $\theta_{y1} = 14.5°$, $\theta_{y2} = -8°$, $d' = 76$ mm, $d = 47.7$ mm.

2.3 Test setups and measurement equipment

The test setup used for the experiments carried out on prototypes TM1 and TM2 is shown in figs. 10 a and 10 b respectively. The ends of prototype TM1 are clamped on the floor by means of the steel connector shown in fig. 6. A hydraulic actuator from Walter & Bai Company with a maximum force capacity of ±500 kN is used to load the prototype at mid-span. An additional V-shaped element (made of hardwood) is attached to the actuator for transferring the load on each panel, as indicated in fig. 10 a and fig. 11 a. During the test, the load is applied up to the ultimate limit state under displacement control with a constant displacement rate of 6 mm/min.

Prototype TM2 and its wooden base are placed on two concrete blocks, as shown in fig. 7. In this case, a hydraulic actuator with a maximum force capacity of ±300 kN is used to apply the bending load at the prototype mid-span. For an accurate measurement of the force, as indicated in the insert of fig. 10, a load cell of 5 KN is added. A custom-made wooden element with two parallel steel rods, is fastened to the actuator in order to dis-

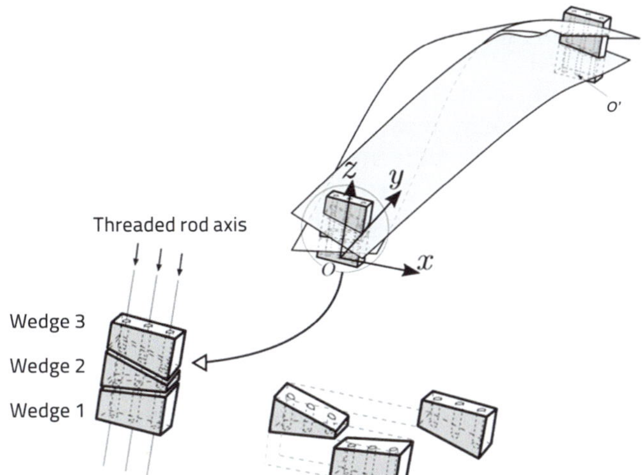

Fig. 7

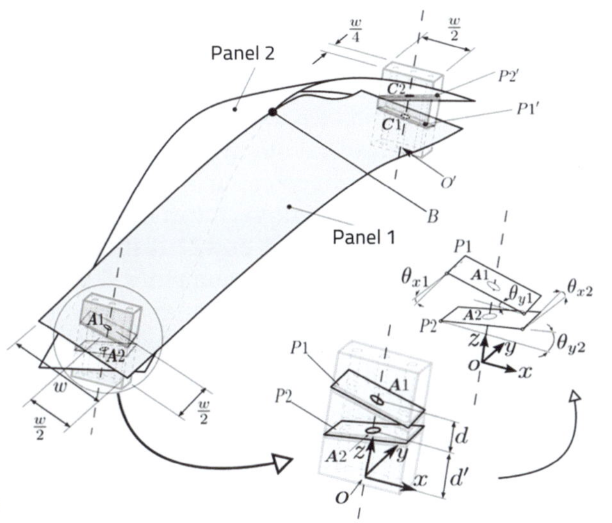

Fig. 8

Fig. 9

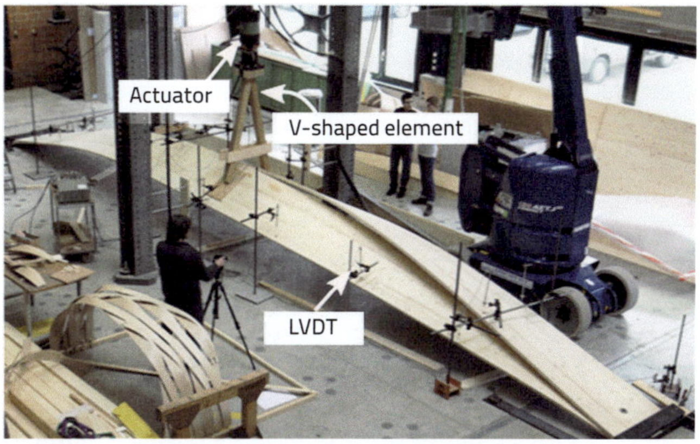

Fig. 10 a

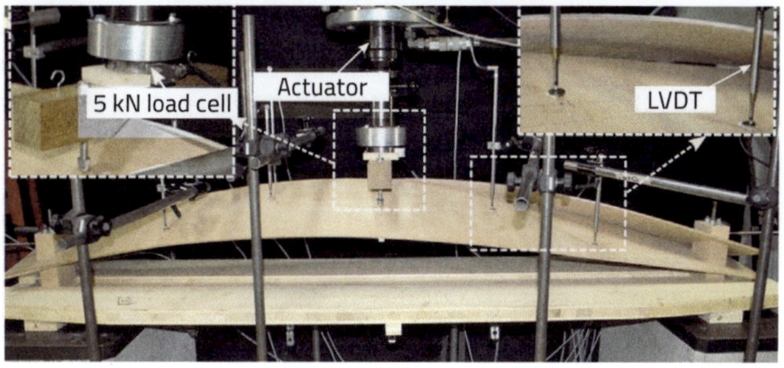

Fig. 10 b

Fig. 7 Textile module assembled with wedge connection

Fig. 8 Definition of the assembly parameters

Fig. 9 Textile module TM2 with wooden base (2) and wedges (1)

Fig. 10 Test setup for a) prototype TM1 and b) prototype TM2

tribute the load on each panel **(figs. 10 b and 11 b)**. Again, the test is run under displacement control (constant rate of 3 mm/min) and the prototype is loaded up to 0.5 kN.

In both cases, the resulting deflections are recorded at several positions along the prototype surface using external transducers mounted on self-supporting vertical steel rods, as shown in figs. 10a and 10b. Variable linear differential transducers (LVDTs) with a measurement range of ±100 mm are employed for measuring the vertical component of the displacement for prototype TM1. Data is typically recorded at three positions (1, 2, 3 in fig. 11a) and the displacement of the actuator is measured by using a LVDT mounted onto the actuator. The positions of LVDTs in the local frame (L,T) are: LVDT1 (L = 6.1875 m, T = −0.4125 m), LVDT2 (L = 8.075 m, T = 0.4125 m), and LVDT3 (L = 1.25 m, T = 0.4125 m).

For prototype TM2, LVDTs with measurement ranges of ±30 mm and ±50 mm are placed at seven positions, as indicated in fig. 11 b. Those with measurement ranges of ±50 mm are only used to record the vertical displacements at points 4 and 5. For positions 6 and 7, the transducers are directed perpendicular to the top wedge connection in order to check its deformation along the horizontal y-axis during loading. As can be observed in the enlarged region of fig. 10 b, small circular pieces of metal are glued onto the panels to provide a flat mounting surface in order to maintain contact with the transducers during the test. A grid mesh composed of 78 × 8 elements of 3 × 3 cm size (not visible in the image presented), is also drawn on the upper surface of each panel. This grid corresponds to the numerical mesh used in the finite element analysis. Finally, the positions of LVDTs in the local frame (L,T) are given by: LVDT1 (L = 0.54 m, T = 0.06 m), LVDT2 (L = 0.84 m, T = −0.03 m), LVDT3 (L = 1.8 m, T = 0.06 m), LVDT4 (L = 1.5 m, T = −0.03 m), and LVDT5 (L = 0.84 m, T = 0.03 m).

3 Numerical modeling

A three-dimensional, geometrically nonlinear, finite element analysis is proposed at this stage to simulate the assembly process of the textile module, including the fixation of the pinned and wedge connections. The aim is to calculate the construction stresses (initial stresses) and to reproduce the structural behavior of the structure under bending static loads.

3.1 Material properties

In this study, GFP and Okoumé plywood are modeled as single-layer orthotropic linear elastic materials, assuming that the layers are rigidly bonded together. When these simplifying assumptions are taken into account, only four independent engineering constants are required to fully characterize the material behavior, namely: the longitudinal E_L and transverse E_T Young's moduli, in-plane shear modulus G_{LT} and Poisson's ratio ν_{LT}.[16,17] The equivalent unidirectional lamina is consequently assumed to be under a state of plane stress with the principal material directions (L,T) indicated in fig. 3. The equivalent material properties given in Table 1 are used for the numerical analysis. In the calculation, the average value $\nu_{LT}=0.3$ will be considered for Poisson's ratio. Volumetric changes (shrinking or swelling of wood panels) due to moisture loss or gain are not taken into account here.

3.2 Finite element model

The commercial FE code ABAQUS is used to build a three-dimensional model of the assembly process for the textile module. Because large rotations and deflections, as well as boundary nonlinearities (i.e. sudden change in contact conditions between the panels) are expected, a geometrically nonlinear analysis is considered, where the specified displacements and loads are applied incrementally.

Each panel is discretized by means of six-node, triangular, thin-shell elements of uniform size (no refinement). Specifically, fully integrated STRI65 elements with five degrees of freedom per node (three displacements u_x, u_y, u_z and two rotations θ_x, θ_y) are used. These elements, adapted for large rotation but small strain, provide accurate solutions in the framework of the classical (Kirchhoff) shell theory, where the shell normal remains perpendicular to the shell reference surface (i.e. negligible transverse shear).[18–20]

An implicit time-integration procedure that relies on the Newton-Raphson iterative scheme[21] is typically used to solve the simultaneous incremental, nonlinear equations. Accordingly, an estimate of the incremental displacement field satisfying the displacement and traction boundary conditions is obtained at the end of a generic time increment. In this analysis, the calculations are achieved with a maximum time frame of 0.1 minute for each increment.

3.3 Boundary and assembly conditions

The simulation of the textile module that is pinned and built by wedge assembly requires the consideration of the construction procedure presented in section 2.2. The same nomenclature is repeated here. Thus, rectangular shell surfaces are used to represent the panels, which are positioned in the global coordinate system (x,y,z) of ABAQUS, as indicated in section 2.2. In particular, they are symmetrically rotated by an angle α (here $\alpha=10°$) about the global z-axis. The assembly process is then simulated by applying appropriate boundary and assembly conditions on the panels, as detailed below for the pinned and wedge connections. The pinned connection (prototype TM1) has a modeling procedure as follows:

- The boundary conditions $u_x=u_y=u_z=0$ (ui is the displacement component in the i-axis direction) are imposed at points A1 (panel 1) and A2 (panel 2). These points are located at the origin A of the frame.
- Only the vertical displacement $u_z=0$ is imposed at points C1 and C2 (Fig. 4). Consequently, these points are forced to remain in the plane (x,y).
- The displacement vectors $\mathbf{u}_{C1}=-u_C\hat{\mathbf{e}}_1$ and $\mathbf{u}_{C2}=u_C\hat{\mathbf{e}}_1$ ($\hat{\mathbf{e}}_1$ is the unit vector in the x-direction) are applied to points C1 and C2 respectively to correspond with the common point C' of the y-axis,

$$u_C=(l-w)\sin(\alpha) \qquad \text{Equ. 1}$$

where u_C is equal to 2005.64 mm and 364.66 mm for TM1 and TM2 respectively.
- The midpoints B1 and B2 of the panel edges are simultaneously shifted to the contact point B by applying the displacement vectors $\mathbf{u}_{B1}=-u_B\hat{\mathbf{e}}_1+w_B\hat{\mathbf{e}}_3$ and $\mathbf{u}_{B2}=u_B\hat{\mathbf{e}}_1+w_B\hat{\mathbf{e}}_3$ ($\hat{\mathbf{e}}_3$ is the unit vector in the z direction) at points B1 and B2 respectively. The expression of u_B for this transformation can be determined as

$$u_B=(l-w)/2\cdot\sin(\alpha)-w/2\cdot\cos(\alpha) \qquad \text{Equ. 2}$$

with respective numerical values of 623.66 mm and 64.15 mm for TM1 and TM2.

The parameter w_B measured by the experimental prototype is imposed, modifying the y-coordinate of point C' that becomes point C (figs. 4 and 5). Note that w_B corresponds to the distance BB' in fig. 5.

For the wedge connection (prototype TM2), the modeling procedure is as follows:

- The boundary conditions $u_x=u_y=0$ are imposed at points A1 and A2.
- $u_z=d'$ is imposed at points A2 and C1, while $u_z=d'+d$ is imposed at points A1 and C2. For the definition of d and d' see fig. 8.

- $\theta_x = \pm\theta_{x1}, \theta_y = \pm\theta_{y1}$ are applied to constrain the nodes on the areas *p1* and *p2'* and $\theta_x = \pm\theta_{x2}, \theta_y = \pm\theta_{y2}$ to constrain the areas *p2* and *p1'*.
- The displacements in *x*-direction are obtained at points C1 and C2 using the displacement vectors $\mathbf{u}_{C1} = -u_C\hat{\mathbf{e}}_1$, $\mathbf{u}_{C2} = u_C\hat{\mathbf{e}}_1$, and equation 1. As previously, the points are free to move in the *y*-direction to the common point C where AC = *s*.
- For the wedge connection, only one displacement component (in the *x*-direction) is needed to position points B1 and B2 at the common point B. Thus, the displacement vectors $\mathbf{u}_{B1} = -u_B\hat{\mathbf{e}}_1$ and $\mathbf{u}_{B2} = u_B\hat{\mathbf{e}}_1$ should be applied to points B1 and B2 respectively. The expression of u_B is given in equation 2.

3.4 Contact interaction

To prevent interpenetration of the two panels, contact regions have to be defined. The general finite-sliding, surface-to-surface algorithm of ABAQUS/Standard is used, allowing arbitrary large sliding, as well as large rotations and deformations of the surfaces. A frictionless, hard-contact, pressure-overclosure relationship is considered, where the penetration of the slave surface into the master surface is minimized and no tensile stress is transferred through the interface. Master (*MS1, MS2*) and slave (*SS1, SS2*) surfaces that can potentially come into contact are specified in fig. 12 by subdividing each panel into two regions. Moreover, surface *SS1* (resp. *SS2*) is located on the lower face of panel 1 (resp. panel 2) while *MS1* (resp. *MS2*) belongs to the upper face of panel 1 (resp. panel 2). Two contact surface pairings (*MS1, SS2*) and (*MS2, SS1*) can then be identified for the tracking contact algorithm. Frictional effects can be included by considering the (simplest) classical, isotropic Coulomb model to describe the frictional behavior between the contacting panels. Representative values of 0.3 and 0.5 will be taken for the (static) coefficient of friction μ in our case.

3.5 Numerical results

Fig. 13 a shows the initial three-dimensional shapes obtained numerically for the large-scale and intermediate-scale textile modules. In each case, the contact opening variable COPEN (in millimeters) corresponds to the clearance between the potentially contacting surfaces. Negative values of COPEN in fig. 13 a indicate small overclosures of the surfaces. For the large-scale geometry, two widths have been considered for the panels (i.e. $w = 1.650$ m and $w = 0.77$ m) that correspond to the prototypes TM0 and TM1. The FE mesh incorporates STRI65 shell elements with an element mesh size of 110 mm. Preliminary calculations were conducted with several element mesh sizes ranging from 80 to 150 mm, revealing a difference of less than 2 % for the in-plane principal stresses. In order to reduce the computational time, a uniform mesh size of 110 mm (resp. 20 mm) has been chosen for the simulation of large (resp. intermediate) scale geometries.

The Tsai-Hill failure criterion is now considered to estimate the macroscopic strength of the textile module. According to the Tsai-Hill failure theory[22] the macromechanical failure criterion for anisotropic materials is given by

$$I_F = \frac{\sigma_{LL}^2}{f_{m,L}^2} - \frac{\sigma_{LL}\sigma_{TT}}{f_{m,L}^2} + \frac{\sigma_{TT}^2}{f_{m,T}^2} + \frac{\sigma_{LT}^2}{f_v^2} \leq 1 \qquad \text{Equ. 3}$$

where σ_{LL}, σ_{TT} and σ_{LT} are the local stresses in the orthotropic material directions. In equation 3, $f_{m,L}$, $f_{m,T}$ and $f_v (= f_{v,L} = f_{v,T})$ are the bending and in-plane shear strengths of laminated wood materials given in Table 1. I_F values greater than 1.0 may lead to failure. The Tsai-Hill criterion I_F is visualized for each module in fig. 13b. Analysis of fig. 13b clearly indicates that maximum values of I_F are located at the contact point B (the most critical point of the textile module during the assembly process). Table 2 gives the numerical values of the stress components at point B for prototypes TM0, TM1, and TM2. According to the Tsai-Hill criterion, a module with the same characteristics as prototype TM0 is predicted as being likely to fail during the assembly process, as was experimentally observed in the laboratory.

	TM0	TM1	TM2
σ_{LL} (MPa)	45.1	18.01	9.61
σ_{TT} (MPa)	−3.3	−0.302	−0.61
σ_{LT} (MPa)	0.65	0.248	0.48

Table 2 Construction stresses values at point B for TM0, TM1, and TM2

Although the application of this criterion also strictly predicts failure for TM1 **(Fig. 13b)**, it should be noted that the characteristic value of $f_{m,L}$ on the fifth percentile ($f_{m,L} = 15.8$ MPa) provided by the manufacturer for GFP material and used to calculate the Tsai-Hill criterion, is rather conservative. Three-point bending tests, carried out on a sample size of 26 GFP plates of size 486 × 50 mm^2 revealed that the median value of the bending resistance of the GFP material is approximately 24 MPa. Nevertheless, due to the presence of knots and other irregularities in the wood, a relatively large variability in the results was observed, with bending strength values ranging from 4.3 to 55 MPa. If one takes the median value of $f_{m,L}$ ($f_{m,L} = 24$ MPa), the maximum value of I_F obtained for TM0 is 5.58, while for TM1 the maximum value of I_F obtained is 0.62. This value corresponds with our experimental observation that the TM1 module systematically withstands the forces applied to it during its construction.

4 Results and discussion

To compare the experimental and simulated initial shapes for TM2, the *z*-coordinates of points 1–5 **(Fig. 11b)** in the global frame (*x,y,z*) are reported in Table 3. It can be seen that the values obtained from FE model and those measured with the prototype correspond well. The greatest discrepancy is found to be 14% at point 4.

Force-displacement curves are presented in fig. 14 for prototype TM1, where the vertical displacements are recorded at locations (points) 1, 2, and 3 **(Fig. 11a)**. Here, compressive loads are given positive values. The upper left figure provides a plot of the recorded load versus actuator displacement. Experimental results indicate a nonlinear response of the structural element with a maximum sustained load of about 13 kN, wich is reached when the actuator moves down to 170 mm. At point 1, this maximum load leads to a deflection of 120 mm for the timber module **(fig. 14)**. Smaller and comparable deflections of about 80 mm are obtained at points 2 and 3, farther away from the loaded surface. Note that the value at point 1 is far more than the maximum deflection (span length/ 300=33.5 mm) of the serviceability limit state, indicating a degree of inherent flexibility of the structure.

Numerically, the module is loaded at midspan such that half of the total (vertical) force is uniformly distributed over a small (600 × 300 mm) rectangular region of each panel **(Fig. 14)**. In ABAQUS, a second "loading" step has been added to the first building step (section 3.3), allowing the bending load to be applied on the simulated (prestressed) textile module. The applied load is linearly ramped over the step up to 13 kN. Simulated force-displacement curves are given in fig. 14 at the corresponding experimental positions. The actuator displacement obtained from the FE model corresponds to the displacement of a node in the center of the rectangular load area. The graphs indicate a nonlinear (elastic) response of the structure to bending load. The experimental decrease of the load (after 13 kN) is clearly not reproduced in the simplified, elastic FE model considered.

It is observed that the numerical model tends to underestimate the rigidity of the structure. The discrepancy between the experimental and simulated curves might be primarily attributable to the material parameters used. Moreover, the assembly conditions considered for the large panels are clearly less controlled than those considered for the intermediate-scale specimen. The idealized "pinned connection" could not accurately represent the more complex (real) assembly conditions in this case. Nevertheless, relatively good correspondence is observed between the experimental and numerical predictions.

The force-displacement curves obtained for the prototype TM2, when loaded up to 0.5 kN, are plotted in fig. 15. Compressive loads are once again given positive values. To verify good reproducibility, the loading procedure was repeated three times and the corresponding curves are referred to as Test 1, Test 2, and Test 3 in fig. 15. A first series of tests was conducted directly after receiving the Okoumé material from the manufacturers. A second series of tests (Test 4 and Test 5) were carried out on samples of the same batch of panels about eight months later, where the same experimental procedure was applied. The experimental results indicate an overall nonlinear response of the structure. Figures labeled "Position *i*" ($i = 1...5$) refer to vertical displacements obtained with transducers positioned at points *i* ($i = 1...5$) as indicated in fig. 11b. Interestingly, it appears that certain regions of the structure flex upward during the loading test. Specifically, we observed that all the considered points shift downwards (towards the ground), except for point 1, which is shifted in the opposite direction (positive direction of *z*-axis). However, for simplicity of graphic representation, all of the displacements have been plotted with positive values in the graphs.

Fig. 15 indicates—particularly in positions 3, 4, and 5—smaller displacements for Tests 4 and 5 than those obtained in the first series of tests with the same loading conditions. This effect might be attributable to a change in environmental conditions. At 0.5 kN, a displacement amplitude of 10 mm is measured at point 1 that is comparable to the displacement recorded at the opposite point 3 (for the same panel). The vertical displacement at point 2 remains surprisingly small (less than 1 mm) compared to the other. Nevertheless, the curves of the second series of tests indicate the same trend, even though the sudden increase of the slope observed for the first series after 0.5 mm no longer appears.

	Point 1	Point 2	Point 3	Point 4	Point 5
Experimental values (mm)	198	271	135	220	221
FE values (mm)	202	293.6	156.4	257	257.5

Table 3 z-coordinate of points 1–5 in global frame (x,y,z)

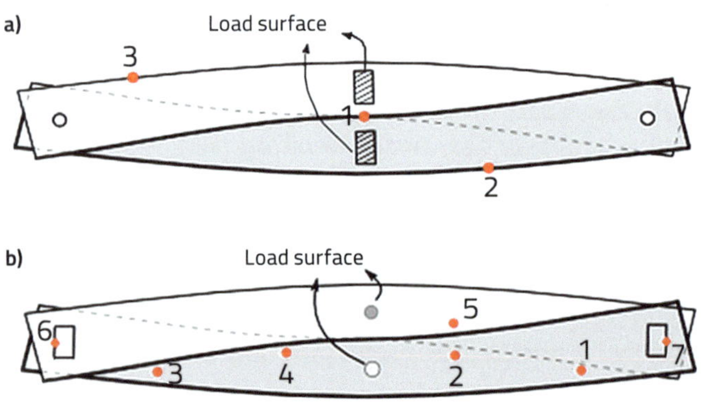

Fig. 11 Location of measurement points for **a)** TM1 and **b)** TM2 (top view)

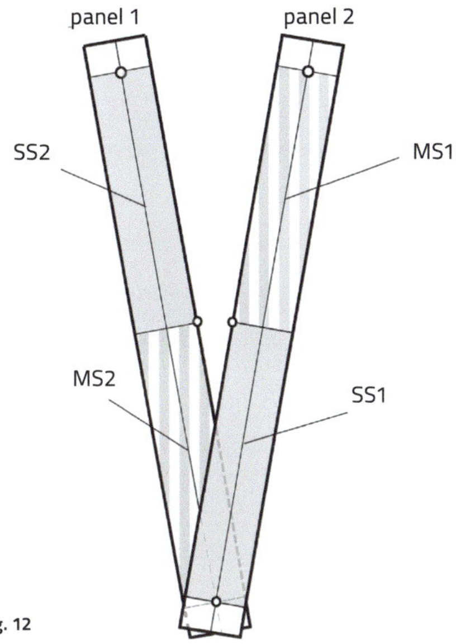

Fig. 12

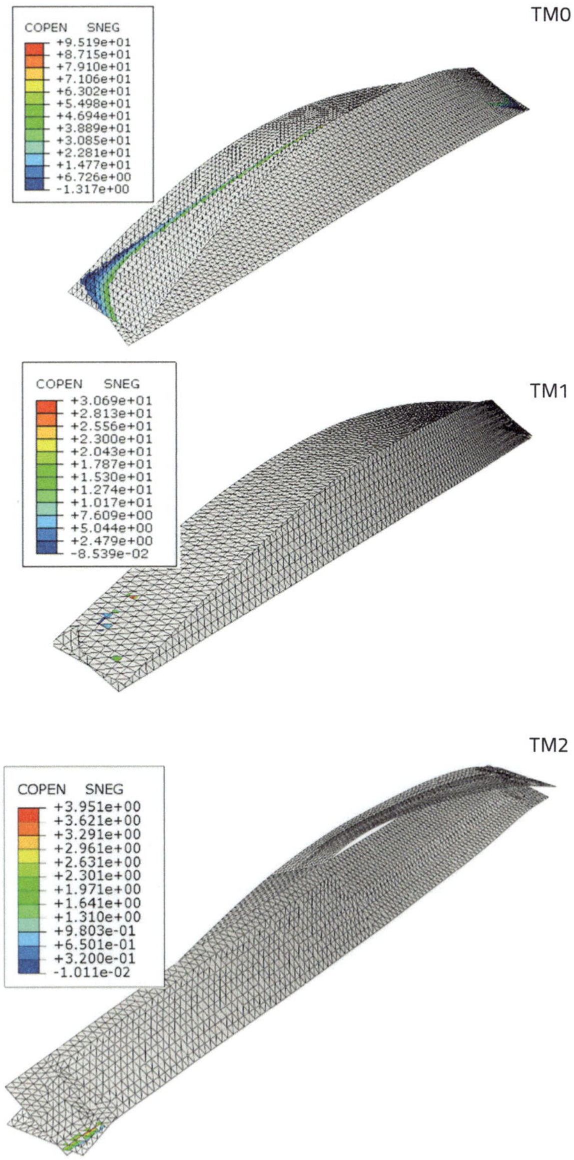

Fig. 13 a

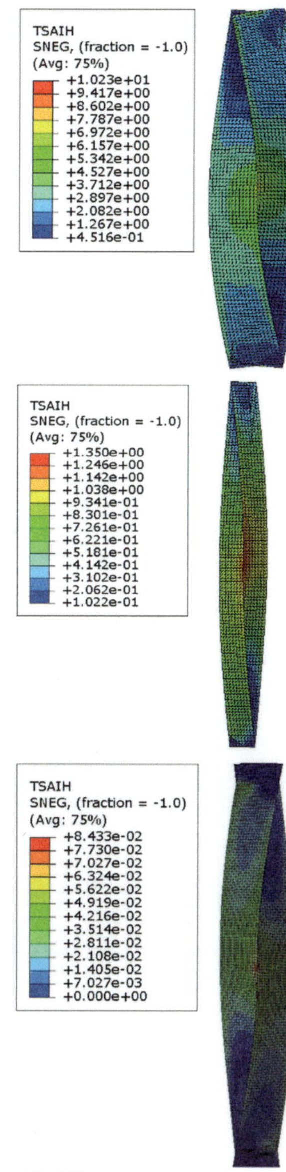

Fig. 13 b

Fig. 12. Definition of contact interaction between the panels

Fig. 13 a) Simulated initial shapes and visualization of clearance values between contacting surfaces of TM0, TM1, and TM2 in millimeters
b) Representation of the Tsai-Hill criterion IF

The largest displacements (i.e. about 15 mm) are encountered at positions 4 and 5. Potential wedge displacements can be recorded with transducers that are positioned horizontally at the two checking positions 6 and 7 **(Fig. 11b)**. A maximum (horizontal) displacement of 0.2 mm is recorded at these points for the wedges. This indicates a relatively small rotation of the wedge elements due to the bending load. Numerically, the wedges are fixed and the displacement is accordingly zero.

Fig. 15 depicts the load-displacement curves obtained with the FE model by considering the manufacturer and laboratory material data. Half of the total (vertical) force is uniformly distributed over a small circular disk with a radius of 15 mm on each panel **(Fig. 15)**. The applied load is ramped linearly over the step up to 0.5 kN. Using the manufacturer's data (labeled "FEM-manufacturer data" in the legend), the numerical model correctly reproduces the experimental behavior of the textile module at positions 3, 4, and 5. The experimental trend is also satisfactorily recovered at position 1. Introducing the elastic moduli from the laboratory measurements in the FE model ("FEM-laboratory data" in the legend), one recovers the experimental load-displacement curves of Tests 4 and 5 at positions 1, 3, and 5. It is worth noting that this material data was obtained during the period corresponding to the second series of tests, wich were conducted under the same environmental conditions. The model does not seem to be able to reproduce the small displacements recorded experimentally at position 2 and significantly overestimates them in the load range considered.

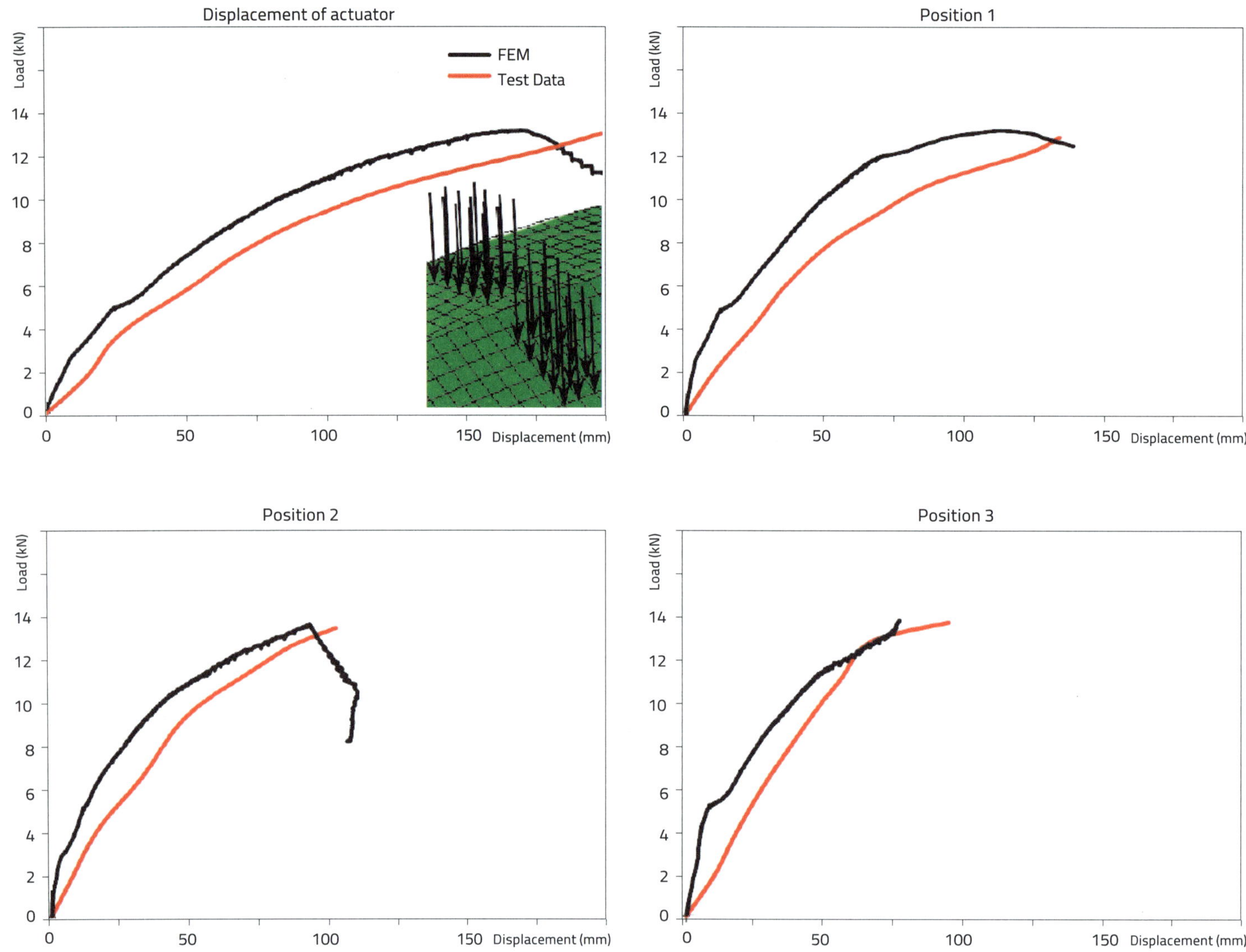

Fig. 14 Experimental and simulated force-displacement curves (TM1 prototype)

The effects of tangential friction between the contacting panels have also been addressed, taking the values $\mu = 0.3$ and $\mu = 0.5$ for the coefficient of friction. The simulations indicate that these effects do not influence the global behavior of the structure in this case. These are not documented here, while the plots are similar to those presented in fig. 15. It is concluded that no sliding occurs between the surfaces of the panels.

Finally, the introduction of wedge elements significantly affects the overall rigidity of the structure. Considering the TM2 geometry once again, fig. 16 compares the simulated displacement-load curves obtained at points 3 and 4 when the structure is assembled with pin or wedge connections. In both cases, the same material (i.e. Okoumé plywood) is considered and the modules are loaded up to 0.5 kN. As expected, the wedge connecting elements provide more rigidity for the structure than the pin connections.

5 Conclusion

This work has involved the investigation of a novel class of timber structures based on the logic and principles of textile techniques. A geometrically nonlinear, finite element model has been developed for the construction of a single textile module, including pinned and so-called "wedge connections" for the assembly conditions. For comparison, large-scale and intermediate-scale experimental prototypes with the previous connections have been constructed.

The proposed analysis first aimed at reproducing the initial shape of the structure and thus evaluating the resulting construction (initial) stresses induced during the assembly process. It was shown that the simulated shape could satisfactorily correspond to the experimental shape at several measurement points. Moreover, the anisotropic Tsai-Hill criterion based on the maximum in-

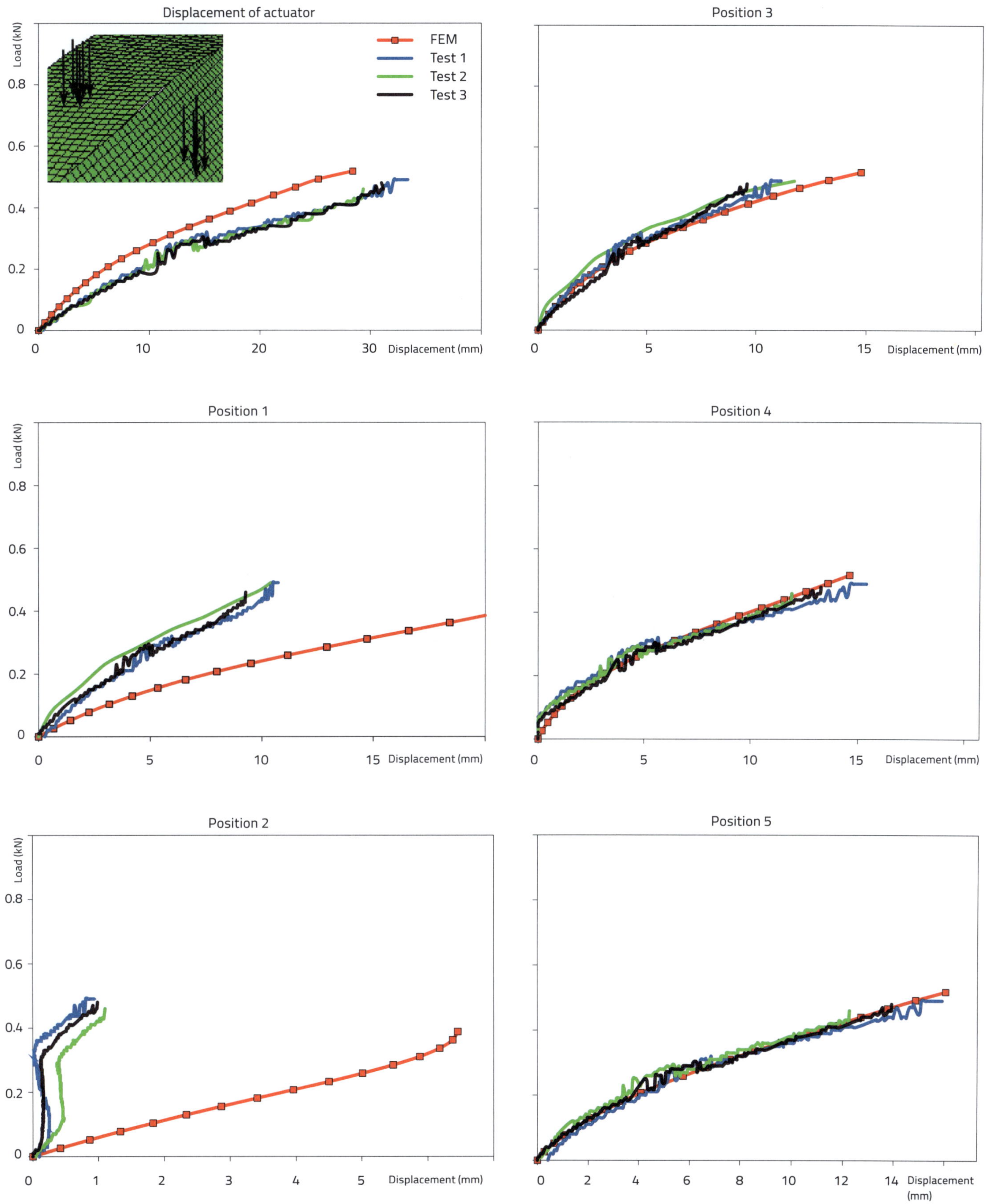

Fig. 15 Experimental and simulated force-displacement curves (TM2 prototype)

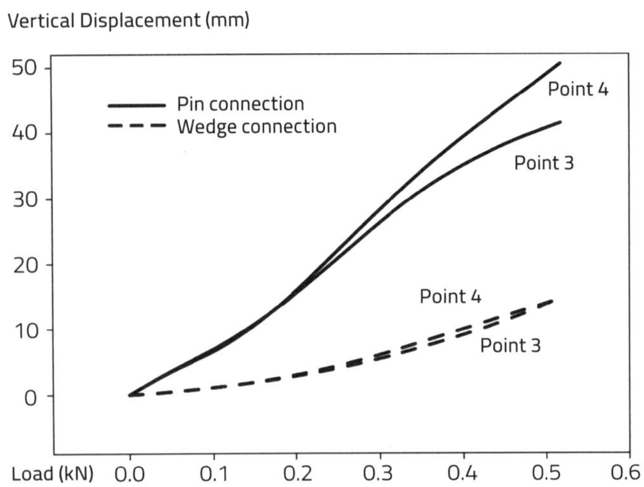

Fig. 16 Simulated displacements at points 3 and 4 for pin and wedge connections (TM2 geometry)

duced stresses allows one to select safe design parameters. It was observed that a length-to-width ratio $l/w = 7.5$ for the large-scale GFP panels led to failure during construction, while $l/w = 15$ was safe. For the intermediate prototype ($l/w = 9.75$), wedge connections comparatively led to lower levels of construction stresses and could be adopted at a larger scale. Second, the structural behavior of the textile module has been examined under bending tests. For the two geometries being considered, the resulting deflections have been measured and calculated at several locations. They highlight a nonlinear bending response of the textile module. Good correspondence is generally observed between the experimental results and the FE predictions at intermediate and large scales. Finally, the introduction of wedge elements was found to significantly improve the overall rigidity of the textile module.

Acknowledgements

The authors would like to thank the Swiss National Science Foundation for supporting this work through contract number 200021–126802. They would also like to express their gratitude to Maria Lindqvist for her participation in the experimental tests.

References

1 Langenbach, R. "Resisting earth's forces: Typologies of timber buildings in history." *Structural Engineering International: Journal of the International Association for Bridge and Structural Engineering (IABSE).* 2008;18:137–40.

2 König, H. *Bauen mit Holz als aktiver Klimaschutz. Bauen mit Holz: Wege in die Zukunft.* München: Prestel Verlag. 2011.

3 Iffland, J.S.B. "Folded plate structures." *ASCE J Struct Div.* 1979;105:111–23.

4 Happold, E. and W.I. Liddell. "Timber lattice roof for the Mannheim Bundesgartenschau." *Structural Engineer.* 1975;53:99–135.

5 Ibid.

6 Kelly, O.J., R.J.L. Harris, M.G.T. Dickson, and J.A. Rowe. "Construction of the downland gridshell." *Structural Engineer.* 2001;79:25–33.

7 Douthe, C. and O. Baverel. "Design of nexorades or reciprocal frame systems with the dynamic relaxation method." *Computers and Structures.* 2009;87:1296-307.

8 Popovic Larsen, O. *Reciprocal frame architecture.* London: Architectural Press; 2007.

9 Weinand, Y. "Innovative timber constructions." *Journal of the International Association for Shell and Spatial Structures.* 2009;50:111–20.

10 Weinand, Y. and M. Hudert. "Timberfabric: Applying textile principles on a building scale." *Architectural Design.* 2010;80:102-7.

11 D'Amato, E. "Finite element modeling of textile composites." *Composite Structures.* 2001;54:467–75.

12 D'Amato, E. "Nonlinearities in mechanical behavior of textile composites." *Composite Structures.* 2005;71:61–7.

13 Kalidindi, S.R. and E. Franco. "Numerical evaluation of isostrain and weighted-average models for elastic moduli of three-dimensional composites." *Composites Science and Technology.* 1997;57:293–305.

14 Page, J. and J. Wang. "Prediction of shear force using 3D non-linear FEM analyses for a plain weave carbon fabric in a bias extension state." *Finite Elements in Analysis and Design.* 2002;38:755–64.

15 Whitcomb, J., K. Woo, and S. Gundapaneni. "Macro finite element for analysis of textile composites." *Journal of Composite Materials.* 1994;28:607–18.

16 Daniel, I. and O. Ishai. *Engineering mechanics of composite materials.* Oxford, UK: Oxford University Press; 1994.

17 Bodig, J. and B.A. Jayne. *Mechanics of Wood and Wood Composites:* Van Nostrand Reinhold Company; 1982.

18 ABAQUS: ABAQUS Theory Manual Version 6.9. 2009.

19 Ertas, A., J.T. Krafcik, and S. Ekwaro-Osire. "Performance of an anisotropic Allman/DKT 3-node thin triangular flat shell element." *Composites Engineering.* 1992;2:269–80.

20 Murthy, S.S. and R.H. Gallagher. "A triangular thin-shell finite element based on discrete Kirchhoff theory." *Computer Methods in Applied Mechanics and Engineering.* 1986;54:197–222.

21 Reddy, J.N. *An introduction to non-linear finite element analysis.* New York: Oxford University Press; 2004.

22 See note 16 above.

Reprinted from *Engineering Structures*, vol. 46, p. 557–568, Oxford: Elsevier, 2013

4.3

Generation process and analysis of innovative timberfabric vaults

Etienne Albenque, Markus Hudert, Laurent Humbert, and Yves Weinand

Modular elements called timberfabric modules are generated by curving and connecting two slender, wooden panels. This paper concerns the study of vaults obtained by assembling several timberfabric modules together. A parametric tool is presented which can automatically generate a three-dimensional, finite element model of a structure for a given set of initial parameters. This tool assists the architectural design of these structures, providing insights into their geometry as well as into the construction stresses and the kinematic constraints between different constitutive elements.

Keywords ***timberfabric vaults, innovative structures, Python script, generative algorithm***

1 Introduction[I]

In recent years, textiles have increasingly served as a reference for architects and civil engineers.[1,2] The research project titled "Timberfabric"[3,4] focuses on the assembly principles and techniques that are used for the production of textile structures. By combining these principles with the particular properties of laminated timber panels, the research sets out to develop a modular construction system for supporting building envelopes. The basic unit for this system was developed during the first stage of the research. This basic unit, which will be referred to as a "timberfabric module," consists of two interlaced, mutually supporting timber panels.[5,6] Based on this unit, a considerable amount of empirical research has been carried out in which various vault-shaped, multi-module configurations were developed.[7] In parallel, numerical models have been developed to generate the deformed shape of several particular configurations of one timberfabric module.[8] Fig. 1 shows a prototype of a vault built by connecting together entire timberfabric modules with half timberfabric modules. The connectors are made from planar wooden panels cut by a numerically controlled machine. The design process of this prototype is based on an empirical approach, the principal steps of which are summarized by a simplified diagram (Fig. 2).

The complexity of the structural and architectural design of a timberfabric vault is a consequence of its spatial configuration, which depends on the geometry of the timberfabric module, which in turn depends on the material properties and assembly conditions of the timber panels. Moreover, construction stresses are generated during the assembly of the timberfabric modules and those stresses will affect the resistance and the behavior of the finished structure.

The objective of the work presented in this paper was to develop a parametric tool to assist in the structural and architectural design of a timberfabric vault. The research is focused on one particular family of timberfabric vaults. However, it aims at establishing an analytical framework for all timberfabric structures and describes the current state of a numeric tool that can be further developed.

The developed tool permits the automatic generation of a three-dimensional, finite element model of a timberfabric vault for a given set of initial parameters. This model can be used to study the spatial quality, as well as the structural behavior and the resistance, of the projected structure. The elaboration of this tool is presented in three steps. Firstly, we present the strategy adopted to generate a three-dimensional, finite element model of one timberfabric module, with given boundary conditions, by simulating the deformation of two flat wooden panels for which the user can specify both the dimensions and the mechanical properties. Secondly, several geometric properties shared by the resulting timberfabric modules are highlighted and "intermediate variables" are introduced to describe the geometry obtained for the given set of initial parameters. Furthermore, the results of a parametric study show how the initial parameters of the flat wooden panels influence the "intermediate variables" that describe the geometry of the corresponding timberfabric module. Finally, we

Fig. 1

Fig. 1 Timberfabric vault prototype

Fig. 2 Construction process of timberfabric vaults

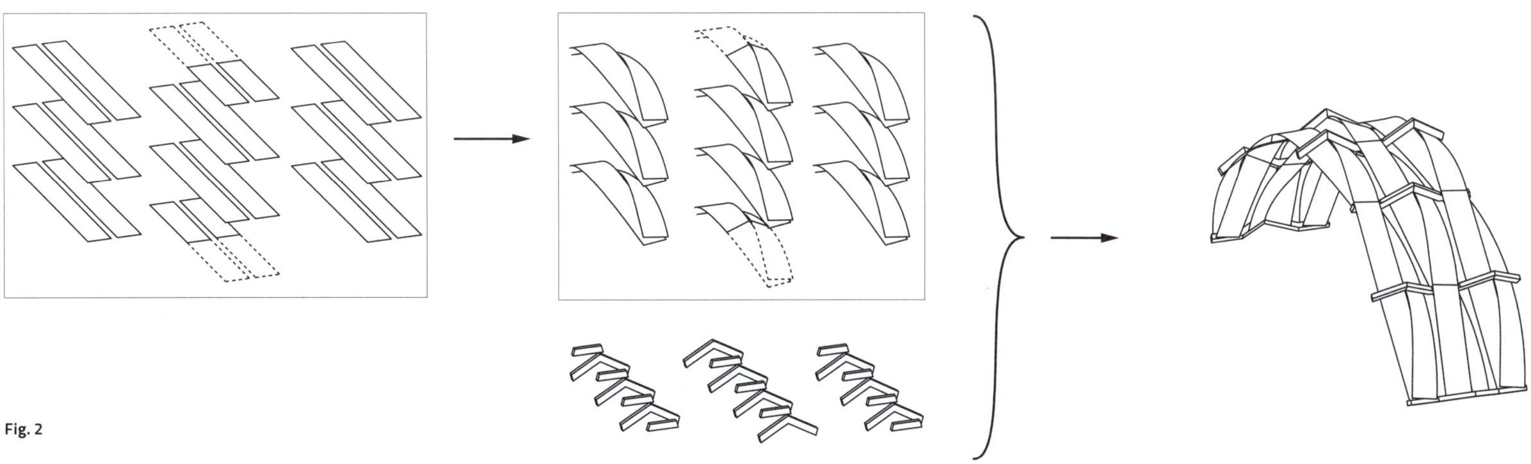

Fig. 2

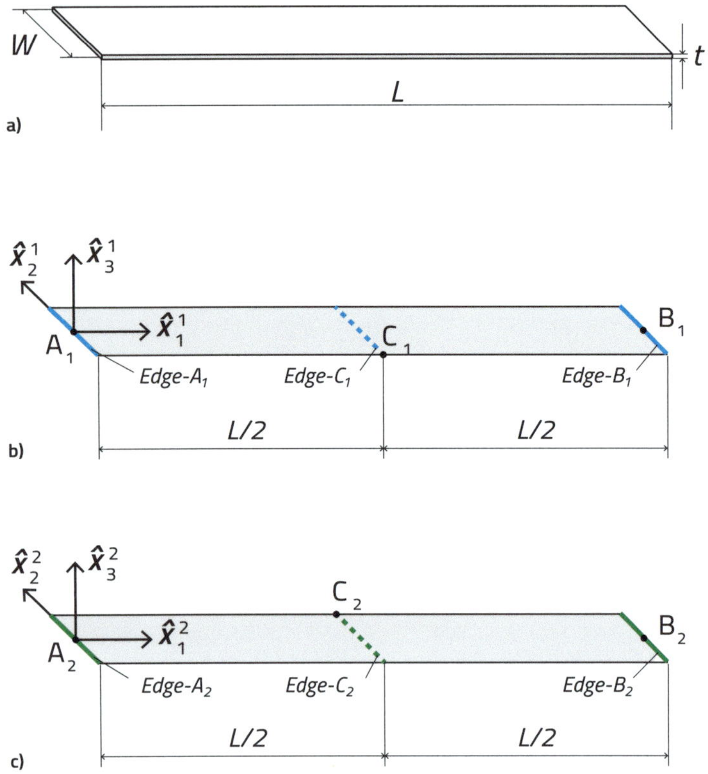

Fig. 3

Fig. 3 a) Parameters of panels
b) and c) Mid-surface of panels 1 and 2, definition of geometric entities

Fig. 4 Calculation steps in the process of generating a timberfabric module

show how those "intermediate variables" can be used to describe the relative positions of timberfabric modules in a timberfabric vault system. The numeric tool of the first part is therefore completed in order to enable the automatic generation of a full vault from one timberfabric module. The geometry of the connection pieces and their behavior are also generated automatically.

The different steps presented in this paper do not follow the same empirical logic developed before. In contrast, several initial hypotheses are based on observations of the built prototypes and their relevance to the modeling of a physical reality is demonstrated. This chronology illustrates the synchronicities between empirical and numerical approaches to design.

2 Parametric model for generating a timberfabric module

Here we describe the methodology employed to generate the geometry and calculate the initial stress state of a timberfabric module for a given set of geometric parameters and material properties of the timber panels. The methodology is based on an editable Python script written under the ABAQUS Scripting Interface environment[9] that builds and runs a non linear, finite element model.

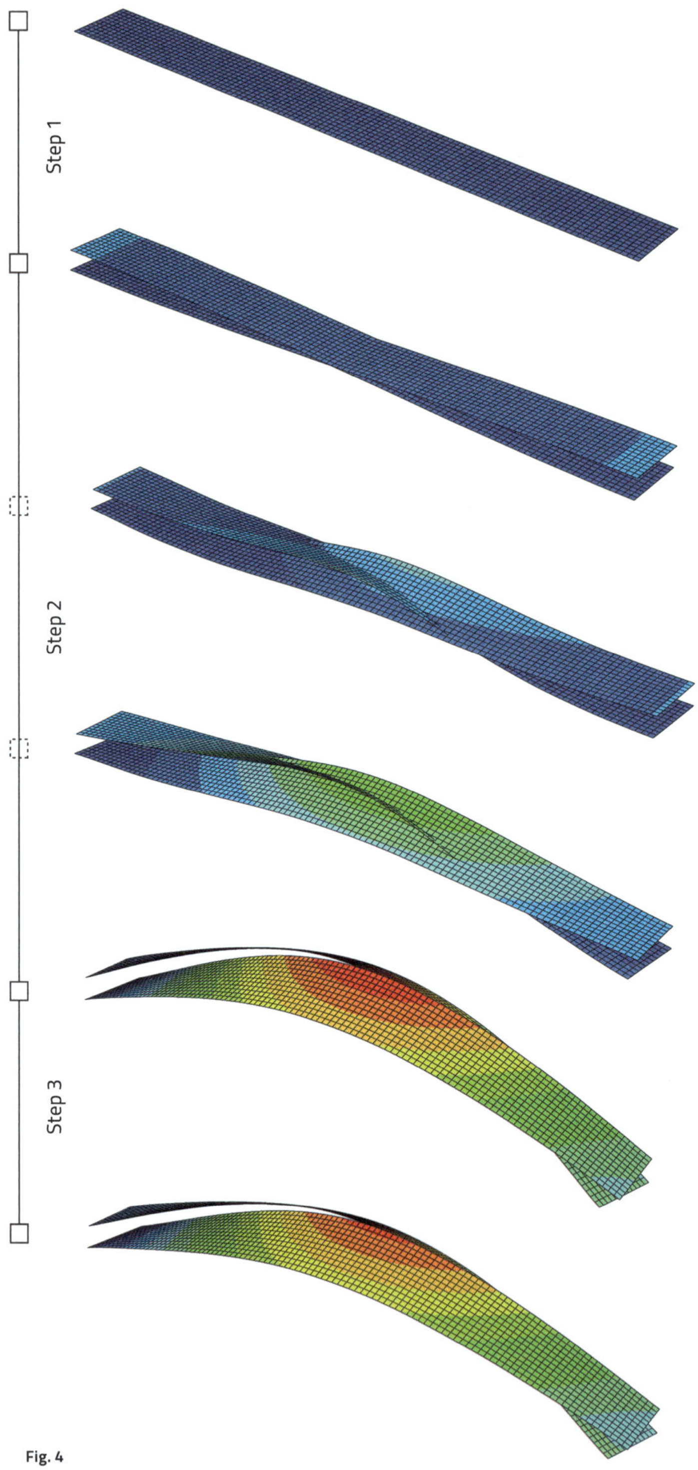

Fig. 4

2.1 Geometric parameters and material properties of the panels

Timberfabric modules (TM) can be regarded as elementary units that are connected together to form a more complex spatial structure, such as the vault prototype depicted in fig. 1. The module is formed by two self-supporting, initially flat, rectangular wooden panels that are deformed and assembled together by connection pieces at their extremities. The two panels that comprise a module come into contact at a single point on one of their longitudinal edges. The width, length, and breadth of the undeformed planar panels are noted respectively W, *L*, and *T* (Fig. 2). Depending on the scale of the timberfabric module, the timber panels are either constructed with plywood panels (for reduced-scale prototype) or laminated wood panels (for full-scale TMs). In both cases, the panels are constituted of orthotropic layers that are considered to be rigidly bonded together.[II]

2.2 Analysis framework

The finite element (FE) code ABAQUS is used to build a three-dimensional model of the deformation process that enables one to generate a timberfabric module from two planar timber panels. Rather than using the graphic user interface, the FE model is generated from the instructions written in a script using the programming language Python.[9] By automating the construction of the FE model, this method allows one to vary parameters with ease.

The initial mathematical settings of the model have been chosen in accordance with the nature of the physical problem.[III] The geometry of the panels is described with reference to their central surface. The position vector of a point/node *M* of the middle surface is denoted by $\mathbf{r}^M$ and the unit vector normal to the middle surface at this point is denoted by $\hat{\mathbf{n}}^M$. Moreover, $\mathbf{u}^M$ and θ^M are respectively the displacement and rotation vectors at *M*, with components and θ_j^M given along a base vector $\hat{\mathbf{e}}_j$. In particular, θ_j^M is the component of the rotation of the normal vector around the base vector $\hat{\mathbf{e}}_j$.

In the following, the same annotation will be used for node sets and for the corresponding geometric points. Depending on the context, the superscript *M* will either refer to a node or to the geometric point of coordinates $\mathbf{r}^M = (r_1^M, r_2^M, r_3^M)^T$.

Let A_1, B_1, A_2, and B_2 be the mid-points of the transverse edges of panels (1) and (2) (Fig. 3). The mid-points of a longitudinal edge of panels (1) and (2) are C_1 and C_2 and *Edge*-A_1, *Edge*-B_1, *Edge*-A_2, *Edge*-B_2, *Edge*-C_1, and *Edge*-C_2 refer to the transverse edges corresponding to the points A_i, B_i, and C_i. Again, the same notation is used both for the edge as a geometric entity and for the edge as a set of node numbers.

2.3 Initial position and boundary conditions

In the initial state of the numerical model, the two rectangular surfaces simulating the panels are laid flat and are placed in the same geometric position.[IV]

Thickness t (mm)	Width W (mm)	Length L (mm)	Connector height h_c (mm)
8.0	280	2450	80

E_1 (MPa)	E_2 (MPa)	ν_{12}	G_{12} (MPa)
4163	5088	0.3	552

Table 1 Geometric and mechanical properties of the panels

To simplify the operations that simulate the interweaving of the panels, it is permitted for the model to run through intermediate steps that would not be possible in the physical world; for example, the intersection of the two panels (Fig. 4). By comparing the initial simulated configuration and the geometry of a real TM, one notices that they share several geometric properties. In the model, these geometric properties are expressed by constraining the possible relative movements (displacements and rotations) of several points.[V] In the virtual space of the model:

- A_1 the central point of *Edge*-A_1 is fixed through all the subsequent transformations.
- All the middle-points of transverse panel edges, A_1, B_1, A_2, *and* B_2 are bound to remain in the same vertical plane.

2.4 Interactions and imposed displacements

Interaction conditions

In the actual process of fabrication, the panels are deformed in several sequential steps with intervening means of temporary fixation. During the final step, the edges of the panels are forced into position in order to adapt to certain geometric conditions imposed by connection pieces. For the numerical simulation, we chose to impose some of those conditions beforehand. In the FE model, imposing interaction conditions consists of the elimination of degrees of freedom (DOF) of a group of nodes and the coupling of their motion to the motion of a single node (called the "master node" in ABAQUS). For example, if all the displacement degrees of freedom of the nodes of a straight edge are coupled with the displacement DOFs of one node of the edge, then the nodes will be aligned for any deformation state of the system. Thus, the effect of a straight, rigid piece of connection pinned continuously along the edge is simulated.[VI] As for the initial boundary conditions, these constraints are defined by close observation of physical mock-ups and reproduce the kinematic constraints imposed by physical connection pieces. It is worth noting that these connection pieces have been developed empirically following construction constraints. Therefore, the abstract geometric model is already informed by the constraints of the physical construction.

Along the extremities of a TM, a rigid connection forces the extreme transverse edges of the panels into two parallel planes. As will be shown in the following section, this allows geometric continuity of surfaces belonging to adjacent TMs forming an arch. In the model, this effect is translated thus:

- The interaction conditions constrain all the edges defined above.
- (*Edge-A_1, Edge-B_1, Edge-A_2, Edge-B_2, Edge-C_1, and Edge-C_2*) remain rectilinear.
- The panels are constrained to remain locally tangential to planes along their extreme transverse edges (*Edge-A_1, Edge-B_1, Edge-A_2, Edge-B_2*), while at each extremity of the timberfabric module, the planes for each panel are constrained to be parallel (they have the same normal vector).

Imposed displacements

With the boundary and interaction conditions defined above, the final step to generate the geometry of the TM is to impose displacements on a set of points. The first displacements simulate the effect of the connection pieces at the extremities of the modules. In the first step, the displacement of node A_2 is constrained along the straight line $D(A_1,\hat{\mathbf{n}}^{A_1})$ and the displacement of node B_1 is constrained along the straight line $D(B_2,\hat{\mathbf{n}}^{B_2})$. The norm of the dispacement, h_c, is also an initial parameter of the design. It specifies the height of the connection pieces at the extremities of the TM, thus, $\mathbf{u}^{A_2}=h_c\,\hat{\mathbf{n}}^{A_1}$ and $\mathbf{u}^{B_1}=h_c\,\hat{\mathbf{n}}^{B_2}$.

In the second step, we impose $\mathbf{u}^{C_1}=\frac{W}{2}\hat{\mathbf{e}}_2+\frac{W}{2}\hat{\mathbf{e}}_3$ and $\mathbf{u}^{C_2}=-\frac{W}{2}\hat{\mathbf{e}}_2+\frac{W}{2}\hat{\mathbf{e}}_3$ The imposed displacements on nodes A_2 and B_1 are maintained.

The displacements $\frac{W}{2}\hat{\mathbf{e}}_2$ and $-\frac{W}{2}\hat{\mathbf{e}}_2$ move respectively the points C_1 and C_2 in the plane, $P(A_1,\hat{\mathbf{e}}_2)$, a plane that already contains the points A_1, A_2, B_1, and B_2. In both cases, the last term $\frac{W}{2}\hat{\mathbf{e}}_3$ is added to force adoption of one of the two equi-probable buckling modes of the panels: indeed, the displacement along $\hat{\mathbf{e}}_2$ induces an important bending moment in the panels along $\hat{\mathbf{e}}_3$ and if no displacement along $\hat{\mathbf{e}}_3$ were specified, the panels would continue to have two equi-probable, lateral buckling modes. For reasons of convergence of the algorithm, the norm of this displacement has the same order of magnitude as the expected displacement. In the final step, the constraints on *Edge-C_1* and *Edge-C_2* are released and the points C_1 and C_2 reach their "natural" post-buckling position.

2.5 Example

For the prototype of the vault described in the introduction, the panels are made of commercial Okoumé plywood.[VII] The deformed shape and displacement fields are reproduced in fig. 4 at the beginning and end of each step of imposed displacements. The objective of this modeling is to calculate the geometry and the stress state of the timber panels that constitute a TM, but is not intended to simulate the real process of fabrication. We can thus allow the pair of panels modeled to interpenetrate each other during the calculation and propose a virtual assembly process that is far simpler than the actual fabrication assembly process of a TM prototype.

2.6 Internal forces and boundary reactions in deformed state

The process described above allows one to generate the deformed shape of two wooden panels with complex boundary and interaction conditions. At this stage, it is legitimate to question whether these particular conditions are relevant to describing the physical reality of a timberfabric module.

If there is a boundary condition on node M, vector $\mathbf{R}^M$ designates the reaction force at the node. In its final deformed shape, the system constituted of the two wooden panels is subjected to several internal forces and to six reaction forces at points A_1, A_2, B_1, B_2, C_1, and C_2. By considering the nature of the boundary conditions and the absence of exterior actions, it can easily be demonstrated that all these forces are collinear to $\hat{\mathbf{e}}_2$ and $\mathbf{R}^M$ ¥ $\hat{\mathbf{e}}_2 = \mathbf{0}$ for M Œ $\{A_1, A_2, B_1, B_2, C_1, C_2\}$.

Furthermore, if we introduce C—the point such that $\mathbf{r}^C=(\mathbf{r}^{C_1}+\mathbf{r}^{C_2})/2$—the deformed shape of the panels has an axial symmetry of axis $\Delta(C,\hat{\mathbf{e}}_3)$. By using the axial symmetry and the absence of external forces, we can show that $\mathbf{R}^{A_1}=-\mathbf{R}^{B_2}$, $\mathbf{R}^{A_2}=-\mathbf{R}^{B_1}$ and $\mathbf{R}^{C_1}=-\mathbf{R}^{C_2}$.

By observing the numerical results, it is possible to assimilate the reaction forces $\mathbf{R}^{C_1}$ and $\mathbf{R}^{C_2}$ to internal forces. Numerically, we observe that the distance between the positions of C_1 and C_2 in the deformed state is inferior to 0.5 % of the length of the panels. Accordingly, $d=\|\mathbf{r}^{C_2}-\mathbf{r}^{C_1}\|<0.005\times L$.

By assuming that this distance is negligible, the two opposed reaction forces can be assimilated to a contact force between the panels. To confirm this hypothesis, the model was repeated by slightly shifting the initial position of the points C_1 and C_2 such that initially

$$\mathbf{r}^{C_1}=\left(\frac{L+d}{2},-\frac{W}{2},0\right)^T,\ \mathbf{r}^{C_2}=\left(\frac{L-d}{2},\frac{W}{2},0\right)^T$$

After this iteration, we oberved $d'=\|\mathbf{r}^{C_2}-\mathbf{r}^{C_1}\|<0.001\times L$. The result of the first round of modeling has thus permitted a more accurate prediction of the coordinates of the contact point.

Furthermore, in part 3, several variables will be introduced to quantitatively describe the model. The numerical results of the variables for the two situations described above is negligible, meaning that the first approximation is acceptable. Finally, the four remaining reactions are consequences of the descision to impose the co-planarity of the points A_1, A_2, B_1, and B_2. It will be demonstrated in part 4 that this geometric condition is necessary for the assembly of several timberfabric modules and that the corresponding reaction forces will be calibrated from one timberfabric module to another.

3 Geometric aspects of timberfabric modules

In order to describe the relative positions of several timberfabric modules that form a timberfabric vault, it is essential to introduce new variables that describe the geometry of a timberfabric module. The definition of such "intermediate variables" necessitates some geometric observations of timberfabric modules.

3.1 Geometric description

The drawings in fig. 5 illustrate a few geometric consequences of the boundary and interaction conditions described in Part 2:

- The points A_1, A_2, B_1, B_2, C_1, and C_2 belong to the plane **P** $(A_1, \hat{\mathbf{e}}_2)$.
- The panels 1 and 2 have tangency planes along their extreme transverse edges and the planes corresponding to each panel are parallel at each side of the timberfabric module. In mathematical terms, P $(A_1, \hat{\mathbf{n}}^{A_1})$ PP $(A_2, \hat{\mathbf{n}}^{A_2})$ and P $(B_1, \hat{\mathbf{n}}^{B_1})$ PP $(B_2, \hat{\mathbf{n}}^{B_2})$.
- The four normal vectors $\hat{\mathbf{n}}^{A_1}, \hat{\mathbf{n}}^{A_2}, \hat{\mathbf{n}}^{B_1}$, and $\hat{\mathbf{n}}^{B_2}$, do not have components along $\hat{\mathbf{e}}_2$. That is $\hat{\mathbf{n}}^M \times \hat{\mathbf{e}}_2 = 0$, M Œ $\{A_1, A_2, B_1, B_2\}$.

3.2 Identification of intermediate geometric parameters

In order to quantitatively describe the geometry of the timberfabric module, four "intermediate variables" are introduced. Those variables are expressed as a function of components of the displacement and rotation vectors of particular points, so that they can be extracted from the numerical results of the finite element model. A geometric interpretation of those variables is given in fig. 6. More specifically,

$a = |q_2^M| = \arccos(\hat{\mathbf{n}}^M \times \hat{\mathbf{e}}_3)$, M Œ $\{A_1, A_2, B_1, B_2\}$ is the absolute value of the *inclination angle* of the tangency planes at the extremities of a timberfabric module.

$Q = |q_1^{C_1} - q_1^{C_2}| = 2|q_1^{C_1}| = 2|q_1^{C_2}|$ is the *angle* between the panels at mid-span.

$L_f = L + u_1^{B_2}$ and $h_m = u_3^{C_1} = u_3^{C_2}$ are respectively the *span* and the *height* of a timberfabric module.

3.3 Parametric study

Thus far, we have introduced:

- The initial parameters L, W, t, and h_c to describe the geometry of the timber panels and the interaction condition at their ends.
- The homogenized mechanical properties of the panels: E_1, E_2, n_{12}, G_{12}.
- The *"intermediate variables"* a, L_f, h_m, and Q to describe the geometry of the timberfabric module.

By adapting the script that calculates the deformed configuration of the panels of a timberfabric module, it is possible to run a parametric study to observe the influence of the geometric parameters and the mechanical properties on the "intermediate variables" that quantitatively describe the geometry of a TM.

The results of this study provide useful information for the design of vaults with imposed dimensions.

$\frac{E_2}{E^{ref}}$	$\frac{G_{12}}{G_{12}^{ref}}$	$\frac{\nu_2}{\nu_{12}^{ref}}$	α (°)	$\frac{L_f}{L^{ref}}$	θ (°)	$\frac{h_m}{L^{ref}}$
0.8	1	1	40.33	0.948	125.89	0.1594
0.9	1	1	40.31	0.948	125.93	0.1594
1	1	1	40.29	0.948	125.96	0.1594
1.1	1	1	40.28	0.948	125.98	0.1593
1.2	1	1	40.27	0.948	126.00	0.1593
1	0.8	1	39.17	0.950	124.90	0.1570
1	0.9	1	39.74	0.949	125.44	0.1582
1	1	1	40.29	0.948	125.96	0.1594
1	1.1	1	40.83	0.947	126.46	0.1605
1	1.2	1	41.36	0.946	126.93	0.1616
1	1	0.8	40.46	0.948	125.99	0.1597
1	1	0.9	40.38	0.948	125.98	0.1596
1	1	1	40.29	0.948	125.96	0.1594
1	1	1.1	40.20	0.948	125.93	0.1592
1	1	1.2	40.11	0.949	125.89	0.1589

$\frac{W_2}{W^{ref}}$	$\frac{h_c}{h_c^{ref}}$	$\frac{t}{t^{ref}}$	α (°)	$\frac{L_f}{L^{ref}}$	θ (°)	$\frac{h_m}{L^{ref}}$
0.8	1	1	32.29	0.964	127.59	0.1333
0.9	1	1	36.04	0.957	126.56	0.1460
1	1	1	40.29	0.948	125.96	0.1594
1.1	1	1	45.03	0.939	125.57	0.1732
1.2	1	1	50.25	0.928	125.21	0.1873
1	0.8	1	40.28	0.951	125.91	0.1564
1	0.9	1	40.29	0.949	125.93	0.1579
1	1	1	40.29	0.948	125.96	0.1594
1	1.1	1	40.30	0.947	125.99	0.1609
1	1.2	1	40.31	0.946	126.02	0.1624
1	1	0.8	42.92	0.943	127.88	0.1651
1	1	0.9	41.48	0.946	126.86	0.1620
1	1	1	40.29	0.948	125.96	0.1594
1	1	1.1	39.29	0.950	125.16	0.1571
1	1	1.2	38.42	0.952	124.44	0.1552

Table 2 Results of the parameter study

For example, if we want timberfabric modules with a given length, rotation radius, and material, the results of the study will help to determine the corresponding length and width of the panels.[VIII]

The most relevant output variables are α and L_f as they will affect the geometry of the vault. The variables h_m and Q will affect the geometry of the connection pieces. For a given length of the panels, we observe that α and L_f are mostly influenced by the width W, as shown in Table 2. The thickness of the panels and height of the connector have a less critical influence on the geometry, but greatly influence the maximum stresses. Fig. 7 provides a graphic illustration of this influence.

4 Parametric model for timber vaults

The tool presented in Part 2 is further developed to enable the generation of a Finite Element model of timberfabric vaults. The script that builds the model uses a function of ABAQUS that enables us to use the result of a previous analysis as the initial state of a new analysis. Therefore, the deformed configuration of the generated timberfabric module can be imported into a new model, copied, rotated, and translated in the three-dimensional space. The timberfabric modules of the new model can then be connected to one another and to the ground in order to form a statically balanced vault.

The geometric configuration of the vault is based on several assumptions and can be described by a new set of user-parameters. Here, the study is restricted to the family of vaults obtained by assembling identical TMs. The logic of the assembly and its implementation in a script are detailed in the following sections.

4.1 Arch from timberfabric modules

Let TMj be a timberfabric module; the notations introduced previously are extended by adding the value "j" of the module in superscript. Accordingly, the local coordinates system associated to TM1 is written $(A_1^1, \hat{\mathbf{e}}_1^1, \hat{\mathbf{e}}_2^1, \hat{\mathbf{e}}_3^1)$. For brevity, a unit vector at a given point M of TMj will be denoted in the subsequent by $\hat{\mathbf{n}}_M^j$.

In the plane $\mathrm{P}(A_1^1, \mathbf{e}_2^1)$, a unique circle exists that is tangential to the plane $\mathrm{P}(A_1^1, \hat{\mathbf{n}}_{A_1}^1)$ and passes through the points A_1 and B_2. By symmetry, the circle is also tangential to the plane $\mathrm{P}(B_2^1, \hat{\mathbf{n}}_{B_2}^1)$. From Section 3 and according to the geometric construction of fig. 8, the radius R of the circle can be expressed as follows:

$$R = \left(\frac{L_f}{2 \times \sin\alpha}\right) \quad \text{Equ. 1}$$

In addition, the height h_2 of the arch $[A_1^1, A_2^1]$ is defined by:

$$h_2 = \mathrm{R} \times (1 - \cos\alpha) \quad \text{Equ. 2}$$

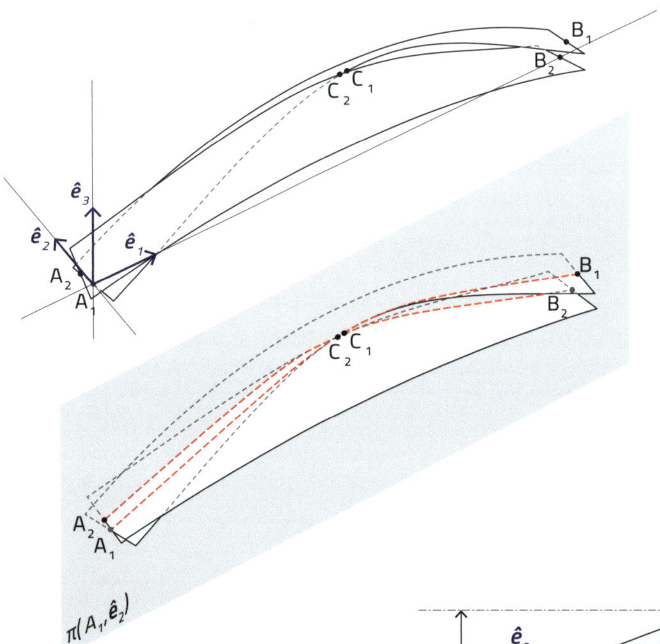

Fig. 5

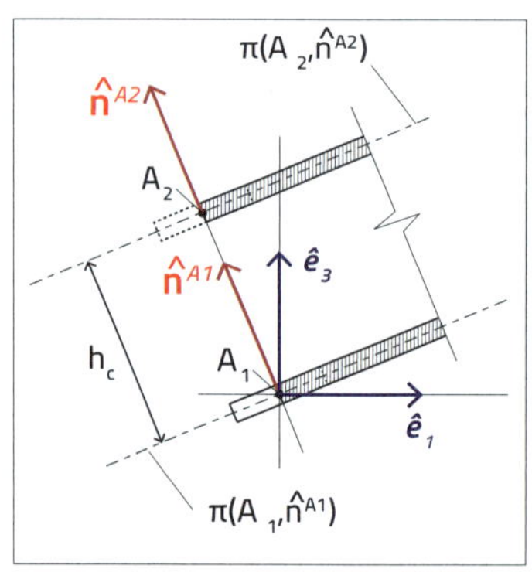

Section along plane P (A1, ê2)

Fig. 5 Geometric characteristics of timberfabric modules

Fig. 6 Definition of "intermediate variables"

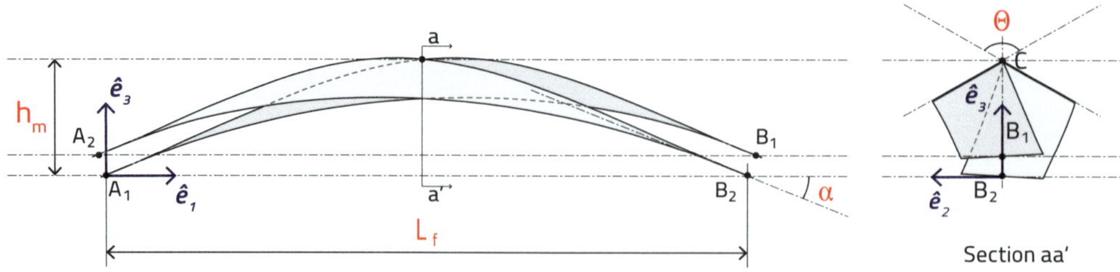

Fig. 6

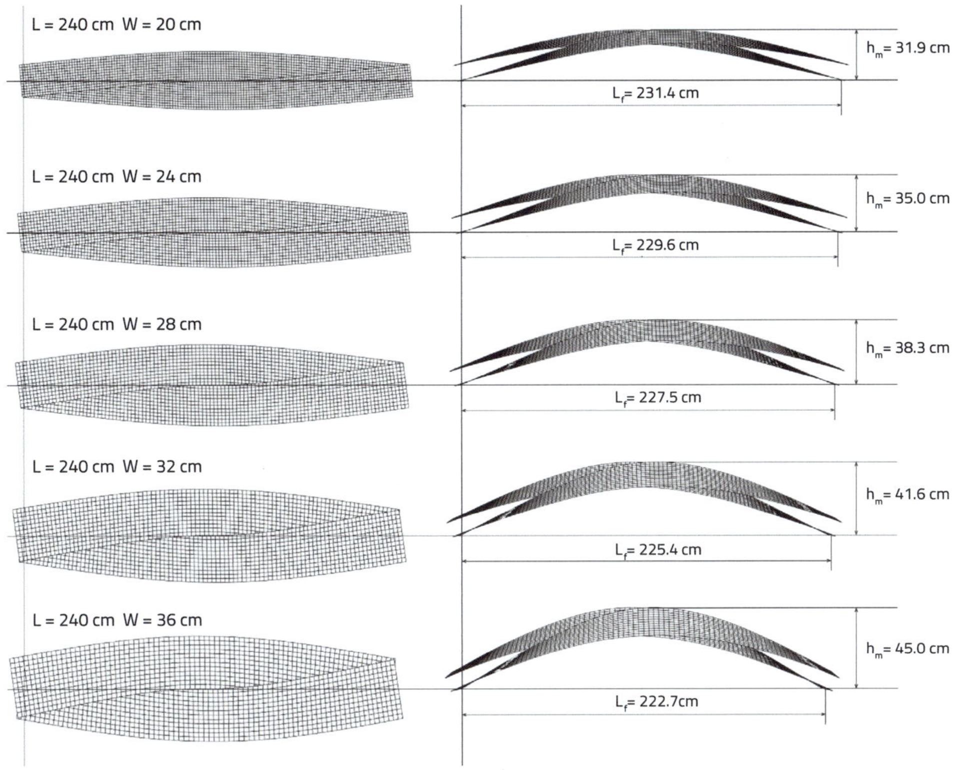

Fig. 7

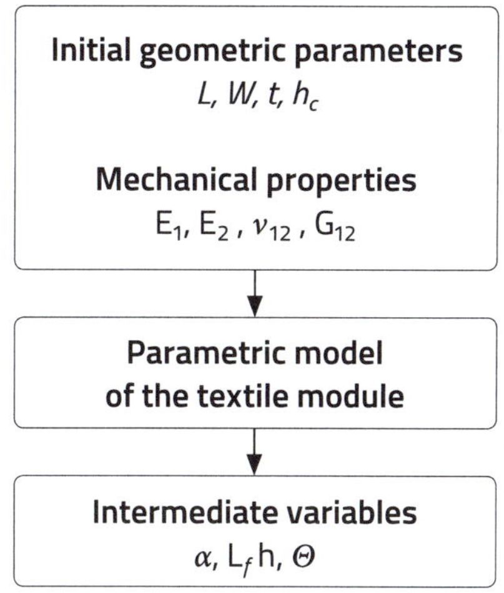

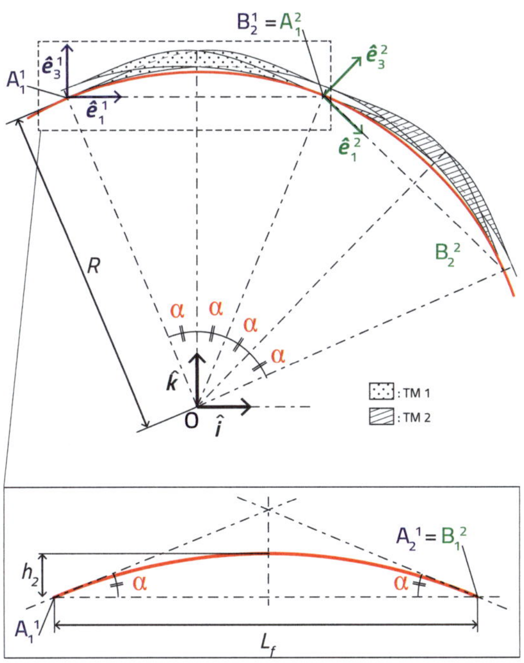

Fig. 8

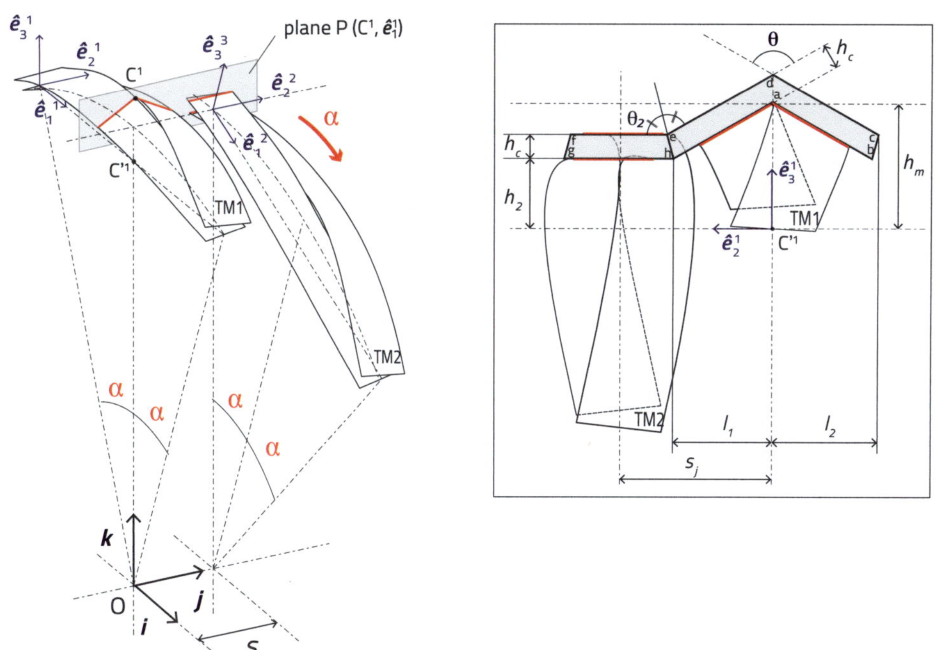

Fig. 9 a

Fig. 9 b

Fig. 7 Graphic results of the parametric study for $t = 8$ mm and $h_c = 80$ mm ($t/t_{ref} = 1$, $h_c/h_{cref} = 1$)

Fig. 8 The assembly of timberfabric modules to form an arch

Fig. 9 Connecting arches together

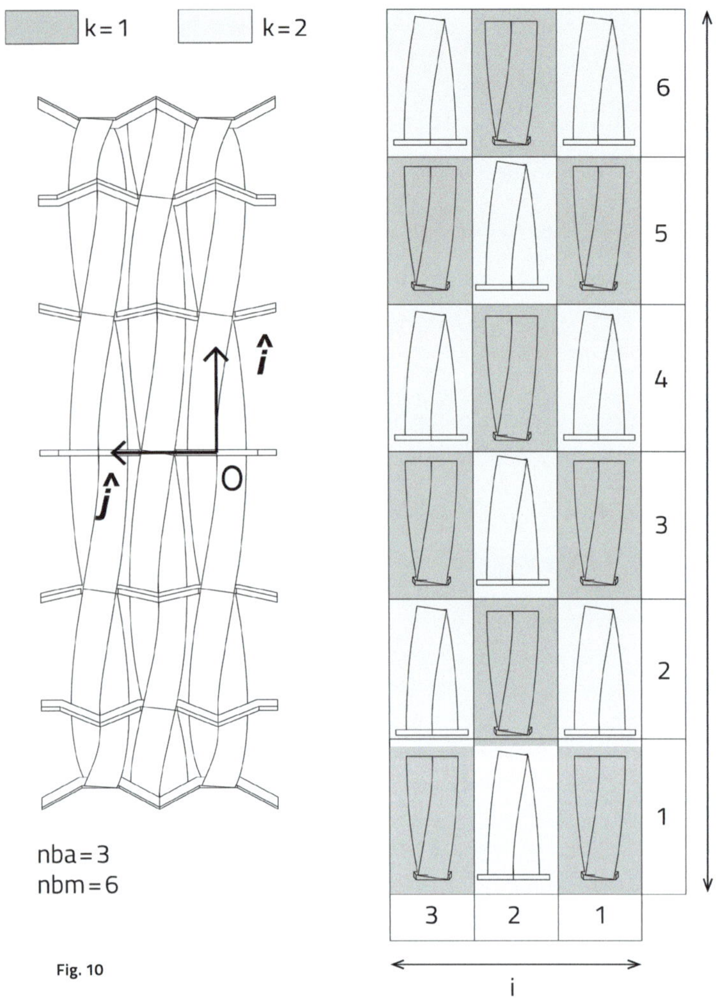

Fig. 10

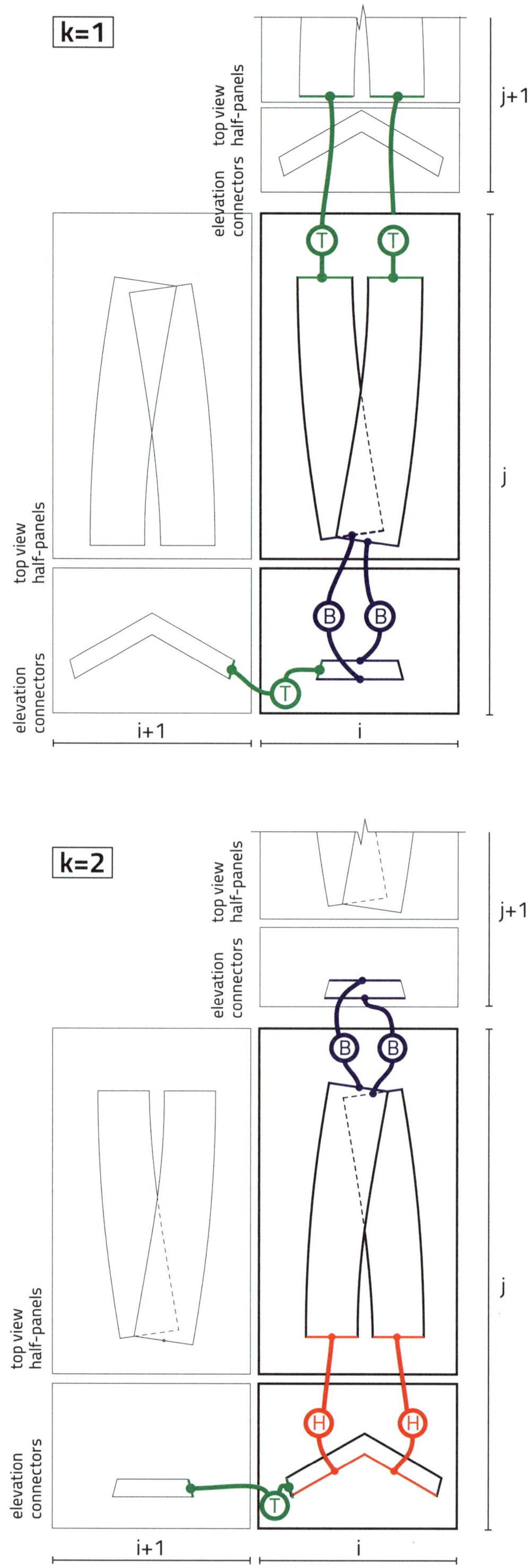

Fig. 11

Fig. 10 "Map" of a timberfabric vault

Fig. 11 Kinematic interactions between constitutive elements of the model

Fig. 12 Geometries generated for several values of the initial and intermediate parameters (hc=80 mm and material properties are given in Table 1)

In the local coordinates system of TM1, the coordinates of the circle center O^1 are $(\frac{L_f}{2}, -h_1, 0)^T$ with $h_1 = R - h_2$.

If we now introduce $(O, \hat{\imath}, \hat{\jmath}, \hat{k})$ as the global Cartesian coordinates for representing the vault system, TM1 is placed in the global system such that O^1 coincides with O and $\hat{\imath}$, $\hat{\jmath}$ and $\hat{k}$ are respectively collinear to $\hat{\mathbf{e}}^1_1$, $\hat{\mathbf{e}}^1_2$, and $\hat{\mathbf{e}}^1_3$. TM2 is obtained by rotating TM1 clockwise about the axis $\hat{\jmath}$ by an angle $2a$, resulting in the following conditions at the ends of the modules:

$$B^1_2 = A^2_1, \quad B^2_1 = A^2_2, \\ \hat{\mathbf{n}}^1_{B_2} = \hat{\mathbf{n}}^2_{A_1}, \; \hat{\mathbf{n}}^1_{B_1} = \hat{\mathbf{n}}^2_{A_2} \qquad \text{Equ. 3}$$

such that $P(B^1_2, \hat{\mathbf{n}}^1_{B_2}) = P(A^2_1, \hat{\mathbf{n}}^2_{A_1})$ and $P(B^2_1, \hat{\mathbf{n}}^1_{B_1}) = P(A^2_2, \hat{\mathbf{n}}^2_{A_2})$.

Equation 3 implies that there is a C^1-continuity between:
- "panel 1" of TM1 and "panel 2" of TM2
- "panel 2" of TM1 and "panel 1" of TM2

We also observe that the boundary reaction force in B^1_2 and B^2_1 are respectively opposed to the boundary reaction force in A^2_1 and A^2_2. These boundary reactions can therefore be replaced by interaction forces between TM1 and TM2. The nature of these interactions and their implementation in the FE model will be detailed in section 4.5. This two-module arch is geometrically smooth and has static equilibrium. Larger arches are obtained by assembling n timberfabric modules in a way that reproduces the geometry that would be obtained with two continuous panels woven together n times.

4.2 Vaults from arches

In this paper, a cylindrical vault is typically obtained by juxtaposing several arches along the axis $(O, \hat{\jmath})$ of the global coordinate system. The distance between two arches is fixed by the parameter s_j as explained below. The modules are rotated from one arch to the other by an angle α around the axis $(O, \hat{\jmath})$ (fig. 9). This implies that the extremity of an arch can be either a full timberfabric module or half a timberfabric module.

Up to this point, describing the kinematic relationship between several nodes has simulated interaction between the panels of a timberfabric module. In order to link the arches together, connection pieces are introduced in the model and referred to as "connectors."

Let TM1 be a timberfabric module placed in the global coordinates system, as in section 4.1 As illustrated in fig. 9, the timberfabric module TM2 is obtained by a translation of TM1 of vector $\mathbf{T} = s_j \hat{\jmath}$ and a rotation of angle α around the axis $(O, \hat{\jmath})$. The mid-surface of the connectors belongs to the plane $P(C^1, \hat{\mathbf{e}}^1_1)$: this plane is perpendicular to the tangency planes $P(A^2_1, \mathbf{n}^2_{A1})$ and $P(A^2_1, \mathbf{n}^2_{A2})$ and includes the points C^1, A^2_1, and A^2_2.

The geometry of a connector can be generated automatically from the previous parameters.[IX]

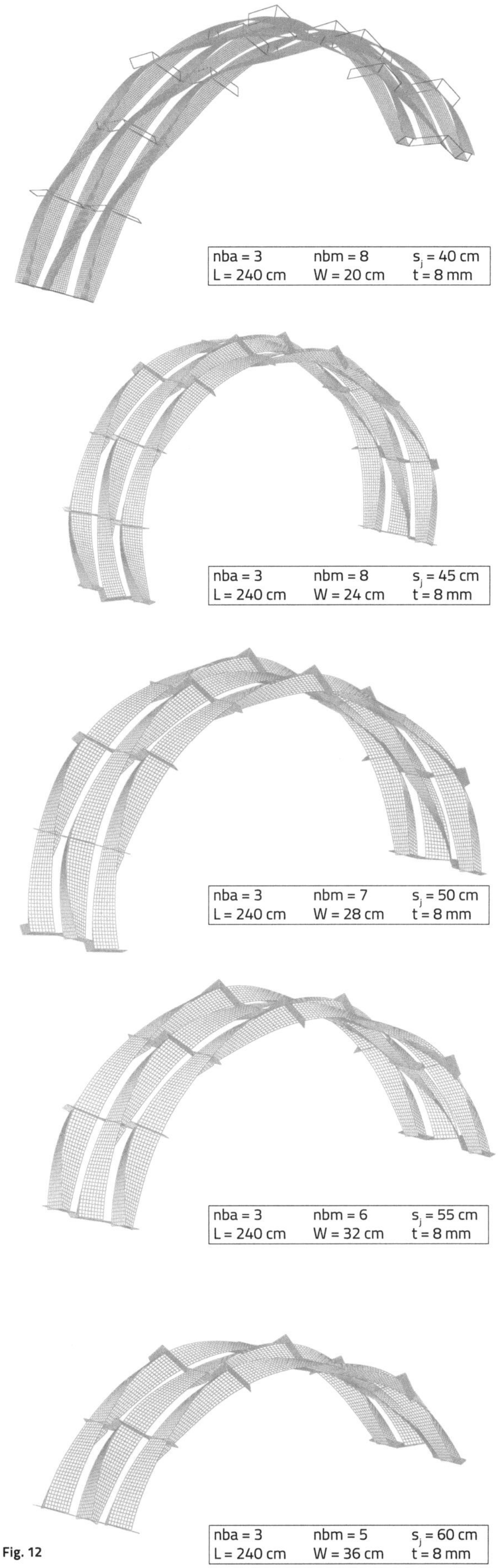

Fig. 12

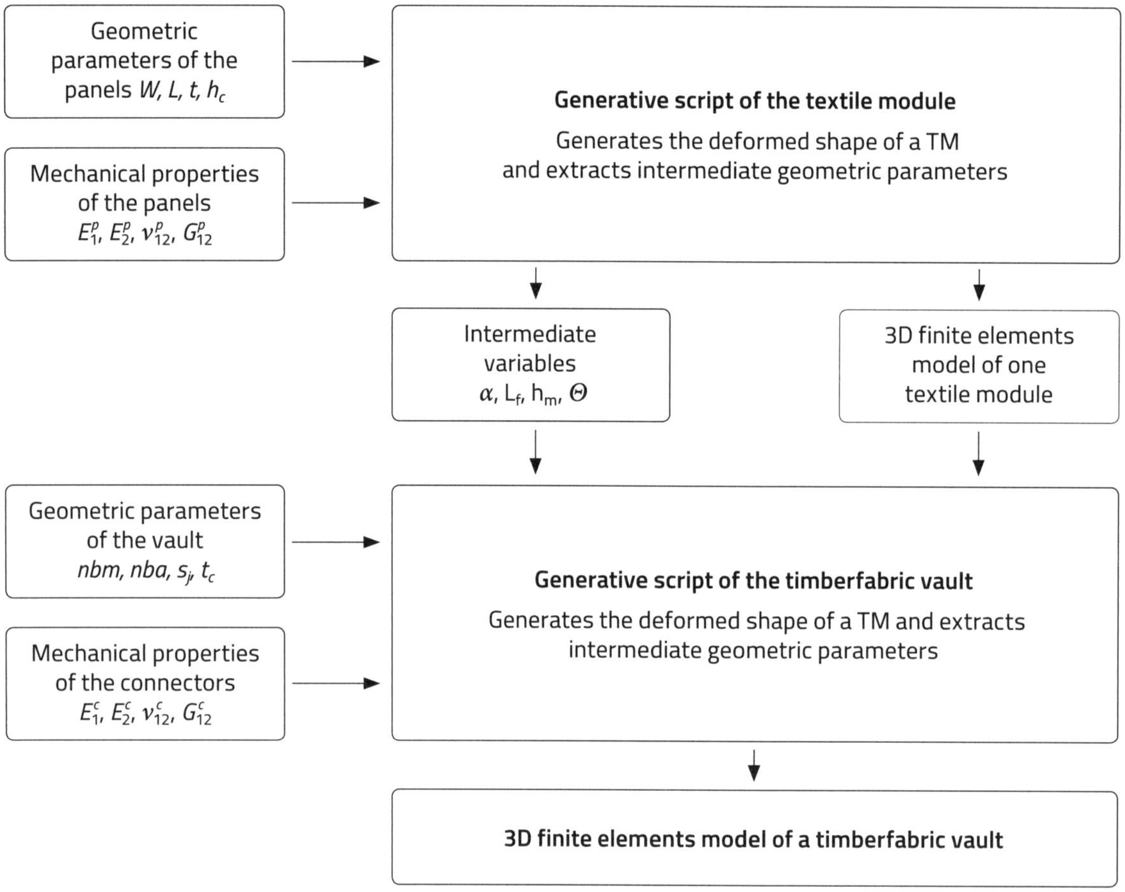

Fig. 13 Synthetic diagram of the complete script

Pieces of wood cut from 25-mm-thick laminated timber panels (class of resistance C30) are assembled to create the connectors. The detailed construction of these connection pieces is not approached in this paper.

4.3 Map of the vault

The α-rotation from one arch to the other implies that some timberfabric modules have to be cut in half so that all the arches can be connected to the ground at the same level. The "half-modules" are shown in fig. 1. In order to give a clear description of the structure, we will generate the vault by only assembling halves of timberfabric modules together. The continuity of the panels of the full timberfabric modules will be restored in the finite element model by specifying appropriate interaction conditions. As represented in fig. 10, the bottom section of the timberfabric module is referred to as "half-module $k=1$" and the top part as "half-module $k=2$": each module consists of two panel halves and one connector. With this virtual cutting of the timberfabric modules, the vault can be described with a chess-like diagram.

When using the script, the user enters a value to the parameters "*nba*" and "*nbm*" to specify the number of arches and the number of modules (in the study of the prototype, we have $nba=3$ and $nbm=6$). The program automatically computes the corresponding modules.[X]

Points	a	b	c	d
Coordinates*	$(0,h_m)$	$(-l_1,h_2)$	$(-l_2,h_2+h_c)$	$(0,h_m+h_3)$

Points	e	r	g	h
Coordinates*	(l_2,h_2+h_c)	$(2s_j-l_2,h_2+h_c)$	$(2s_j-l_1,h_2)$	(l_1,h_2)

* in $(c'^1, \hat{e}_2^1, \hat{e}_3^1)$

Table 3 Coordinates of the points defining the geometry of the connectors

4.4 Geometric assembly process

The geometric assembly is achieved by performing a double loop on i in $[1,2,..,nba]$ and j in $[1,2,..,nbm]$. For each couple (i, j), the three parts corresponding to the half module $k(i, j)$ are imported into the model, and then translated, and rotated. First, the instances of the parts are placed in the global coordinate system $(O,\hat{\imath},\hat{\jmath},\hat{k})$ such that their reference points A_1 (also noted $A_{1\,j=1..nbm}^{j}$ in section 4.1) and point O are coincident and local, and the global axis is parallel. They are then placed in the vault system by applying an appropriate translation and rotation.[XI]

4.5 Kinematical interactions and boundary conditions

Finally, to model the vault, it is necessary to introduce kinematic interactions between the different geometric entities of the model in order to:

- restore the material continuity of the elements that have been divided (recall that this division was undertaken to facilitate the description of the virtual assembly process)
- simulate the kinematic relationship between the panels and the connectors

The definition of these interactions is also generated automatically by looping over *i* in [1, 2, ..., *nba*] and *j* in [1, 2, ..., *nbm*–1]. Fig. 11 summarizes the interactions that are defined, depending on the value of *k(i, j)*, and the letters refer to the nature of the interaction between the two connected edges. For all the elements that have been virtually "cut" in the model, the corresponding meshes have to be tied back together. In fig. 11, the letter "T" stands for "tie." Accordingly, the DOFs of each node of one edge are constrained to the DOFs of the corresponding node of the other edge.

For the interactions between the panels and the connectors, there are two objectives: ensuring the static equilibrium of the vault and imposing geometric conditions as described in Section 3.1 In practice, the panels are fixed onto the connectors by punctual screws with given characteristics and a given repartition. To model the global effect on the structure's behavior of these locally complex connections, all the nodes of an edge are constrained to the central node of this edge and simple kinematic interactions are defined between pairs of central nodes. Two different connections are implemented, as represented in fig. 11 where "B" stands for "beam" and "H" stands for "hinge." The beam constrains all the DOFs of one node to the DOFs of the connected node, whereas the hinge constrains all the DOFs except the rotational DOF along the axis of the edge. Finally, the boundary conditions consist of blocking all the DOFs of the connectors of the extremities (corresponding to the couples *(i,* 1) and *(i, nbm* + 1) for *i* in *[1, 2, ..., nba])*.

4.6 Application

As was previously argued, the parametric model is a useful tool for modifying the architectural design of the initial structure. Fig. 12 illustrates how we can generate a very different geometry by varying the initial and intermediate parameters of the model. This will be of particular interest during the conception phase since architects, planners, and engineers can discuss and jointly visualize the variation of geometric parameters, such as the curvature radius shown here. The visualization of the global form is crucial for the interdisciplinary understanding necessary to achieve harmony of structure-function.

5 Conclusion

The numeric tool described in this paper can be specifically useful for the architectural and structural design of timberfabric structures. Indeed, the complexity of such structures resides essentially in the interdependence of the geometrical and mechanical aspects. The primary challenge was therefore to find a way to quantitatively describe the geometry of one constitutive element (a TM) with a small set of variables and to elaborate a procedure to calculate them by generating the entire deformed geometry.

Rather than trying to reproduce the precise steps of deformation necessary to assemble flat panels into a timberfabric module and to assemble several TMs together, the strategy was to directly reproduce the final geometry. These objectives came from the observation of the results of the empirical approach for the prototype shown in fig. 1. This also means that the geometry of the TMs that had been calculated was not the geometry of self-equilibrated TMs, but rather a geometry "extracted virtually" from the vault and having more complex boundary conditions.

The diagram shown in fig. 13 synthesizes the actions performed automatically by the scripts and highlights the input parameters that need to be specified by the user, as well as the "intermediate variables" that need to be introduced in order to enable the automatic generation of the vault after having calculated one timberfabric module. It is important to note that this strategy of generating modules from flat panels allows one to calculate the stress state of the panels that constitute a timberfabric module. This information is essential to study the long-term behavior and the resistance of such structures. Overall, the complexity of timberfabric structures calls for the elaboration of powerful tools. The work presented in this paper gives a strategy that generates finite element models for a rather simple family of timberfabric vaults, but the structural analysis still has to be performed manually. Further elaboration of the tool should allow the evaluation of structures made from timberfabric modules of various sizes and the ability to automatically perform service and ultimate limit state verifications.

References

1 Garcia, M. (ed.) "Architextiles." *Architectural Design*, 2006;76(6).

2 Tsukui, N. (ed.) "Cecil Balmond." *A+U Special Issue*, 2006.

3 Weinand, Y. "Innovative timber constructions." *Journal of the International Association for Shell and Spatial Structures*, 2009; 50(161): 111–120.

4 Hudert, M. "Timberfabric—Applying Textile Assembly Principles for Wood-Construction in Architecture." PhD Thesis, EPFL, 2013.

5 Weinand, Y. and M. Hudert. "Timberfabric: Applying textile principles on a building scale." *Architectural Design*, 2010; 80(4): 102–107.

6 Hudert, M., and Y. Weinand. "Structural Timber Fabric: Applying Textile Principles on Building Scale." *ICSA2010 1st International Conference on Structures & Architecture*, Guimaraes, Portugal, 2010.

7 Hudert, M., and Y. Weinand. "Timberfabric: Innovative Lightweight Structures." *IABSE-IASS Symposium*, London, U.K. , 2011.

8 Sistaninia, M., M. Hudert, L. Humbert, and Y. Weinand. "Experimental and numerical study on structural behavior of a single Textile Module." In: *Engineering Structures*, 2013; 46: 557-568.

9 ABAQUS Version 6.11 Documentation: *ABAQUS Scripting User's Manual and Reference Manual.* Dassault Systems; 2011.

10 Daniel, I., and O. Ishai. *Engineering mechanics of composite materials.* 1994, Oxford, UK: Oxford University Press.

11 Bodig, J. and B.A. Jayne. *Mechanics of wood and wood composites.* 1982: Van Nostrand Reinhold Company.

12 Reddy, J.N. *An introduction to non-linear finite element analysis.* 2004, New York: Oxford University Press.

13 Teboply. *Okoumé plywood datasheet*. www.tebopano.com

Footnotes

I The following text was initially written for publication in the scientific journal *Structural Engineering International.* In the adapted version presented here, several mathematical notations and some highly specific scientific commentaries have been replaced or completed by simpler textual descriptions in the core of the text. The original parts that have been modified are reproduced as end notes to this document.

II The behavior of a panel at a macroscopic scale is viewed as being equivalent to the behavior of a single-layer, homogeneous, orthotropic material with principal (material) axes along the longitudinal and transverse geometric directions of the panel. With those assumptions and considering a linear elastic constitutive law, only four independent constants are required to fully characterize the material's behavior: Young's moduli in the two principal directions E1 and E2, the in-plane shear modulus G12 and the Poisson's ratio $\nu 12$.[10,11] Moisture content and time dependence of the material properties are not taken into account.

III The timber panels are meshed using four-node, quadrilateral shell elements S4R with reduced integration and six active degrees of freedom per node. These general-purpose elements are suitable for the analysis of (doubly) curved thick or thin shells, allowing transverse shear deformation. Because large rotations and displacements are expected, we run a geometrically non linear analysis for which the imposed displacements are applied incrementally.[12]

IV Initial position: $(\mathbf{O}, \hat{\mathbf{e}}_1, \hat{\mathbf{e}}_2, \hat{\mathbf{e}}_3)$ is an orthonormal Cartesian coordinate system. Initially, the panels are positioned in the (χ_1, χ_2) plane with their longitudinal axes along $\hat{\mathbf{e}}_1$. For each panel, the initial position of the nodes introduced previously are such that $\mathbf{r}^{A_1} = \mathbf{r}^{A_2} = \mathbf{O}$, $\mathbf{r}^{B_1} = \mathbf{r}^{B_2} = (L, 0, 0)^T$, $\mathbf{r}^{C_1} = (\frac{L}{2}, -\frac{W}{2}, 0)^T$, $\mathbf{r}^{C_2} = (\frac{L}{2}, \frac{W}{2}, 0)^T$.
In the subsequent section, D $(M, \hat{\mathbf{n}})$ will stand for the straight line through a point M, directed by a (unit) vector $\hat{\mathbf{n}}$ and P $(M, \hat{\mathbf{n}})$ for the plane of normal $\hat{\mathbf{n}}$ at M.

V The following boundary conditions are imposed to block certain displacement degrees of freedom of the nodes A_1, A_2, B_1, and B_2:
- A_1 is fixed: $u_1^{A_1} = u_2^{A_1} = u_3^{A_1} = 0$
- B_2 is constrained on line $\Delta(A_1, \hat{\mathbf{e}}_1)$: $u_2^{B_2} = u_3^{B_2} = 0$
- A_2 and B_1 are constrained in plane $P(A_1, \hat{\mathbf{e}}_2)$: $u_2^{A_2} = u_2^{B_1} = 0$

VI The following interacting conditions impose kinematic constraints respectively between the degrees of freedom of the node sets *Edge-A*$_1$, *Edge-B*$_1$, *Edge-C*$_1$, *Edge-A*$_2$, *Edge-B*$_2$, *Edge-C*$_2$, and the nodes A_1, B_1, C_1, A_2, B_2, C_2. To simplify the expressions, these constraints are written using the position and normal vector of the nodes.
The first set of constraints forces the panels to be tangential to planes along their extreme transverse edges (*Edge-A*$_1$, *Edge-B*$_1$, *Edge-A*$_2$, *Edge-B*$_2$), while at each extremity of the timberfabric module the planes for each panel are constrained to be parallel (they have the same normal vector):
- Panel 1 has to remain tangential to the plane $P(A_1, \hat{\mathbf{n}}^{A_1})$ along *Edge-A*$_1$ and to the plane $P(B_1, \hat{\mathbf{n}}^{B1})$ along *Edge-B*$_1$:

$$(\mathbf{r}^M - \mathbf{r}^{A_1}) \times \hat{\mathbf{n}}^{A_1} = 0, \hat{\mathbf{n}}^M \text{ ¥ } \hat{\mathbf{n}}^{A_1} = 0, M \text{ Œ } Edge\text{-}A_1$$
$$(\mathbf{r}^M - \mathbf{r}^{B_1}) \times \hat{\mathbf{n}}^{B_1} = 0, \hat{\mathbf{n}}^M \text{ ¥ } \hat{\mathbf{n}}^{B_1} = 0, M \text{ Œ } Edge\text{-}B_1$$

- Panel 2 has to remain tangential to the plane $P(A_2, \hat{\mathbf{n}}^{A1})$ along *Edge-A*$_2$ and to the plane $P(B_2, \hat{\mathbf{n}}^{B1})$ along *Edge-B*$_2$:

$$(\mathbf{r}^M - \mathbf{r}^{A2}) \times \hat{\mathbf{n}}^{A_1} = 0, \hat{\mathbf{n}}^M \text{ ¥ } \hat{\mathbf{n}}^{A_1} = 0, M \text{ Œ } Edge\text{-}A_2$$
$$(\mathbf{r}^M - \mathbf{r}^{B2}) \times \hat{\mathbf{n}}^{B1} = 0, \hat{\mathbf{n}}^M \text{ ¥ } \hat{\mathbf{n}}^{B1} = 0, M \text{ Œ } Edge\text{-}B_2$$

The second set of constraints is imposed on the central transverse edges of the panels (*Edge-C*$_1$, *Edge-C*$_2$).
The nodes of these edges should remain aligned:

$$(\mathbf{r}^M - \mathbf{r}^{C_1}) \text{ ¥ } (\mathbf{r}^{M'} - \mathbf{r}^{C_1}) = 0, M, M' \text{ Œ } Edge\text{-}C_1$$
$$(\mathbf{r}^M - \mathbf{r}^{C_2}) \text{ ¥ } (\mathbf{r}^{M'} - \mathbf{r}^{C_2}) = 0, M, M' \text{ Œ } Edge\text{-}C_2$$

VII The mechanical characteristics of the Okoumé Panel are E1 = 4163 MPa, E2 = 5088 MPa for the Young's moduli (in traction/compression) and G12 = 552 MPa for the in-plane shear modulus.[13] For convenience, the geometrical and material properties of the panels are listed in Table 1.

VIII To choose the size of the mesh we used for this study, we first ran a parametric study on the number of elements per width of the panels. By accepting an error of 0.5 % on the results of the "intermediate variables," we chose a mesh size corresponding to 12 elements per width of the panels.

IX Considering the sections of TM1 and TM2 by this plane, the geometry of the connectors is given by the points {a,b,c,d,e,f,g,h}. The coordinates of the points in the system $(C'^1, \hat{\mathbf{e}}_2^1, \hat{\mathbf{e}}_3^1)$ are given in Table 3 **(Fig. 9)**. The parameters introduced in Table 3 are given by:

$$l_1 = (h_m - h_2) \times \text{tanQ}/2$$
$$l_2 = l_1 + h_c/\text{tanQ}_2$$
$$h_3 = h_c \times \left(1 + \frac{1}{\text{tanQ}/2 \times \text{tanQ}_2}\right), \text{Q}_2 = \frac{p + \text{Q}}{4} \qquad \text{equ. 4}$$

In the FE model, those connectors are modeled as two planar shells rigidly bonded along the edge [e,h]: "connector 1" and "connector 2" will respectively refer to the planar surfaces defined by the sets of points {e,f,g,h} and {a,b,c,d,e,h}.

x The nature of the half-module corresponding to a module "*j*" of the arch "*i*" is described by the variable *k* for which possible values are 1 or 2. The choice of the value of *k* for $i = 1$ and $j = 1$, i.e. $k(1,1) \int k_{11} = 1$ or 2, determines the value of *k* for any other integer couple (*i, j*) as

$$k(i,j) = \text{mod}\,(i + j + (k_{11} - 1), 2) + 1, \; 1 \pounds i \pounds nba, \; 1 \pounds j \pounds nbm \qquad \text{equ. 5}$$

mod (*a* ,*b*) being the positive remainder in the division of *a* by *b*, where *a* and *b* are positive integers.

xi The translation vector **T** and rotation angle *b* about the global axis $\hat{\mathbf{j}}$ are defined below:

$$\mathbf{T} = -\frac{L_f}{2}\,\hat{\mathbf{i}} + (i-1) \times s_j \times \hat{\mathbf{j}} + (R - h_2) \times \hat{\mathbf{k}}$$

$$b = (j + k(i,j) - 1 - \frac{nbm}{2}) \times a \qquad \text{equ. 6}$$

Note that by taking nba = 1 (i.e. one arch) and nbm = 2 (i.e. two modules) in equ. 6, the positioning of the two modules TM1 and TM2 in Section 4.1 in the global coordinate system is recovered.

Reprinted from *Structural Engineering International*, vol. 24, num. 2, p. 254-264.

4.4

Mechanical form-finding of the timber fabric structures with dynamic relaxation method

Seyed Sina Nabaei, Olivier Baverel, and Yves Weinand

Timber fabric structures (TFS) combine textile principles with recent industrial developments in producing cross-laminated timber panels. Several individual timber strips are woven according to a pattern, thereby creating innovative spatial structures. The three-dimensional geometry obtained can be regarded as the stress-free configuration of deformed panels under the imposed boundary conditions. We thus propose a form-finding procedure that reproduces this deformed configuration as the steady state of a pseudo-transient, constrained, dynamic problem. The corresponding nonlinear problem involves a finite rotation regime and contact handling through the cross-section on both panel faces. To effectively deal with nonlinear constraints, a new modified dynamic relaxation method is utilized, which combines elastic material behavior with a fictitious stiffness proportional damping into an equivalent fictitious, viscous material model. The procedure is implemented as an ABAQUS/Explicit user subroutine VUMAT and the overall accuracy of the numerical results has been studied for a number of geometrically nonlinear, shell benchmark problems. This numerical approach is then employed to simulate the assembly process for a timber fabric module (TFM), which is an interlaced assembly of two timber strips. The simulated geometry for the deformed surfaces is then extracted and is compared with a three-dimensional processed surface mesh obtained from scanning a built prototype with a non-contact laser scanner arm to validate the simulation procedure.

Keywords ***dynamic relaxation, dynamic explicit, timber fabric structures, three-dimensional finite elements method, three-dimensional mesh processing***

1 Introduction

1.1 Background

A timber fabric module (TFM) **(fig 1 b)** was developed through a research project at the EPFL-IBOIS by applying textile principles at an architectural scale in the particular context of timber structures.[1,2] This new concept of structures is derived from the combination of fabric production techniques—the micro-scale structure of textiles, weaving, braiding, knitting, etc.—and lightweight architecture using timber panels.

The common denominator of different interlacing techniques is the knot **(fig. 1 a)**. The key to the approach employed in the timber fabric project is to reinterpret the "knot" by replacing yarns with panels to reproduce a more complex overlap **(fig. 1 b)**. Generalization of the weaving concept is straightforward: continued weaving with two strips leads to a braided arch. Fig. 1 c shows three parallel braided arches.

Markus Hudert[1] has also investigated the spatial extension of braided arch structures, both in-depth, as double-layer, braided arches, in transverse direction, and with multiple parallel arches connected together. Various different double-layering and transverse multiplication techniques were proposed based on the analysis of the local connection technology and the structural performance. Double-layer, braided arches have different weaving patterns for the upper and lower layers to improve structural performance and reinforce weak areas. Markus Hudert[1] has also examined connection pieces that integrate transverse with lateral load patching. He conducted prototype fabrication in medium and large scales. This morphological investigation has mainly been based on an investigative approach with prototypes built in increasing scales. Although it is helpful to create prototypes to understand and resolve conceptual issues (mainly related to local assembly aspects), the newly discovered morphology can hinder the designer's creativity and his or her handicraft skills.

Broadly speaking, the fabric structure concept is situated in an interdisciplinary context between topology/knot theory, structural engineering, and wood tech-

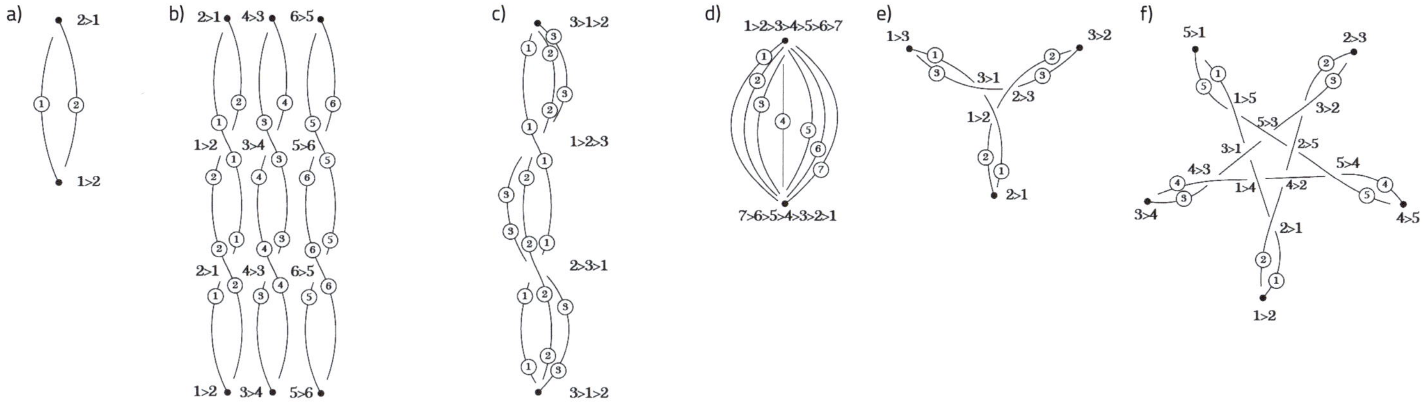

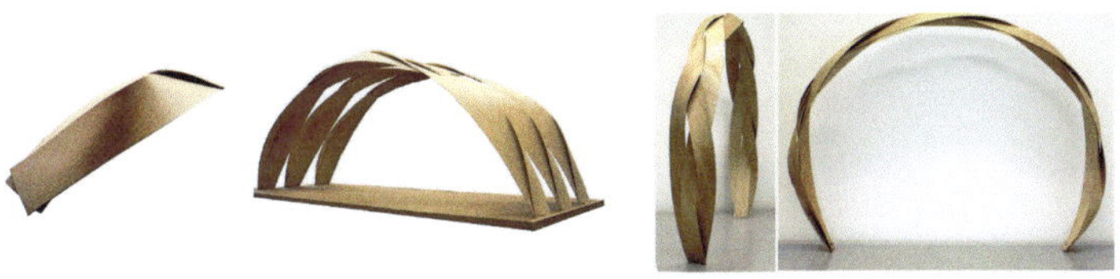

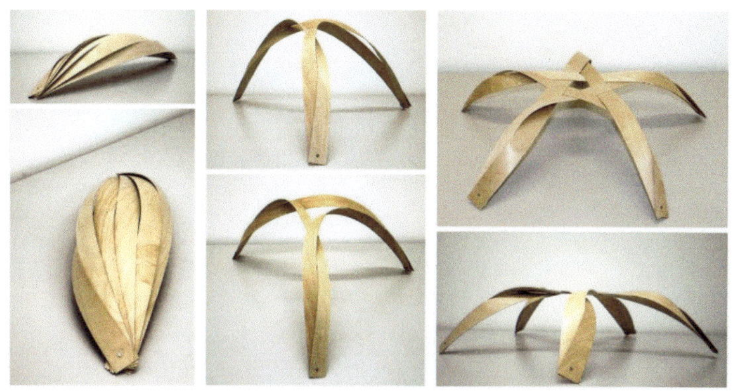

Fig. 1 2-D representation of TFS:
a) A simple two-strand knot, TFM
b) Two-strand braid: Three braids placed side by side
c) Three-strand braid
d) Multi-strand knot
e) Three-strand fan
f) Five-strand fan

nology. In this context, further conceptual structures can be introduced with multi-thread braids (not only limited to two as in Hudert[1]) and with a greater number of overlaps. However, these complex structures quickly become too complex for an empirical approach; the remaining conceptual potential indicates a need for a form-finding/tool procedure.

1.2 Mechanical form-finding

The first analysis regarding the structural behavior of TFS was undertaken by Masoud Sistaninia and his colleagues[3] in order to propose a simulation procedure for the TFM form-finding problem through a nonlinear, static analysis. In order to validate the simulation, the experiment proceeds with two vertical uniformly-distributed loads applied to the mid-span. The deformation is then measured over a number of marked points on panels, in order to compare them with the simulation results. The main drawbacks of the approach taken in Sistaninia[3] is that the edge-edge contact is not treated in the static, nonlinear, implicit solver employed, and the boundary condition at the mid-contact point is a simple symmetry condition imposed to keep panels "connected" (without collisions) at this particular point. As a consequence, we are limited to reproducing interlaced shapes for which we ascertain the exact position of edge-edge contact points from a previous experiment with prototypes. Furthermore, the position of this edge-edge contact point is supposed to be given and has not been treated as an unknown, thus limiting the validity of the approach. Thus, we propose two primary contributions to the TFS form-finding problem:

- *Reformulating the TFS form-finding problem:* The deformed state of the timber fabric structures can be regarded as the stress-free configuration of a flat, initial state under a set of imposed boundary conditions (BC). These constraints can be either i) the order of panels or their offset at a particular overlap, or ii) an imposed displacement/rotation on a degree of freedom for a structural node. Interlacing and then connecting panels together makes them deform and initiates a pre-stressed mod-

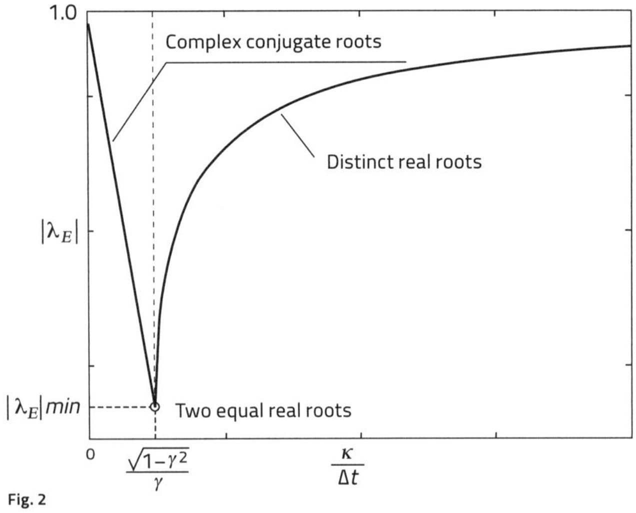

Fig. 2

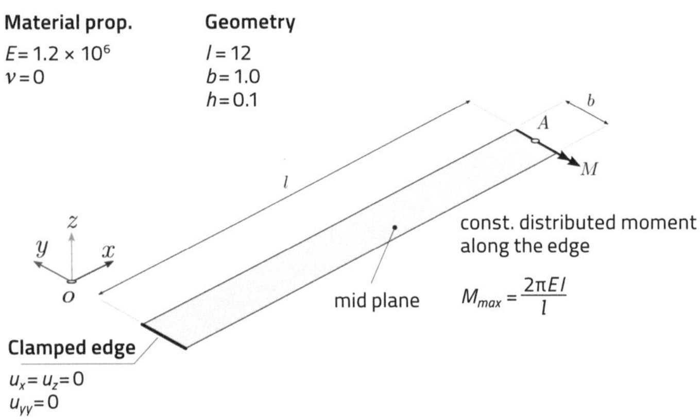

Fig. 3

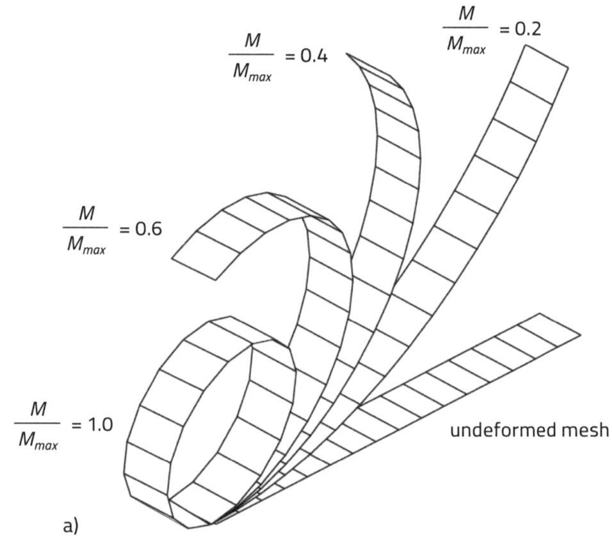

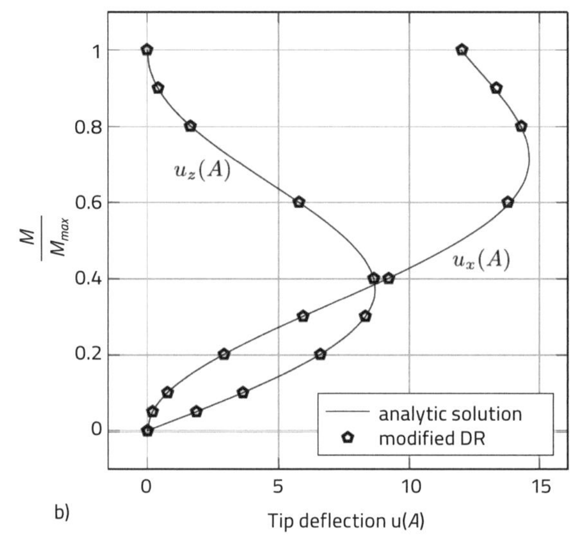

Fig. 4

Fig. 2 Relation between $\frac{\kappa}{\Delta t}$ and $|\lambda_E|$.

Fig. 3 Clamped strip problem, section 3.2.1

Fig. 4 Clamped strip simulation results:
a) Deformed mesh for loadcases
b) Tip deflection $u_x(A)$, $u_z(A)$ versus end moment ratio M/M_{max}.

ule with active bending behavior. We postulate that knowing the flat configuration and the overlaps interpreted as mechanically meaningful BCs suffices to numerically calculate the stress-free form of TSF. In order to distinguish different interlacing patterns, a new graph representation is introduced. As seen in two-dimensional knot graphs in knot theory, it helps to distinguish strands and also to take into account a notation to represent the order of overlapping strands in a particular interlace node. Each strand/panel is represented by a two-dimensional smooth curve and the lower strands are disconnected locally at each overlap. The continuous curve passes on top of the others in this particular overlap node. Each strand is numbered with positive integers starting from 1, once where they start and once where they end. For intermediate nodes, the order of overlapping strands is noted using the > notation: 1 > 2 > 3 stands for a three-strand overlap where strand 1 is woven above strand 2 and strand 3 passes underneath. Examples of this structural typology with more complex nodes are illustrated along with their corresponding graph representation in fig. 2.

The two-dimensional graph enables an effective description and illustrates a braided configuration in an abstract topological representation, regardless of its dual three-dimensional, stress-free shape. The first step toward a numerical approach to studying TFS morphologies is to be able to differentiate one from the other.

- *Use of explicit procedure to effectively deal with complex edge-edge contact handling:*
  Once considered as a mechanical simulation, the interlacing/weaving of panels involves shell, nonlinear mechanics in finite deformation. The appropriate finite element framework and the nonlinear analysis method have to be determined in order to proceed with the simulation. The complex assembly of the panels calls for the use of an existing finite element package. In terms of the numerical method, pseudo-transient, explicit dynamic analysis has been efficiently used to find a static solution for complex, nonlinear, static shell problems. In such an analysis, the initial static problem is considered as being pseudo-transient dynamic by adding fictitious, dissipative features, such as damping and mass scaling. The static solution will be the steady state of the equivalent pseudo-dynamic problem. However, evaluating fictitious energy dissipative parameters is a demanding task, as different parameters are usually interrelated, thus making the manual calibration difficult. The main goal of this study is to establish a valid simulation procedure which can handle the complex modeling requirements of braided structures such as these, while also enhancing an automatically adapted evolution of fictitious dissipative parameters for pseudo-transient analysis.

The paper is structured in the following manner: the dynamic relaxation method (DR) is briefly reviewed and modified in section 2, and the dynamic relaxation method is presented in section 3, where the DR method is reformulated as a fictitious, viscous, elastic material model. This fictitious material model is implemented as an ABAQUS user subroutine material called VUMAT. Results are cross-referenced with a number of nonlinear shell benchmark problems for validation. Regarding the TFS form-finding problem, in section 4 the proposed modified DR is used to simulate an arbitrarily chosen configuration of timber fabric module (TFM). Finally, deformed surfaces from simulation and prototype are processed and analyzed in section 5.

2 The dynamic relaxation method

2.1 A review of dynamic relaxation

Dynamic relaxation (DR) is a numerical iterative method to find the solution to a system of nonlinear equations. It is also referred to as "pseudo-transient dynamic analysis" in the relevant literature. This method is used in structural mechanics to solve the static equilibrium of the system by integrating the damped wave equation to find the steady state of the equivalent dynamic problem. Interestingly, the method was introduced during the 1960s by Otter and Day[4–7] and has its origins in the "second order Richardson method," developed by Frankel[8] in 1950.

Since 1970, DR has been used to solve numerous engineering problems: nonlinear analysis of plates and shells,[9–13] buckling and post buckling analysis,[14–16] membrane and cable-net form-finding and analysis,[17–19] form-finding and analysis of tensegrity structures,[20] inflatable structures,[21] reciprocal frame structures,[22] and medical applications,[23] among others. In addition, several publications have focused on the convergence and the stability of the method and proposed accelerated rates through an improved viscous damping matrix, improved conditions for set-up of the fictitious mass matrix, or by introducing kinetic damping.[24–35]

The governing discretized equation for structural static equilibrium can be written as $\mathbf{Kx}=\mathbf{F}$, where the solution is $\mathbf{x}^*=\mathbf{K\text{-}1F}$. In material/geometrically nonlinear problems, the estimation of the tangent stiffness matrix $\mathbf{K}$ can be costly computationally. In order to obtain the static solution using the DR method, fictitious mass and damping is introduced and the equivalent dynamic equation is integrated using the central difference explicit technique. Let the incremental displacement vector be $\mathbf{x}^n=\mathbf{x}(t_n)$, the dynamic equilibrium equation at n^{th} time step will be

$$\mathbf{M}\ddot{\mathbf{x}}^n + \mathbf{C}\dot{\mathbf{x}}^n + \mathbf{K}\mathbf{x}^n = \mathbf{F}^n \quad \text{equ. 1}$$

According to the central difference integration and supposing a constant time step, mid-step velocity vector, step velocity vector, and step acceleration vector can be written as equation 2.

$$\dot{\mathbf{x}}^{n+\frac{1}{2}} = \frac{1}{\Delta t}\left(\mathbf{x}^n-\mathbf{x}^{n-1}\right), \dot{\mathbf{x}}^n = \frac{1}{2}\left(\dot{\mathbf{x}}^{n+\frac{1}{2}}+\dot{\mathbf{x}}^{n-\frac{1}{2}}\right)$$
$$\ddot{\mathbf{x}}^n = \frac{1}{\Delta t}\left(\dot{\mathbf{x}}^{n+\frac{1}{2}}-\dot{\mathbf{x}}^{n-\frac{1}{2}}\right) \quad \text{equ. 2}$$

Substituting equation 2 values into Equation 1, gives an iterative linearized equation for the mid step velocity vector (equ. 3). The DR algorithm can then be resumed in algorithm 1.

$$\dot{\mathbf{x}}^{n+\frac{1}{2}} = \left(\frac{\mathbf{M}}{\Delta t}+\frac{\mathbf{C}}{2}\right)^{-1}\left(\left(\frac{\mathbf{M}}{\Delta t}+\frac{\mathbf{C}}{2}\right)\dot{\mathbf{x}}^{n-\frac{1}{2}}+\mathbf{F}^n-\mathbf{K}\mathbf{x}^n\right)$$
$$\mathbf{x}^{n+1} = \mathbf{x}^n + \dot{\mathbf{x}}^{n+\frac{1}{2}}\Delta t \quad \text{equ. 3}$$

Algorithm 1 DR method

Require: $\mathbf{C}$, $\mathbf{M}$ and δ as admissible error

```
Initialize with n = 0, x^0 and ẋ^0 = 0
while (r^n = F^n − Kx^n) ≥ δ do
    if n ≥ 0 then
        ẋ^(1/2) ← (Δt/2) M^-1 r^0
    else
        ẋ^(n+1/2) ← (M/Δt + C/2)^-1 ((M/Δt + C/2)) ẋ^(n−1/2) + F^n − Kx^n
    end if
    x^(n+1) ← x^n + ẋ^(n+1/2) Δt
    n < n+1
end while
return x
```

$$\dot{\mathbf{x}}^{\frac{1}{2}} \leftarrow \frac{\Delta t}{2}\mathbf{M}^{-1}\mathbf{r}^0$$

$$\dot{\mathbf{x}}^{n+\frac{1}{2}} \leftarrow \left(\frac{\mathbf{M}}{\Delta t}+\frac{\mathbf{C}}{2}\right)^{-1} \left(\left(\frac{\mathbf{M}}{\Delta t}+\frac{\mathbf{C}}{2}\right)\right)\dot{\mathbf{x}}^{n-\frac{1}{2}}+\mathbf{F}^n-\mathbf{K}\mathbf{x}^n$$

$$\mathbf{x}^{n+1} \leftarrow \mathbf{x}^n + \dot{\mathbf{x}}^{n+\frac{1}{2}}\Delta t$$

2.2 Fictitious mass and damping

Different choices for the damping $\mathbf{C}$ and mass $\mathbf{M}$ matrices lead to distinct DR-derived variants where the rates of convergence and stability conditions will be different. The mass matrix $\mathbf{M}$ is usually scaled in order to increase the time-step size, although it can also be left as a lumped mass diagonal matrix. Some examples of the fictitious mass proposed in scientific literature are as follows: a mass matrix computed equation from different directional densities,[36] a diagonal mass matrix consisting of the diagonal elements of the stiffness matrix,[26] a mass matrix based on the stiffness matrix proposed with an insight on the Gerschgorin theorem,[27] a mass/scale factor determined by the incremental rate of change of the kinetic energy upon internal energy for the structure[37] and, finally, a diagonal lumped mass matrix that is scaled at each node to align with the maximum values of its elements.[23] A typical equation approximation for $\mathbf{C}$ is the Rayleigh damping, which presumes the damping to be a linear combination of mass and stiffness matrices, $\mathbf{C}=c\mathbf{M}+\kappa\mathbf{K}$. The other main alternative is to introduce kinetic damping, as introduced by Cassell et al.[38] The idea behind kinetic damping is to observe the evolution of the kinetic energy of the system and to neutralize the velocity components as soon as a local kinetic energy peak is observed. Supposing the Rayleigh damping, the main approach has been to suppose a particular construction for damping matrix C and then to adapt the damping coefficient to a critical damping value at each iteration via the eigenvalue analysis of the iterative error vector. (Similar to Lynch[39] and Papadrakakis[26] in several successive publications.) This approach is also called "adaptive damping."

3 Modified dynamic relaxation method

3.1 Mathematical formulation

The idea here is to add the fictitious damping contribution to the material elastic response and to treat the dynamic relaxation concept as a fictitious viscous material model. Revisiting the Rayleigh damping equation, $\mathbf{C}=c\mathbf{M}+\kappa\mathbf{K}$, a stiffness proportional damping matrix is supposed as $\mathbf{C}=\kappa\mathbf{K}$ and by following the eigenvalue analysis of the error vector, a near critical damping coefficient is obtained.

Let $\mathbf{M}$ be a diagonal lumped mass matrix with diagonal elements m_{ii}, $\mathbf{D}=\mathbf{M}^{-1}\mathbf{K}$ and $\mathbf{C}=\kappa\mathbf{K}$. Substituting these into equation 3 and rewriting the whole with regard to the step displacement vector $\mathbf{x}$, results in equation 4.

$$\alpha_1\mathbf{x}^{n+1}+\alpha_2\mathbf{x}^n+\alpha_3\mathbf{x}^{n-1}-\mathbf{F}^n=0$$
$$\alpha_1=\frac{\mathbf{M}}{\Delta r^2}\left(\mathbf{I}+\frac{x\Delta t}{2}\mathbf{D}\right)$$
$$\alpha_2=\frac{\mathbf{M}}{\Delta r^2}\left(\Delta t^2\mathbf{D}-2\mathbf{I}\right)$$
$$\alpha_3=\frac{\mathbf{M}}{\Delta r^2}\left(\mathbf{I}-\frac{x\Delta t}{2}\mathbf{D}\right)$$

equ. 4

Following Papadrakakis[39] and Lynch[26], if $\mathbf{x}^*$ is the solution for the static equilibrium equation $\mathbf{K}_x=\mathbf{F}$, then the incremental displacement error vector can be introduced as $\varepsilon^n=\mathbf{x}^n-\mathbf{x}^*$. Furthermore, successive error vectors are supposed to be linearly dependent via a matrix $\mathbf{E}$ as $\varepsilon^{n+1}=\mathbf{E}\varepsilon^n$. Let λ_E be the largest eigenvalue of $\mathbf{E}$ (including the non-real ones), then $\varepsilon^{n+1}=\lambda_E\varepsilon^n=\lambda_E^2\varepsilon^{n-1}$. In order to have the iterative method converge, λ_E should have its complex norm less than one, $|\lambda_E|<1$. Substituting an incremental error vector into equation 4, λ_E can be calculated as the solution of a quadratic equation, where m, λ_D are respectively any eigenvalue of $\mathbf{M}$ and $\mathbf{D}$.

$$\lambda_1\lambda_E^2+\lambda_2\lambda_E+\lambda_3=0\Rightarrow$$
$$\lambda_E=\frac{-\lambda_2\pm\sqrt{\lambda_2^2-4\lambda_1\lambda_3}}{2\lambda_1}$$
$$\lambda_1=\frac{m}{\Delta t^2}\left(1+\frac{\varkappa\Delta t}{2}\lambda_D\right)$$
$$\lambda_2=\frac{m}{\Delta t^2}\left(\Delta t^2\lambda_D-2\right)$$
$$\lambda_3=\frac{m}{\Delta t^2}\left(1-\frac{\varkappa\Delta t}{2}\lambda_D\right)$$

equ. 5

It can be shown that the minimum $|\lambda_E|$ occurs when the quadratic equation 5 has two equal real roots that correspond to a near critical damped system. **(fig. 2)**

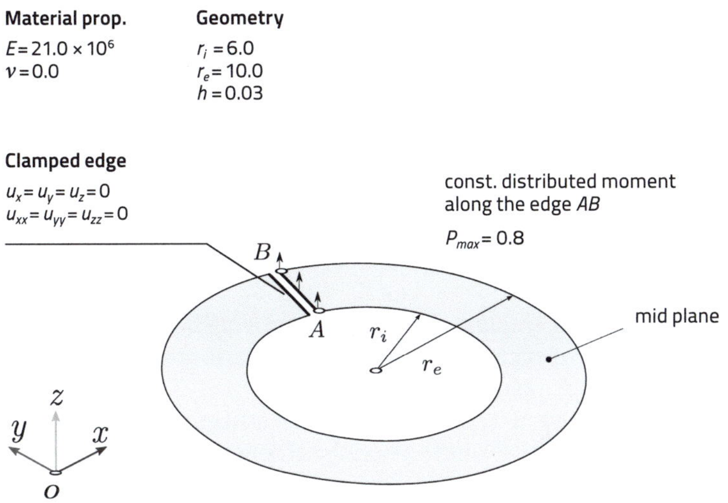

Fig. 5 Annular ring problem, section 3.2.2

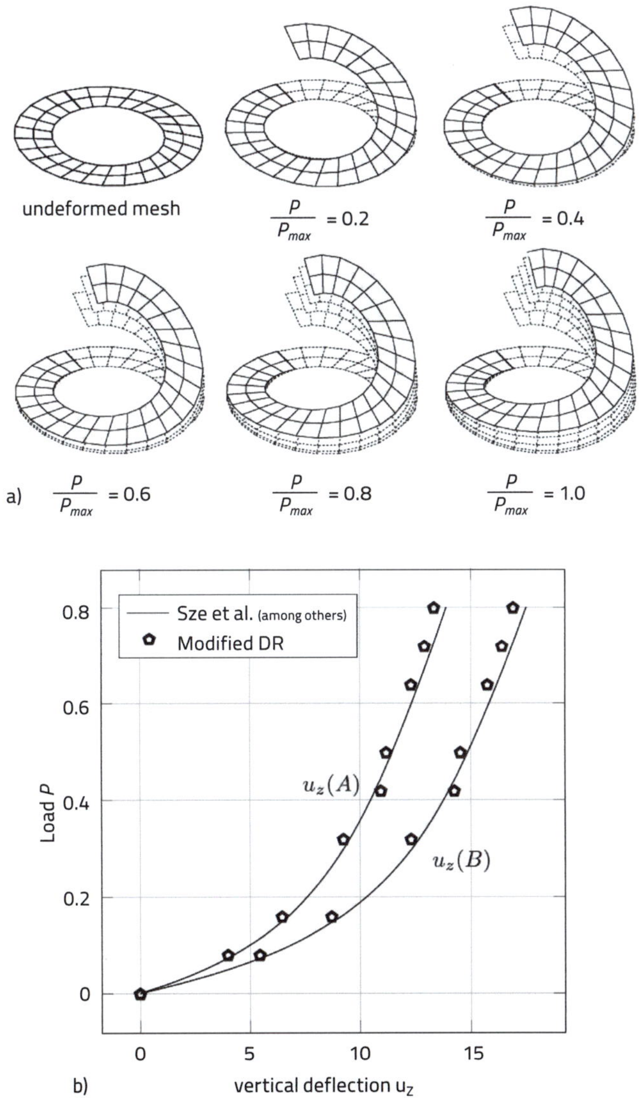

Fig. 6 Annular ring plate: a) Superimposed deformed mesh for load cases, b) Vertical deflection u_Z versus free edge shear force P

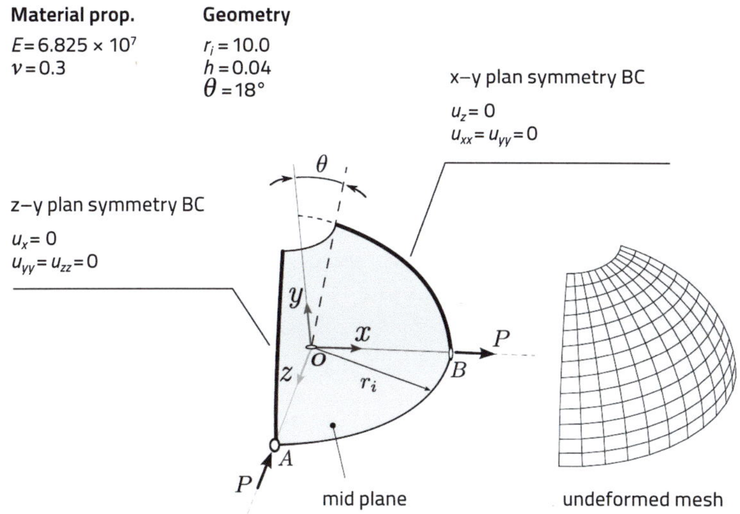

Fig. 7 Buckling of a pinched hemisphere problem, section 3.2.3

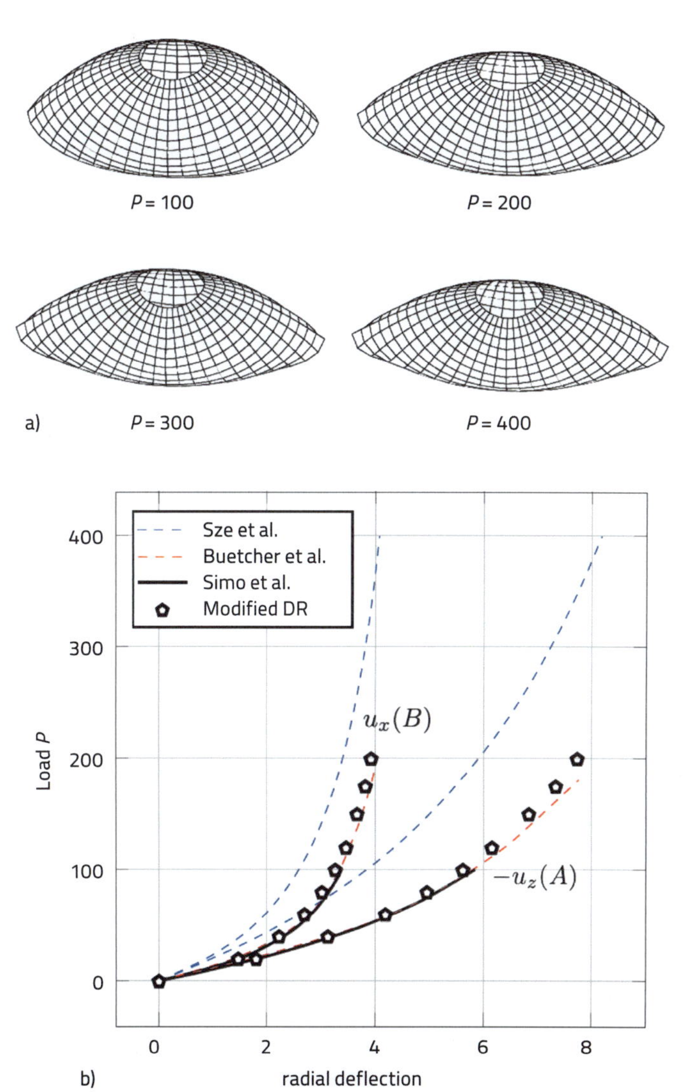

Fig. 8 Buckling of a pinched hemisphere:
a) Deformed mesh configuration for P = 100; 200; 300; 400
b) Radial deflection versus radial load P

In the case of explicit time integration, maximum stable time step must satisfy the condition

$$\Delta t_{cr} \leq \frac{2}{\omega_{\max}}$$

for an undamped system with ω_{max} as the maximum frequency of the system and

$$\Delta t_{cr} \leq \frac{2}{\omega_{\max}} \leq \frac{2}{\omega_{\max}}(\sqrt{\xi^2+1}-\xi)$$

in case of damping, where $0.0 \leq \xi \leq 1.0$ is the fraction of critical damping in the mode with the highest frequency.[40] The relation for the maximum time step size can also be written as a function of $\lambda_{max} = \omega^2_{max}$ and the maximum eigenvalue of the system, as

$$\Delta t_{cr} = \frac{2}{\sqrt{\lambda_{\max}}}$$

Following the notation used in equation 5, it is convenient to recall that $\lambda_{\max}$ is related to the maximum eigenvalue of $\mathbf{D} = \mathbf{M}^{-1}\mathbf{K}$, $(\lambda_D)_{\max}$[40,41] and since equation 5 is true for any eigenvalue, one can suitably replace λ_D as a function of maximum time step size Δt_{cr}. If we then assume that the actual increment time step Δt is also proportional to this stable time step size by a real coefficient $0.0 \leq \gamma \leq 1.0 \in$ as $\Delta t = \gamma \Delta t_{cr}$, then all terms of the discriminant in equation 5 can be rewritten with respect to γ, Δt and κ. If then, it is set to zero to obtain the fastest convergence condition, the following equation can be established between the stiffness proportional viscous damping coefficient κ and the time step:

$$\frac{\kappa}{\Delta t} = \frac{\sqrt{1-\gamma^2}}{\gamma} \qquad \text{equ. 6}$$

Regarding the discretized system, in explicit central difference FEM, the maximum eigenvalue of the assembled system λ_{max} is bound with sufficient accuracy, by the maximum eigenvalue of any element

$$\lambda^{(e)}_{\max} \text{ or } \Delta t_{cr} = \frac{2}{\sqrt{\lambda^{(e)}_{\max}}} \leq \Delta t_{cr} \leq \frac{2}{\sqrt{\lambda_{\max}}}$$

where $\Delta t^{(e)}_{cr}$ is the element's stable time step size.[41] On the other hand, the maximum stable time step for any element must satisfy the Courant-Friedrichs-Lewy condition,[42] which is based on the smallest transit time of a dilatational wave across any of the elements in the mesh. The Courant-Friedrichs-Lewy condition in our context of structural mechanics can be taken as

$$\Delta t^{(e)}_{cr} \leq \frac{L^{(e)}}{c} \leq \Delta t_{cr}$$

Here, $L^{(e)}$ is the element characteristic length (or if simplified, the "smallest" dimension of the element),

$$c = \sqrt{\left(\frac{\lambda + 2\mu}{\rho}\right)}$$

and λ, μ are the Lamé constants, ρ being the density.

In the numerical implementation of explicit FEM solvers, the estimation of system frequencies at each increment to determine the maximum stable time step size is a costly task, so the Courant-Friedrichs-Lewy condition is practically verified for the smallest element in order to determine the Δt_{cr}.

This, along with the fact that we have related the stiffness proportional damping coefficient κ to the time step size with equation 6, enables us to adapt the fictitious stiffness proportional damping contribution to a near critical value at each time step, based entirely on the time step size estimation. In fact, the ratio

$$\frac{\min \Delta t^{(e)}_{cr}}{\Delta t_{cr}}$$

is equivalent to the coefficient γ in equation 6, which should be selected deliberately by the user in our algorithm. This ratio can be calculated knowing the finite element's mass and stiffness matrices and the problem configuration. The finite element's mass and stiffness matrices give us the maximum eigenvalue of the element, and the maximum stable time step size chosen for the element is

$$\Delta t^{(e)}_{cr} = \frac{2}{\sqrt{\lambda^{(e)}_{\max}}}$$

(see Tablei[43] for a demonstration of thin-shell problems). Here we take the approximate value of $\gamma = 0.9$ for our numerical simulation. The main interest of a stiffness proportional matrix over the other proposed options is that it can be implemented as a viscous elastic material model and used as an existing time step estimation algorithm inside an explicit procedure, without explicitly having access to assembled mass and stiffness matrices.

The viscous stress is generated proportionally to the strain rate in the form of $\sigma^{vis} = \kappa \mathbb{C}\dot{\varepsilon}$. The total damped stress $\sigma = \mathbb{C}\varepsilon + \kappa\mathbb{C}\dot{\varepsilon}$ is then used to evaluate the amount of internal force necessary to compute the accelerations for velocity update. The above-mentioned idea is resumed under the following algorithm:

Algorithm 2 Incremental total stress and internal force update algorithm

Require: γ, $\mathbb{C}$, t, $\Delta t^{(n)}$, $\sigma_p^{(n)}$, $\delta\varepsilon_p^{(n)}$, $f_{int}^{(n)}$

if $t = 0$ **then**

  for all *integration points p* **do**

    $\sigma_p^{(0)} \leftarrow \mathbb{C}\,\delta\varepsilon_p^{(0)}$

  end for

else

  for all *integration points p* **do**

    $\varepsilon_p^{(n+1)} \leftarrow \varepsilon_p^{(n)} \leftarrow \delta\varepsilon_p^{(n)}$

    $\sigma_p^{(n+1)} \leftarrow \mathbb{C}\varepsilon_p^{(n+1)}\left(\frac{\sqrt{1-\gamma^2}}{\gamma}\right)\mathbb{C}\delta\varepsilon_p^{(n)}$

    $f_{int}^{(n+1)} \leftarrow f_{int}^{(n)}\left(\frac{\sigma_p^{(n+1)}+\sigma_p^{(n)}}{2}\right)\delta\varepsilon_p^{(n)}$

  end for

end if

return $\delta\varepsilon_p^{(n+1)}$, $f_{int}^{(n+1)}$

γ	*Viscous damping parameter,* $\frac{x}{\Delta t} = \frac{\sqrt{1-\gamma^2}}{\gamma}$
$\mathbb{C}$	*Material stiffness matrix*
$\Delta t^{(n)}$	*Time step size in n^{th} increment*
$\sigma^{(n)}$	*Cauchy stress vector in n^{th} time increment*
$\delta\varepsilon^{(n)}$	*Incremental Strain vector in n^{th} increment*
$\varepsilon^{(n)}$	*Green Strain vector in n^{th} increment*
$f_{int}^{(n)}$	*Internal force in n^{th} increment*

3.2 Numerical implementation

In this section numerical results obtained from the simulation with the VUMAT user subroutine are examined for accuracy. In particular, verification is realized using results presented in Sze,[44] Arciniega,[45] Buechter,[46] and Simo,[47] among others.

Four problems are tackled here: The isotropic elastic behavior is assumed throughout section 3.2.1 and section 3.2.3, and an orthotropic laminate case is studied in section 3.2.4 The case study in section 3.2.1 is a classical benchmark problem with an analytical solution to test the large rotation of shells. The bending/torsional coupling is tested in section 3.2.2, and inextensible bending behavior is examined with the buckling hemisphere problem in section 3.2.3 This problem is considered as being both a linear and a nonlinear benchmark problem. Here loads are increased until they reach the large deformation state. Finally, the post-buckling behavior of an orthotropic laminate cylindrical roof is studied in section 3.2.4 for double lamination, in order to demonstrate the generalizability of the proposed approach.

The ABAQUS S4, 4-node, quadrilateral shell finite element with explicit time integration scheme and second-order accuracy is used for all simulations. The VUMAT user subroutine describes the fictitious viscous material behavior and replaces the software's default elastic

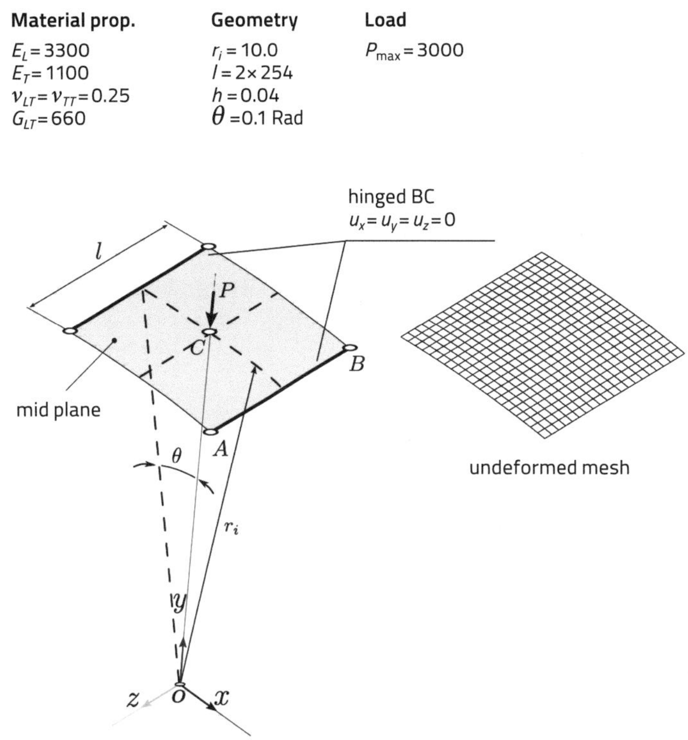

Fig. 9 Hinged cylindrical laminate problem, section 3.2.4

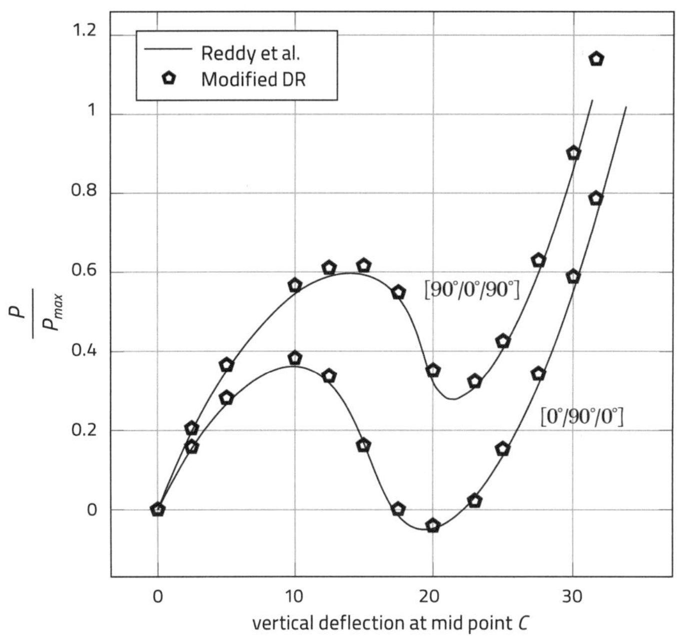

Fig. 10 Hinged cylindrical laminate problem: Vertical deflection of the mid point versus relaxed reaction force

behavior in order to determine the stress and internal energy update for each increment. According to the communicated syntax, VUMAT must define the Cauchy stress and the rotations are applied by the framework with the co-rotational formulation based on the Green-Naghdi rate. The rotation, however, should be implemented if a rate other than the default rate is required.

The analysis is run over time until it reaches a steady state for the desired degree of freedom. The only dissipative feature is the fictitious stiffness proportional damping with algorithm $\gamma=0.9$, as detailed in algorithm 2. The default linear, bulk viscosity parameter is thus turned off. The proposed approach is based on a pseudo-transient dynamic analysis, so wherever the density is not specified in the case studies, we assumed the deliberately chosen value of 500 kg/m^3.

3.2.1 *Roll up a clamped plate strip*

A distributed moment is applied on one end of a plate strip, while the other end is clamped. This problem has an analytical solution for the tip deflection in 2D case (in xz plan):

$$u_1(A) = \frac{EI}{M} sin\left(\frac{MI}{EI}\right) - 1$$
$$u_3(A) = \frac{EI}{M}\left(1 - \cos\left(\frac{MI}{EI}\right)\right) \qquad \text{equ. 7}$$

If , $M = M_{\max} = \frac{2\pi EI}{I}$ the timber strip is rolled into a full circle. The problem configuration is shown in fig. 3 and the material properties are set as $E=1.2\times10^6$, $\nu=0.0$. The results obtained for simulations with $M/M_{max}=1.0$, 0.9, 0.8, 0.6, 0.4, 0.3, 0.2, 0.15, 0.1 are compared with the computed analytical solutions. The deformed configurations are also extracted for a number of the load cases in fig. 4.

3.2.2 *Annular ring plate under transverse shear edge force*

An isotropic open ring is clamped on one edge and a uniformly distributed edge shear force P is applied to the other edge along the z-axis. **(fig. 5)** The strip undergoes a spring-like oscillation under bending and torsion caused by the sheared edge force until it reaches the equilibrium steady state. The plate dimensions are specified in fig. 5 and the thickness is intentionally $h=0.03$. Material elastic properties are $E=21\times10^6$, $\nu=0.0$. The strip is simulated under $P/P_{max}=0.1$, 0.2 the two-dimensional 0.4, 0.5, 0.6, 0.8, 0.9, 1.0 load cases and vertical deflection of points A and B (two ends of the loaded edge) are extracted to plot the load-deflection curve.

The final load case ($P=P_{max}=0.8$) involves a deformation of $\approx 3\times r_i$. Deformed configurations are plotted without magnification in fig. 4 and the simulated load-deflection curve for both end points are compared with results presented in Sze,[44] and Arciniega,[45] among others.

3.2.3 *Buckling of a pinched hemisphere*

Two equal point loads P are applied to a pinched hemispherical shell with an opening on top. The problem is simulated in one quadrant due to the symmetry boundary conditions at opposite ends **(fig. 7)**. This problem is a test to examine the method's ability to represent inextensible bending behavior and rigid body modes and is used as both a linear and nonlinear benchmark, depending on the applied load level. Here, the load is increased to cause a deformation level up to ≈ 60% of the radius to be placed in a large deformation regime. The material is assumed to be linear elastic with $E=6.825\times10^7$, $\nu=0.3$ and shell thickness is $h=0.04$.

Radial deflections under the applied load curves are extracted for load cases $P=20$, 40, 60, 80, 100, 120, 140, 160, 180, 200, and are compared to transcribed results in the literature. Particularly, modified DR results agree

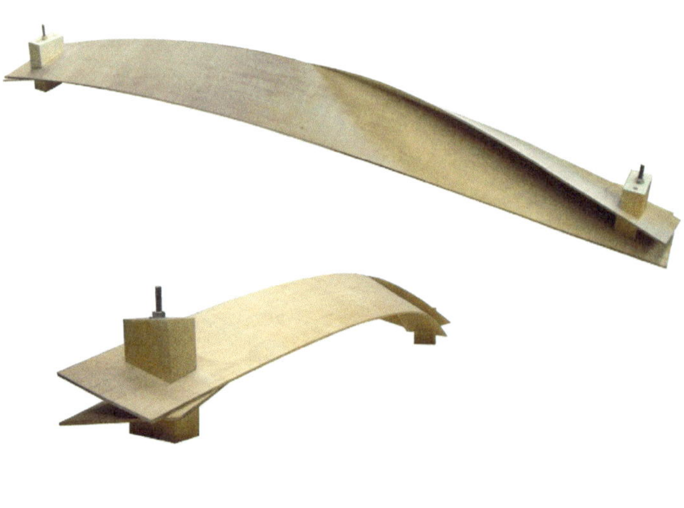

Fig. 11

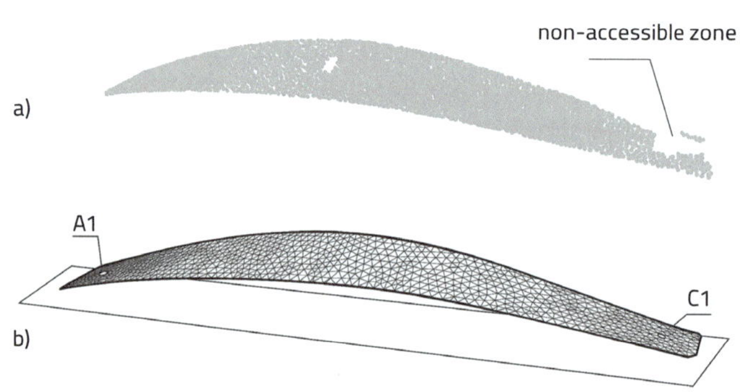

Fig. 12

Fig. 11 TFM built prototype

Fig. 12 Reconstruction of 3-D surface mesh corresponding to the top-surface of Panel 1
a) The scanned point cloud of the prototype after registration
b) The 3-D mesh surface reconstruction of the point cloud (see Fig. 13 a)

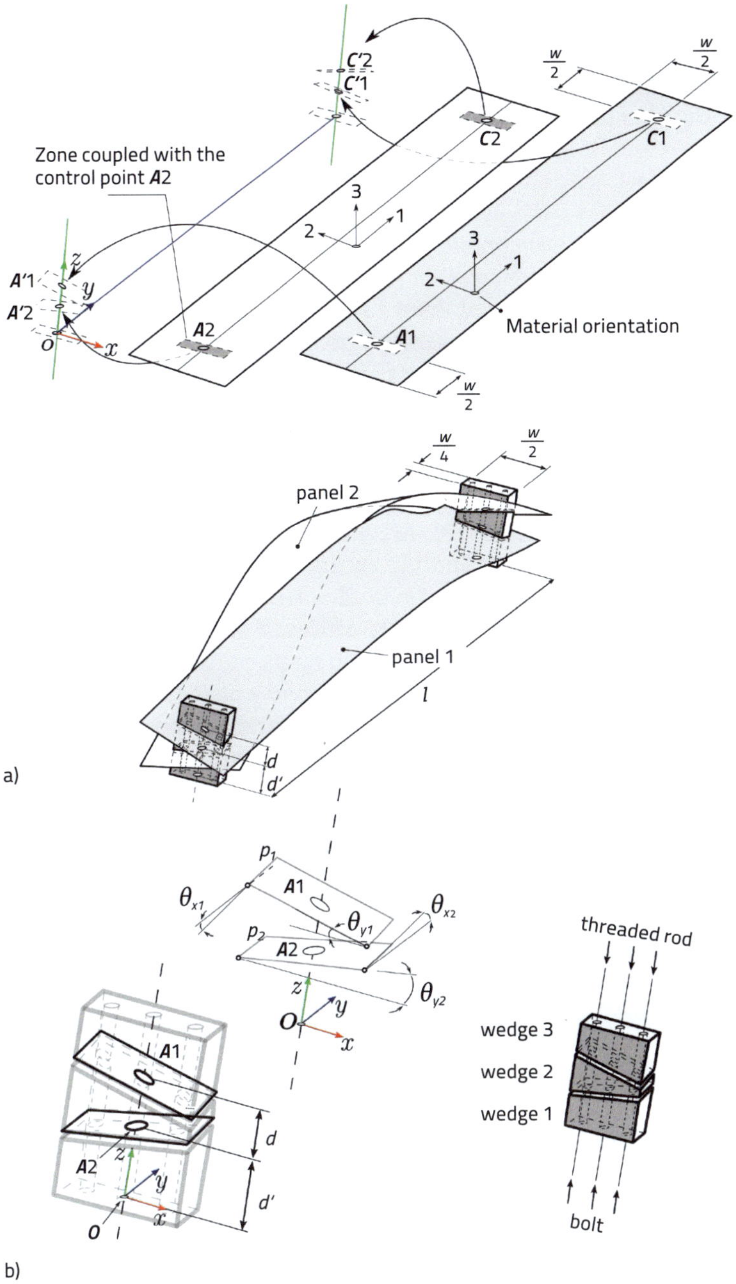

Fig. 13

Fig. 14 a) $0.0 \leq t \leq 2.0$ b) $2.0 \leq t \leq 4.0$

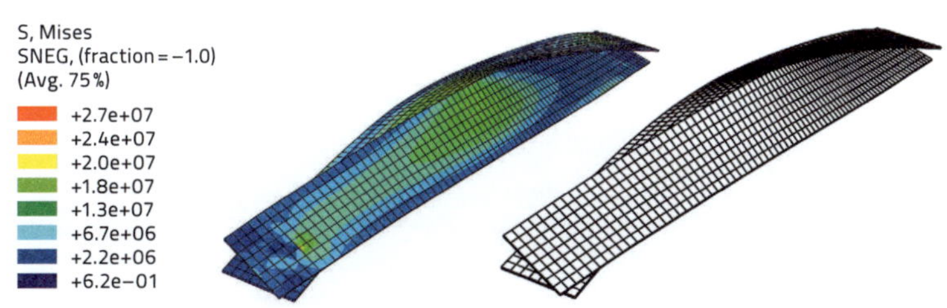

Fig. 15 a) b)

Fig. 13 a) TFM prototype geometry and assembly,
b) Connection relative angles and offset, wedge connection detail

Fig. 14 Deformed snapshots for simulation Step 1 $0.0 \leq t \leq 2.0$, see column **a)**, view from top to bottom and Step 2 $2.0 \leq t \leq 4.0$, see column **b)** view from top to bottom

Fig. 15 Simulation results at the end of Step 3
a) The von Mises stress on mid-surface panels [Pa]
b) Deformed snapshot for t = 14.0

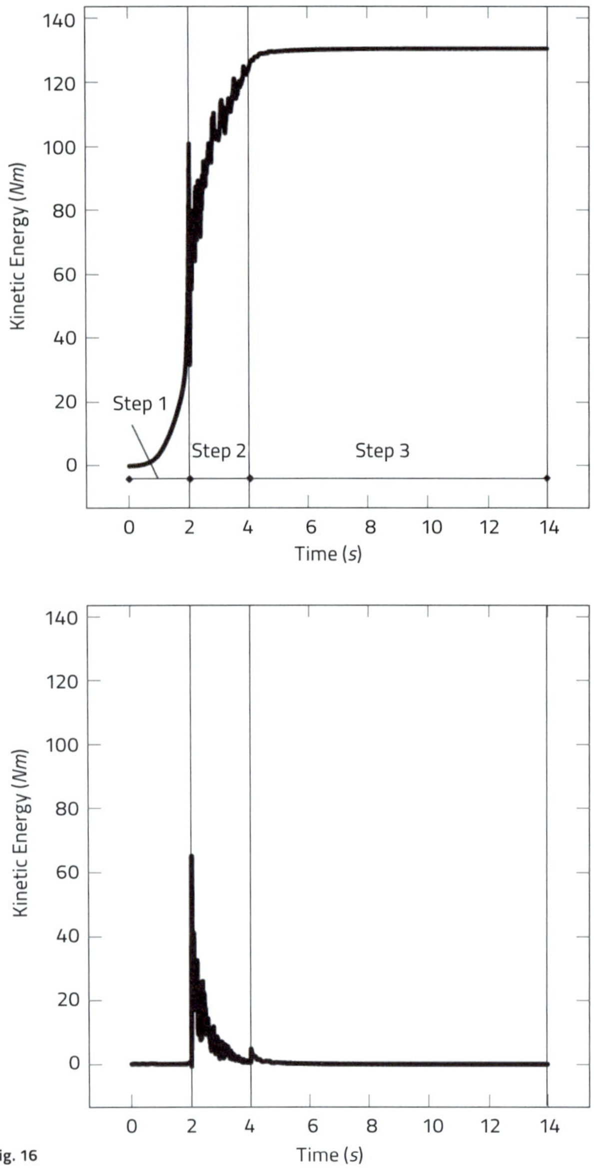

Fig. 16

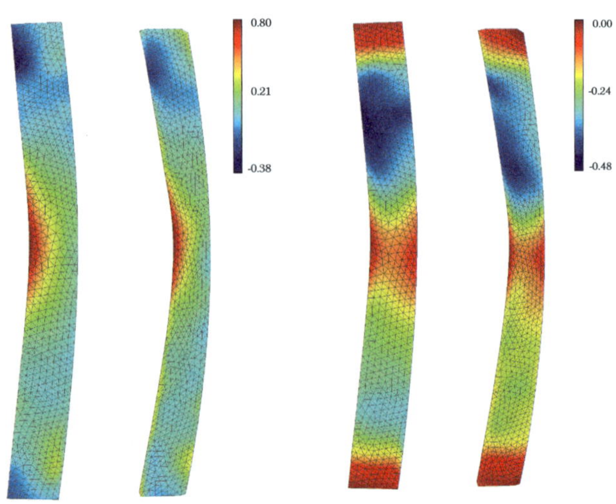

Fig. 17 a) Mean curvature (m^{-1}) b) Gaussian curvature (m^{-1})

with values ratified by Simo et al.[47] which are also reported by Buechter and Ramm,[46] and Jiang and Chernuka.[48] These curves are different; their values are shown in Sze[44] **(fig. 8b)**.

3.2.4 *Hinged cylinder roof: post-buckling behavior of an orthotropic laminate*

A vertical point load "P" is applied to the mid point of a cylindrical shell with hinged boundary condition assumed for lateral edges. The other two transverse curved edges remain free. The problem involves buckling and post-buckling of a lightly curved laminate shell (see fig. 9 for problem configuration).

Two laminate configurations are considered: [0°/90°/0°] and [90°/0°/90°]. All lamination plies have an equal thickness of $h/3$ and ply material assumed properties are assumed as $E_L=3300$, $E_T=1100$, $G_{LT}=660$, and $\nu_{LT}=\nu_{TT}=0.25$. Ply direction 0° is attained once the fibers run along the longitudinal edge. Viscous material definition can be used with both imposed load and/or imposed displacement regimes. Here, the problem is simulated with the imposition of the various displacement levels at the midpoint. Instead of the point load, the vertical displacement is imposed at the midpoint and vertical reaction force is extracted once steady state is reached for the reaction force. The load-deflection curve is extracted for the midpoint displacement $-u_Z(C)$2.5, 5, 10, 12.5, 15, 17.5, 20, 23, 25, 27.5, 30, 31.66, and is compared with the results reported by Sze et al.[44] and Arciniega and Reddy[45] **(fig. 10)**.

4 TFM prototype simulation

4.1 Prototype geometry and assembly process

A medium-scale prototype has been built from 2340 × 240 × 6.7 mm laminated wood panels to verify simulation results **(fig. 11)**. In order to ensure that relative panel offset and spatial angles are respected at their correspondent connections, three wooden wedges per connection point are exploited. The module is fixed with three bolts per wedge into a rigid panel. The set of three wedges at one end is mirrored on the other set, corresponding to the mid-plan normal to the y-axis. The geometric parameters for the prototype are as following (notations illustrated in **fig. 13**).

Fig. 16 Energy evaluation during the TFM simulation steps

Fig. 17 Deformed surface curvature comparison:
a) Simulation extracted surface
b) the prototype reconstructed-scanned surface

l	2060 mm	θ_{x1}	17°
w	240 mm	θ_{x2}	11.5°
d	47.7 mm	θ_{y1}	14.5°
d'	76 mm	θ_{y2}	-8°

In order to assemble the woven timber module from flat panels, first **A**1 and **A**2 are positioned respectively at **A**'1 and **A**'2 and are fixed. Subsequently, **C**1 is positioned at **C**'1 and fixed, and finally **C**2 is brought to position, **C**'2 and the entire structure is connected to the base steel profile **(fig. 13a)**.

4.2 Material properties

Cross-laminated *TeboPly™* Okumé plywood panels, supplied by Thebault (France) are used to build the prototype. The panels are 6.7 mm thick. It is a 7-mm, four-ply $[0/90]_s$ laminate. Orthotropic material orientations are illustrated in fig. 13 a and homogenized material constants are $E_1 = 4163Mpa$, $E_2 = 5088Mpa$, $\nu_{12} = 0.3$, $G_{12} = 552MPa$, and $G_{13} = G_{23} = 91MPa$, which correspond to the values obtained from the examples tested in ambient temperature and humidity.

4.3 Surface measurement using laser scanner FARO arm and mesh processing

In order to compare simulation results and validation, the top surface of panel 1 **(fig. 13a for panel 1 at RHS)** from the prototype is scanned using a three-dimensional, non-contact scanning technique with a FARO laser arm scanner. Depending on the size of the prototype, scanning is performed through six snapshots with sufficient superposition to ease the registration. However, due to the tight over-closure between panels along the connection zone between **C**'1 and **C**'2 handle points, the reconstructed panel is slightly smaller than the flat initial rectangle.

After every six snapshots, the point cloud is "cleaned" to contain only the vertices corresponding to the top surface of panel 1, and filtered to merge all the vertices together that are smaller than a 10-mm threshold. Duplicate vertices are also removed. These point clouds are then aligned and merged using the three-dimensional shape registration technique based on the ICP algorithm.[49] Subsequently, a three-dimensional, triangular surface mesh is reconstructed to match the aligned and merged point cloud using the Poisson surface reconstruction technique.[50] The reconstructed mesh, output of Poisson reconstruction, extends beyond the initial point cloud boundary and thus has to be split with poly-lines that correspond to approximate surface edges. This trimmed and reconstructed surface mesh is also re-meshed with uniform 30mm mesh size target length, using an isotropic, surface-based incremental remeshing technique.[51] This remeshed, polygonal, three-dimensional mesh **(fig. 13)** is used for comparison with simulation results in section 5.

4.4 Finite element simulation

The mid-surfaces for each panel are simulated using quadrilateral shell finite element, S4R ABAQUS/Explicit, with reduced integration scheme and enhanced hourglass control. Each FE node has six degrees of freedom: three translational and three rotational. ABAQUS/Explicit user subroutine VUMAT described in Section 3 is exploited for this simulation with constant $\gamma = 0.9$.

Four fixed points, **A'1, A'2, C'1, C'2,** are defined in three-dimensional space, which represent where corresponding points for each panel should be positioned. Two nodes are marked on each panel as positioning points to deform the panel, with **A1**, **C1** for panel 1 and **A2**, **C2** for panel 2. These positioning points help to impose translational/rotational boundary conditions forced by connector wedge blocks.

A rectangular zone (corresponding to the wedge cross-section) is selected per positioning point and is rigidly coupled with it over all six degrees of freedom. ABAQUS/Explicit general contact feature is used to handle face and edge contact during all simulation steps. Only non-penetrative, hard normal contact is considered, and the tangential component is ignored. Form-finding simulation is realized in three steps. Since the analysis is pseudo-dynamic, a notion of time length is assigned to each analysis step, but the duration of each step is chosen to be long enough for imposed boundary conditions to be applied smoothly **(see figs. 13 a and 13 b for notations)**.

- **Step 1:** ($0.0 \leq t \leq 2.0$)
  The panels are positioned with exact relative offset d, d' and their left-hand side handles **A1**, **A2** are angled into fixed points **A'1**, **A'2** respectively. θ_{x1}, θ_{y1} and θ_{x2}, θ_{y2} are imposed respectively for **A1**, **A2**. All degrees of freedom for these points remain fixed during this first step. Keeping **A1**, **A2** fixed during a dynamic/explicit step, the handle point **C1** is brought to position **C'1** and spatial angles $-\theta_{x2}$, $-\theta_{y2}$ are imposed. (Wedges are mirrored.) Rotational and displacement BCs are imposed simultaneously. The left midpoint of panel 1, noted as **B1,** is also fixed in the x-axis in order to ease the overlap that will occur in the following steps.

- **Step 2:** ($2.0 \leq t \leq 4.0$)
  Positioning points **A1**, **A2**, and **C1** are kept at their deformed positions **A'1**, **A'2**, and **C'1** respectively. All degrees of freedom for these three positioning points are fixed during this step except θ_{zz} to let panels reposition respective to each other. The fixed boundary condition for positioning point **B1** is also released. Starting with Step 2, the point **C2** is brought to position **C'2** and spatial angles $-\theta_{x1}$, $-\theta_{y1}$ are imposed. During this step, panels will intersect over the

cross-section and will reposition themselves into the relaxed configuration.

- **Step 3:** ($4.0 \leq t \leq 14.0$)
  Having four positioning points at deformed position and exact angles, all degrees of freedom for these four positioning points are fixed during this step and the module is left to dissipate extra kinetic energy.

Deformed snapshots extracted at equal time intervals illustrate the deformation of panels during Step 1 and Step 2. During Step 3 deformed geometry does not vary much from the deformed state at the end of Step 2, but it is essential to ensure that the system has reached the stationary state to have an accurate stress state. At the end of Step 3, deformed coordinates of nodes for the top surface of Panel 1 are extracted and reported for surface reconstruction and comparison with the scanned prototype geometry (Section 4.3).

5 Results and comparison

With our primary objective being form-finding of TFS, we progress with a comparison of deformed three-dimensional shapes, issue of modeling, and prototype scanning. Here, the main focus will be on the geometric comparison of these surfaces, rather than on the more classic approach of loading the prototype and deformation measurements as tracked in Sistaninia[3].

5.1 Simulation surface processing

Deformed coordinates of nodes of the upper surface of Panel 1 are ascertained from the simulation result database for the last frame of the simulation, Step 3. A three-dimensional, triangular mesh is patched onto these vertices and remeshed based, on the same technique and criteria described in section 4.3.

5.2 Surface comparison

Surfaces are aligned with the help of the ICP registration algorithm, in order to obtain an optimum fit between two, three-dimensional meshes. A maximum normal distance of 9 mm is found between two surfaces. Further inquiry on correspondence between surfaces would be to inspect their local curvature properties, the mean, and Gaussian curvature. These local properties on three-dimensional mesh surfaces can be estimated using the Algebraic Point Set Surface (APSS) approximated curvatures of the best fitting spheres to the mesh point set.[52]

Results for theses approximated curvatures are respectively shown in figs. 17 a and 17 b and indicate a good correlation between numerical and experimental results for both absolute value and distribution. The estimated value for the Gaussian curvature takes non-zero values, particularly in surface zones extending between the mid-contact point and the connection wedge, which means that the TFM deformed surface cannot be developed in its precise definition. Furthermore, the deformation map from the flat panel onto the deformed surface of the TFM is not isometric. Meanwhile, the absolute value for the Gaussian curvature remains relatively small. The distribution for Gaussian curvature follows the intuition that flat, cut wedges impose planarity (zero Gaussian curvature) on a locally restrained zone at both ends. Analysis of the mean curvature leads us to the important impact of the mid-contact point between panel edges, which imposes the local flexural deformation in this area. Transverse stability of each panel in the TFM assembly relies on in-plane stiffness at this mid-contact point. Distribution of both Gaussian and mean curvature are not symmetrical, which limits the simplification of the simulation.

6 Conclusion and future work

Pseudo-dynamic, explicit finite element methods, such as dynamic relaxation, can effectively deal with geometrically nonlinear problems of form-finding and initial stress prediction for space structures, where the objective is the free manipulation of thin shells undergoing finite rotation.

Timber fabric structures (TFS) are novel conceptual structures that originate from a reinterpretation of knot and braided structures with thin timber panels that can be generalized into a wide family of spatial structures where the form and structural behavior are strongly interrelated.

Mesh processing pipeline (from shape laser scanning up to three-dimensional mesh surface reconstruction and discrete curvature estimation techniques) seems to represent a promising approach for complex geometry, structural form investigation and may efficiently replace traditional loading experiments and reshape the special focus on the deformed shape comparison.

The simulation procedure with geometrically nonlinear, shell FE, although precise, is expensive and less responsive to design changes. Consequently, we are currently working on the application of physically based modeling techniques for discrete thin shell and elastic rod simulation, to approach a more interactive, finite strip simulator with an aim to integrate it into a physically based NURBS/CAD tool for active bending form exploration.

Acknowledgments

This research work was sponsored by the Swiss National Science Foundation (SNSF) under grant No. 20002_137884/1. This support is gratefully acknowledged.

References

1 Hudert, M. "Structural Timber Fabric: Applying Textile Principles in Building Scale." PhD Thesis, IBOIS, EPFL, Lausanne, 2012.

2 Weinand, Y., and M. Hudert "Timberfabric: Applying Textile Principles on a Building Scale." In *Architectural Design*, 80(4), 102–107, 2010.

3 Sistaninia, M., M. Hudert, L. Humbert, and Y. Weinand. "Experimental and Numerical Study on Structural Behavior of a Single Timber Textile Module." In *Engineering Structures*, 46(0), 557–568, 2013.

4 Otter, J. and A. Day. "Tidal Computations." In *The Engineer*, 1960.

5 Day, A. "An Introduction to Dynamic Relaxation." In *The Engineer*, 219, 218–221, 1965.

6 Otter, J., A. Cassell, R. Hobbs, et al. "Dynamic Relaxation." In *Inst. Civ. Engrs. Proceedings*, vol. 35, 633–656, 1966.

7 Otter, J. "Computations for Pre-Stressed Concrete Reactor Pressure Vessels using Dynamic Relaxation." In *Nuclear Structural Engineering*, (1), 61–75, 1965.

8 Frankel, SP. "Convergence Rates of Iterative Treatments of Partial Differential Equations." In *Mathematical Tables and Other Aids to Computation*, 4(30), 65–75, 1950.

9 Frieze, P., R. Hobbs, and P. Dowling. "Application of Dynamic Relaxation to the Large Deflection Elasto-Plastic Analysis of Plates." In *Computers & Structures*, 8(2), 301–310, 1978.

10 Pica, A. and E. Hinton. "Transient and Pseudo-Transient Analysis of Mindlin Plates." In *International Journal for Numerical Methods in Engineering*, 15(2),189–208, 1980.

11 Pica, A. and E. Hinton. "Further Developments in Transient and Pseudo Transient Analysis of Mindlin Plates." In *International Journal for Numerical Methods in Engineering*, 17(12), 1749–1761, 1981.

12 Ramesh, G. and C. Krishnamoorthy. "Geometrically Nonlinear Analysis of Plates and Shallow Shells by Dynamic Relaxation." In *Computer Methods in Applied Mechanics and Engineering*, 123(1), 15–32, 1995.

13 Kommineni, J. and T. Kant. "Pseudo-Transient Large Deflection Analysis of Composite and Sandwich Shells with a Refined Theory." In *Computer Methods in Applied Mechanics and Engineering*, 123(1), 1–13, 1995.

14 Ramesh, G. and C. Krishnamoorthy. "Postbuckling Analysis of Structures by Dynamic Relaxation." In *International Journal for Numerical Methods in Engineering*, 36(8), 1339–1364, 1993.

15 Papadrakakis, M. "Postbuckling Analysis of Spatial Structures by Vector Iteration Methods." In *Computers & Structures*, 14(5), 393–402, 1981.

16 Alamatian, J. "Displacement-based Methods for Calculating the Buckling Load and Tracing the Postbuckling Regions with Dynamic Relaxation Method." In *Computers and Structures*, 114115, 84–97, 2013.

17 Barnes, M. "Form Finding and Analysis of Tension Structures by Dynamic Relaxation." In *International Journal of Space Structures*, 14(2), 89–104, 1999.

18 Adriaenssens, S. and M. Barnes. "Tensegrity Spline Beam and Grid Shell Structures." In *Engineering Structures*, 23(1), 29–36, 2001.

19 Ye, J., R. Feng, and S. Zhou. "The Modified Dynamic Relaxation Method for the Form-Finding of Membrane Structures." In *Advanced Science Letters*, 4(810), 2845–2853, 2011.

20 Bel Hadj Ali, N., L. Rhode Barbarigos, and I. Smith. "Analysis of Clustered Tensegrity Structures using a Modified Dynamic Relaxation Algorithm." In *International Journal of Solids and Structures*, 48(5), 637–647, 2011.

21 Rodriguez, J., G. Rio, J. Cadou, and J. Troufflard. "Numerical Study of Dynamic Relaxation with Kinetic Damping Applied to Inflatable Fabric Structures with Extensions for 3-D Solid Element and Nonlinear Behavior." In *Thin Walled Structures*, 49(11), 1468–1474, 2011.

22 Douthe, C. and O. Baverel. "Design of Nexorades or Reciprocal Frame Systems with the Dynamic Relaxation Method." In *Computers & Structures*, 87(21), 1296–1307, 2009.

23 Joldes, G., A. Wittek, and K. Miller. "An Adaptive Dynamic Relaxation Method for Solving Nonlinear Finite Element Problems. Application to Brain Shift Estimation." In *International Journal for Numerical Methods in Biomedical Engineering*, 27(2), 173–185, 2011.

24 Brew, J. and D. Brotton. "Nonlinear Structural Analysis by Dynamic Relaxation." In *International Journal for Numerical Methods in Engineering*, 3(4), 463–483, 1971.

25 Wood, WL. "Note on Dynamic Relaxation." In *International Journal for Numerical Methods in Engineering*, 3(1), 145–147, 1971.

26 Papadrakakis, M. "A Method for the Automatic Evaluation of the Dynamic Relaxation Parameters." In *Computer Methods in Applied Mechanics and Engineering*, 25(1), 35–48, 1981.

27 Underwood, P. "Dynamic Relaxation (in Structural Transient Analysis)." In *Computational Methods for Transient Analysis*, Amsterdam, 245–265, 1983.

28 Cassell, A. and R. Hobbs. "Numerical Stability of Dynamic Relaxation Analysis of Nonlinear Structures." In *International Journal for Numerical Methods in Engineering*, 10(6), 1407–1410, 1976.

29 Alwar, R., N. Rao, and M. Rao. "An Alternative Procedure in Dynamic Relaxation." In *Computers & Structures*, 5, 271–274, 1975.

30 Tong, P. and Ty Tsui. "Stability of Transient Solution of Moderately Thick Plate by Finite Difference Method." In *AIAA Journal*, 9(10), 2062–2063, 1971.

31 Rezaiee-Pajand, M., M. Kadkhodayan, and J. Alamatian. "Timestep Selection for Dynamic Relaxation Method." In *Mechanics Based Design of Structures and Machines*, 40(1), 42–72, 2012.

32 Alamatian, J. "A New Formulation for Fictitious Mass of the Dynamic Relaxation Method with Kinetic Damping." In *Computers and Structures*, 9091(1), 42–54, 2012.

33 Rezaiee-Pajand, M., S. Sarafrazi, and H. Rezaiee. "Efficiency of Dynamic Relaxation Methods in Nonlinear Analysis of Truss and Frame Structures." In *Computers and Structures*, 112113, 295–310, 2012.

34 Rezaiee-Pajand, M., M. Kadkhodayan, J. Alamatian, and L. Zhang. "A New Method of Fictitious Viscous Damping Determination for the Dynamic Relaxation Method." In *Computers and Structures*, 89(910), 783–794, 2011.

35 Rezaiee-Pajand, M. and S. Sarafrazi. "Nonlinear Dynamic Structural Analysis using Dynamic Relaxation with Zero Damping." In *Computers and Structures,* 89(1314), 1274–1285, 2011.

36 Rushton, K. "Large Deflection of Variable Thickness Plates." In *International Journal of Mechanical Sciences,* 10(9), 723 –735, 1968.

37 Chung, W., J. Cho, and T. Belytschko. "On the Dynamic Effects of Explicit Fem in Sheet Metal Forming Analysis. Engineering Computations." In *Journal for Computer Aided Engineering,* 15(6), 750–776, 1998.

38 Cassell, A., P. Kinsey, and D. Sefton. "Cylindrical Shell Analysis by Dynamic Relaxation." In: *Inst. Civ. Engrs. Proceedings,* vol. 39. 75–84, 1968.

39 Lynch, R., S. Kelsey, and H. Saxe. "The Application of Dynamic Relaxation to the Finite Element Method of Structural Analysis." Tech. Rep.; Technical report No. THEMISUND681, University of Notre Dame, 1968.

40 Hughes, T. "Analysis of Transient Algorithms with Particular Reference to Stability Behavior." In *Computational Methods for Transient Analysis*, Amsterdam, 67–155, 1983.

41 Belytschko, T. "An Overview of Semi-Discretization and Time Integration Procedures." In *Computational Methods for Transient Analysis,* Amsterdam, 1–65, 1983.

42 Courant, R., K. Friedrichs, H. Lewy. "über die Partiellen Differenzengleichungen der Mathematischen Physik." In *Mathematische Annalen,* 100(1), 32–74, 1928.

43 Tabiei, A. and R. Tanov. "A Nonlinear Higher Order Shear Deformation Shell Element for Dynamic Explicit Analysis." In *Finite Elements in Analysis and Design,* 36(1), 17–37, 2000.

44 Sze, K., X. Liu, and S. Lo. "Popular Benchmark Problems for Geometric Nonlinear Analysis of Shells." In *Finite Elements in Analysis and Design,* 40(11), 1551–1569, 2004.

45 Arciniega, R., and J. Reddy. "Tensor-Based Finite Element Formulation for Geometrically Nonlinear Analysis of Shell Structures." In *Computer Methods in Applied Mechanics and Engineering,* 196(4), 1048–1073, 2007.

46 Buechter, N. and E. Ramm. "Shell Theory Versus Degeneration a Comparison in Large Rotation Finite Element Analysis." In *International Journal for Numerical Methods in Engineering,* 34(1), 39–59, 1992.

47 Simo, J., D. Fox, and M. Rifai. "On a Stress Resultant Geometrically Exact Shell Model." *In Computer Methods in Applied Mechanics and Engineering,* 79(1), 21–70, 1990.

48 Jiang, L. and M. Chernuka. "A Simple Four-Noded Co-Rotational Shell Element for Arbitrarily Large Rotations." In *Computers and Structures,* 53(5), 1123 –32, 1994.

49 Besl, P. and N. McKay "A Method for Registration of 3-D Shapes." In *IEEE Transactions on Pattern Analysis and Machine Intelligence,* 14(2), 239–256, 1992.

50 Kazhdan, M., M. Bolitho, and H. Hoppe. "Poisson Surface Reconstruction." In *Proceedings of the Fourth Euro-graphics Symposium on Geometry processing.* Euro-graphics Association, 61–70, 2006.

51 Botsch, M. and L. Kobbelt. "A Remeshing Approach to Multi-Resolution Modeling." In *Proceedings of the 2004 Eurographics/ACM SIGGRAPH symposium on Geometry Processing.* ACM, 185–192, 2004.

52 Guennebaud, G. and M. Gross. *Algebraic Point Set Surfaces.* ACM Trans Graph, 26(3), 2007.

Reprinted from *International Journal of Sapce Structures,* vol. 28, No. 3 & 4, 2013.

4.5

„These programs don't provide easy, off-the-shelf solutions"

Interview between Olivier Baverel (Laboratoire Navier, Ecole nationale des Ponts et Chaussées, Paris) and Yves Weinand
May 26, 2015

Yves Weinand: Olivier, we both come from the academic world and we're particularly interested in the practical and applied potential of our research. However, I am under the impression that our activities have not garnered the interest of a large audience. We remain marginal in combining research with practical applications. Together, we have already had the opportunity to address the question of active bending. In the classes I give at EPFL,[1] I try to clarify the notion of active bending by presenting built examples that have left their mark on the history of construction. Such as the structures that Colonel Emy[2] designed for the French military, or the creation of glue-laminated wood.[3] I've noticed that there are few applications of these types of structures. What is your opinion about this?

Olivier Baverel: Most likely it's because we've not yet perfected our tools, and so long as they're not user-friendly, we won't have many users. It's important to remember that the truss beam already appeared 500 or 600 years ago, but no one used it until the moment when mathematical tools allowed for it to be better understood.
It was only around 1820 that we began to understand and began to be capable of predicting the behavior of a truss beam: fifty years later, the Eiffel Tower was erected. There's a similar story in relation to Frei Otto and metallic-textile structures, which until the 1970s, existed solely to cover small spaces. From the moment we understood the theory and created tools to predict forms and their applications, large-spanning projects appeared, such as the Munich stadium in 1972.
As far as we're concerned, we're close to having a final and predictive tool, but there's still a lot of work to be done. Once these tools can easily be used by a substantial group of engineers and architects, that's when I believe these structural typologies will fully emerge. Until now, if not for the last ten years or so, it was not at all conceivable. Neither the digital tools nor the calculation capacity were sufficiently developed, from either a theoretical or mechanical point of view. We're not there yet, but we're getting closer.

Y. W.: Rib shells are actively bending structures that fascinate us. We share this common passion, but I'm under the impression that there's a large step to be made in terms of convincing firms to share our passion for bending structures. Shigeru Ban, for example, proposed this type of structure, but was unable to construct works that are efficient or maximized from a structural perspective. These structures seem to evolve from a more formal conception, rather than from a structural perspective. In that sense, his work remains modernist. It does not correspond exactly to what we are attempting to discuss in the fourth chapter of this book, namely: stronger harmony between form and structure, and about structures that are actively flexed. Do you think there's a way to motivate people around these types of structures, to convince them through objective arguments—for example, in relation to flexibility and in relation to seismic activity? Do you think it's possible to convince people in a pragmatic manner, and not from a kind of researcher's idealistic motivation?

O. B.: Yes, that would be possible the day a designer has a tool in his or her hands that is easy to use. The relevance of these extremely slim structures is to take exposed beams made from small wooden sections, assemble them on the ground and shape them to create a three-dimensional form. These structures have a low environmental impact, as they don't need a complex transformation of wood, no bending or bonding, and little or no material waste. There is currently nothing simpler today than to make a lattice and set out the lattices one next to the other. Why look for more structural optimization when latticework already has tremendous capacities? Lattice will consume a bit more material than a ribbed shell, but today, in relation to the overall cost of things, it's barely significant. If we begin to consider environmental costs more seriously, and not solely financial costs, these slim, optimized structures that don't involve excessive transformations and important losses would become truly worthy of interest.

In parallel, these last fifteen years have seen the rise of digital tools like Archicad or Vectorworks in architecture. De Casteljau and Bézier, engineers at Citroën and Renault, developed algorithms called NURBS for the shaping of car bodies. With embossing techniques, using a thirty-ton press, any kind of metal plate can be shaped. However, these tools have not yet been adapted to our industry, which is primarily based on the assembly of different metals and not by embossing a shape out of a forty-by-forty-meter piece of metal. We have spent the last fifteen years using a tool that is a poor example of what we should be fabricating. This use of this inappropriate tool for the task means a fairly significant over-consumption of material in many contemporary examples. This over-consumption is testimony to a lack of capacity in representing the elements that need to be fabricated. Thanks to tools like Rhino[4] and Grasshopper,[5] industry has progressed from that perspective. But there's still a lot of work to do in order to have tools that are completely adapted to our industry. The representational capacity means that structures such as ribbed shells will no doubt be far more interesting than classical structures in the coming years. Remember that to build an airplane, Dassault Systèmes proposes a suite of more than 150 programs; but only five years ago only two or three programs were available to help in the construction of a building. We were lagging very far behind. Today, diversification is taking place on many levels thanks to programs like Rhino and Grasshopper, and associated plug-ins.

Y. W.: I feel the time is ripe for considering applications; the tool developed by Sina Nabaei[6] for example, but also for the application of actively flexed structures. The NCCR[7]—where we are active—could allow for it, though we lack engineers to be able to take Sina Nabaei's work further. His work only includes one practical application to date (a spider form). This sole example is not sufficient for illustrating the tremendous potential of this tool. I'm under the impression that, at that level, very few people would be capable of helping us at the moment.

O. B.: There are two answers to your question: the first is that we need to employ young people who will use Sina Nabaei's tool, take it further, and potentially create a large-scale model at the EPFL or at the Ècole des Ponts to demonstrate the capacity of these algorithms.
You also need to have some luck and find the opportunity. If we look at the gridshells created by Frei Otto, it was pure chance that a flower exhibition at Mannheim necessitated a construction over a sixty-meter range. There was sufficient intellectual impetus for it to be produced, and enough luck to make it happen. Insofar as the Sina Nabaei tool is concerned, we are not mature enough yet. First and foremost, Sina Nabaei designed a complex engineering tool of a computer scientist. We would need to build a prototype at the university level with a range of about ten meters. Then, we would need the opportunity to test it in a real project. We are still missing these two stages. We had a few opportunities with our gridshell structures, which we constructed out of composite materials at the Ècole des Ponts. This really showcased our representational capacities and know-how in terms of calculations. We had two concrete opportunities, one for a festival and the other for an ephemeral cathedral (a temporary structure), using composites. We had to deal with the French planning office for this large-scale construction, who are very competent, but to whom we had to explain our structure. At that time there was no official Eurocode for composite materials. We therefore took the double risk of designing an atypical structure with a material that had never before been used in our industry. Thankfully, the contractor and auditor were supportive and it was a success.

Y. W.: Many people seem to be fascinated by braided structures, but this fascination seems limited to the aesthetic perspective. The structural advantages of this type of construction are not perceived, nor are they commented on. We have not yet been able to create a wider acceptance for this type of structure. It seems to me that the practical application proposed by Sina Nabaei is still somewhat awkward or unfinished in its global form, and that it would be necessary to add an architectural aspect to it. I believe that we should deploy this type of structure quickly and efficiently in earthquake-prone areas, or in response to natural disasters, for example, rather than aiming for a specific commission. Rather than building a pavilion, a pragmatic application would no doubt be very useful towards bringing these structures into the mainstream, in terms of how they are understood in relation to their structural efficiency.

O. B.: One has to pay attention to the goal—with the concept of a pavilion: some are more focused on representation, while others are more interested in the structure. Certainly if we move towards pure aesthetics, interest will quickly wane. You have to clearly demonstrate the structural advantage—or at least the combination of technological and structural advantages—as these are always the two pillars on which we base our constructions. This work remains to be undertaken. It's about creating a prototype that has an architectural function to be determined, an interesting rendering, but above all a true mechanical capacity, whether at the level of the assembly typology chosen, or for the zones with high risk of seismic activity, or simply thanks to the mesh effect that is placed between the slat layers. With Sina Nabaei's thesis, we have a tool and a working method, but we have not developed further in demonstrating how this type of typology is relevant in terms of mechanical behavior. With a good design team, it would be possible to have an extraordinary result, with true added value, where there is a reasonable consumption of materials and an implementation of technology that would no doubt be fairly straightforward.

Y. W.: Are there any industrial partners in France for developing this technology?

O. B.: As you know, France has an annual production rate of around 65 million cubic meters of timber. That's more than a cubic meter per person, but unfortunately the system is not well organized. There are 3.5 million privately-owned forests that are on average only 2.6 hectares in size. Currently, French timber does not have a powerful lobby, when compared to the producers of other materials where the lobby is well organized. There is true production potential, but there are few large sawmills, compared to Austria, for example. Small enterprises like Simonin do their best and are keen to innovate, but they don't have the same capacity as the large concrete or steel factories.

Y. W.: From an academic perspective, I have tried to motivate engineering students to be interested in the global form. Should we consider trying to attract these types of students, specifically to further develop Sina Nabaei's tool and to make it easier to understand? Do you have any ideas or proposals in relation to this?

O. B.: Engineers lack a culture of structures and often remain far from the line of research we seek to advance. They can't understand because they're not taught these aspects during their training. In general, the engineer calculates the structure and justifies it through the Eurocode, but doesn't question the form. I believe this is a serious error in relation to how work is organized. At the EPFL, at the École des Ponts, and at other universities, there are courses on conceptual design and structures that should be further developed and taken up by other institutions. It's also about sensibilities: for an engineer, there's nothing more comforting than making calculations and applying rules such as Eurocode. Of course taking things further, questioning and innovating, is bound to be more complicated. Our engineering students are primarily recruited for their mathematical skills. An emphasis on curiosity and innovation should also be integrated into their curriculum.

Y. W.: But if we consider that the IASS[8] is organizing an event in Amsterdam this summer, where pavilions will be built, I'm under the impression that these pavilions will be consistent with the representation of form, rather than remaining a purely mechanical problem. In terms of final form-finding, I believe Sina Nabaei's tool is very strong. The depth achieved by his tool is not rendered solely through the construction of a pavilion. Freestanding structures are regularly published, but rarely present their mechanical characteristics and performances. Aspects like: Is there friction or not? What was the original situation of equilibrium? What does friction bring as added value? Is there friction? We easily talk about a synthesis between form and structure—morphogenesis—but we don't detail the mechanical performance. It therefore becomes auto representation.

O. B.: Yes, I have exactly the same impression. But this also comes from the fact that understanding structures is complex, even with a simple, static diagram. People prefer to create an aesthetic form, but to move from there and to take the effort to understand the structural behavior, to rationalize the construction, to limit the losses...there are not very many people who think along these lines. I saw this among my architectural students. There's a plug-in in Grasshopper called Karamba,[9] which enables you to do structural analysis calculations.
All of the students were enthusiastic, but as soon as I told them it was about making mechanical structures, evaluating the supporting conditions, what is the load, what are the sections, what types of calculations should be undertaken, etc., all of these questions are immediately less "glamorous." You have to ask yourself many questions, you have to anticipate, examine, and these programs don't provide easy, off-the-shelf solutions. What many people tend to forget is that the mechanics of structures only began to be mastered during the past 180 years or so. Galileo posed the question: how to find the stresses in a cantilever beam? But he didn't manage to find an answer. Galileo was a pioneering thinker who initiated an intellectual revolution throughout Europe, and yet he was unable to solve the simple question of the cantilever beam! This clearly proves that it's complicated and not necessarily accessible to everybody. I'm not surprised by what you describe; we see many things, but we don't explain them sufficiently. In any case, it's a problem; we sell the image without really understanding the content.

Y. W.: When we met up at the IASS conference in Venice in 2009, we noted that engineers are starting to embrace forms. Indeed, the engineer should not be indifferent or neutral to the form proposed by the architect. It is necessary for the engineer to take a stance in relation to this form. He should be interested in this global form from his own point of view. Certain engineers claim that they can calculate any forms! This may be true, but today I consider this stance with some skepticism. I prefer to look for work where there is a strong coherence between form and structure, as with Félix Candela's work, for example, originally from the IASS. Certain new trends at IASS, which I see as weaknesses, don't seek this synthesis that I aspire to, and to, which I would like to attract more adherents—engineers in particular.

O. B.: Yes, I agree. But then, let's be prudent with our comparisons. If, for instance, we consider the IASS in the 1970s, history has forgotten mediocre achievements: we only retain the significant ones. When it comes to the capacity to calculate all forms without limit, there's certainly a lack of culture on the part of the engineer on the critical question of form. As far as I'm concerned, when I discuss this issue with an architect, I try to understand the substance of the discourse around his or her intentions, rather than solely the form. It is work that architects and engineers have to collaborate on. What's important is to convey the architect's concept, rather than the formal object he has rendered, recognizing that the tools he is using are not adapted for presenting the form that he wants.
Incidentally, regarding representational tools, with Nurbs, it is practically impossible to make folds or to create discontinuities in curves, or local singularities on surfaces. In fact, architecture today is a prisoner of a formal universe, which does not represent all of the possible forms, but only those of Nurbs. Today, there are many other solutions, which we don't allow ourselves to create, because we don't have the proper representational tools. Those who believe they can represent everything with Nurbs are fooling themselves. There's a lot of work to be done by the entire community to

develop new representational tools. Furthermore, we also began working on a thesis along this theme two years ago. We are advancing well and are capable of describing formal realms with particular, "non-Nurb" structural typologies. (...)
We can continue our discussion with the massive impact of robots in architectural production. I would like your opinion on this. You have several ongoing theses on this topic, is that correct?

Y. W.: Yes, but I'm not convinced that all systems have to be additive systems. We also know that robots are imprecise. There are sequential aspects that must be respected. I'm interested in advanced building processes in architecture. These processes do not only relate to robotics, but to all technologies that are available to serve the construction and architecture markets. We would like to actively fold panels by using several robots, but I'm not convinced that should be done by robots. We can test it, but we'll eventually find other means. The history of ribbed shells demonstrates that there are other ways to approach this. What interests me is the following question: How to interest industrial partners in this type of structure, generally speaking, in order to advance further?

O. B.: In my opinion, for that you would need to look for a long-term partner. Apropos robots, I'm lucky enough to be mentoring a young researcher who is developing a robot manipulation tool from Grasshopper. The robots will help us to think about how to produce complex structures today. A new technological tool will appear that will be taken up by some like a religion, but will be considered an enemy by others. The thesis is about creating concrete formwork systems with a very particular robot technology, which would not be feasible without a robot. I think there are many techniques that have been abandoned, as they were too time-consuming and expensive, but with robotics, we can once again ask questions about their feasibility. For example, we can revisit Japanese wood joinery and build with wood without using any metallic connectors at a more reasonable cost.

Y. W.: This is precisely what we have defined in our upcoming research topics. I'm fascinated by the fact that reinterpretation of Japanese joinery may have an effect on the way future logistical concepts in the wood industry evolve.

References

1 "Conception de structures" (Structure conceptions), 3rd-year bachelor, architecture section, EPFL.

2 Traité de l'art de la charpenterie (Treatise on the art of carpentry), Colonel Emy, 1878.

3 Otto Hetzer, 1906.

4 Rhino is a 3-D drawing program for graphic design and modeling.

5 Grasshopper® is a graphic editor of 3-D algorithms, entirely integrated with Rhino modeling tools. Unlike RhinoScript, Grasshopper does not require programming or scripting knowledge, but allows designers to build form generators from the simple to the grandiose.

6 Sina Nabaei. "Mechanical Form-Finding of Timber Fabric Structures." Thesis n° 6436, EPFL, Lausanne, 2014.

7 The National Centre of Competence in Research (NCCR) Digital Fabrication—Advanced Building Processes in Architecture is hosted at the ETH Zurich, directed by Matthias Kohler, in collaboration with 13 laboratories—8 laboratories from ETHZ, 2 laboratories from EPFL, 2 from EMPA, and 1 from BUAS. Included is the Laboratory for Timber Construction, IBOIS, directed by Prof. Yves Weinand.

8 IASS 2015 Annual International Symposium on Future Visions, 17–20 August, 2015: The continuous development of design, analysis, and construction techniques for the built environment and shell and spatial structures. http://www.iass2015.org

9 Developed by Bollinger + Grohmann engineers.

4.6

Braided structures: applying textile principles at an architectural scale

Marielle Savoyat

Design IBOIS—Laboratory for Timber Constructions/ EPFL, Swiss Federal Institute of Technology, Lausanne, Switzerland
Prof. Yves Weinand and Dr. Markus Hudert (researcher)

Research and completion 2007–2013

The research undertaken between 2007 and 2013 by Markus Hudert within IBOIS, the Laboratory for Timber Constructions at the Swiss Federal Institute of Technology, analyzes the use of textile techniques at an architectural scale. It soon became apparent that principles of knitting, braiding, and weaving offer great potential for varying structural possibilities when applied to the scale of architecture.

The common denominator between all of these textile techniques is one basic element: that of a thread interlaced with another thread. This starting principle can be transferred onto two interlaced planks of wood. To put this concept into practice, a first prototype called a textile module was created, which demonstrated how the application of a textile technique, when combined with the properties of wooden material, could lead to a particularly efficient freestanding structure.

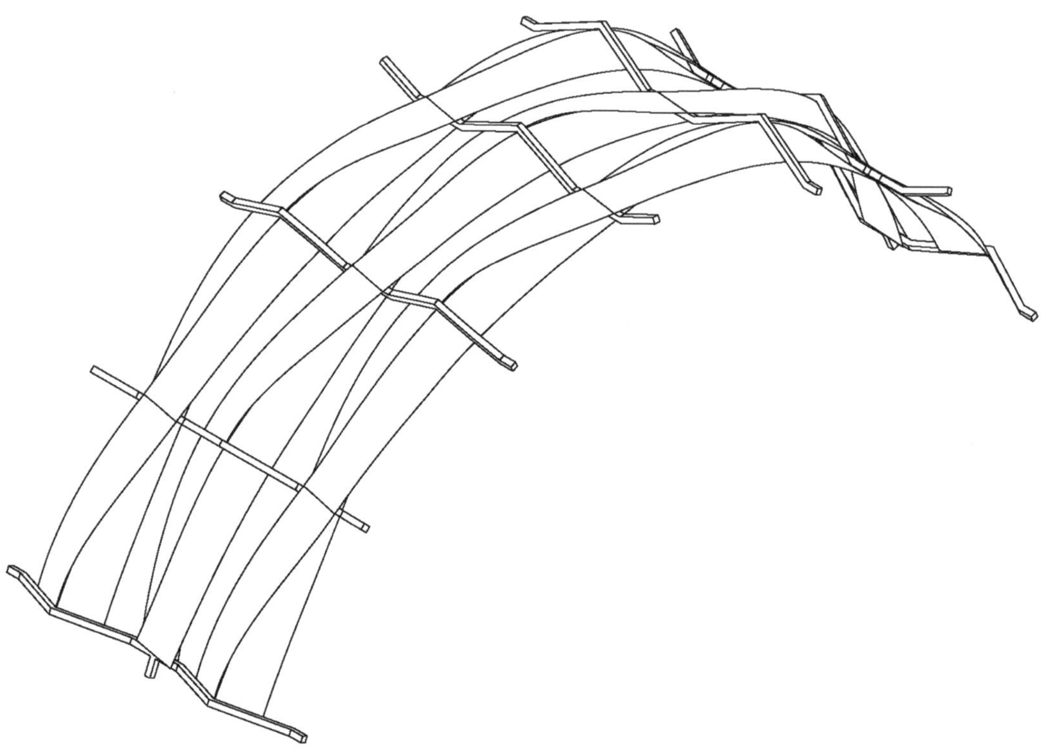

Fig. 1 Axonometry of a braided arch

Fig. 2 Underview of the model

The geometry of the whole was automatically conditioned by the assembly techniques. What's more, when pressure is exerted on the structure,[1] it lengthens and becomes narrower, while the middle section of the basic module gains in height, thus making the structure more rigid. Indeed, superimposing two thin wooden planks and vertically aligning the screw holes cutout at each extremity automatically generates the curvature without relying on a mold. A structural resistance becomes apparent. To obtain structural capacity, the finer layers have to be fixed together.

One of the structural advantages of textile is that it is composed of a great number of simple, interconnected elements, which work together to form a whole. Thus, a weakness in one of the elements does not affect the structure in its entirety in any way. In order to create the same effect at an architectural scale, the creation of a stable structure composed of a multitude of elements also becomes essential, so that the deformation of certain elements would not compromise the entire building.

Markus Hudert's research was applied to exploring the ways that the textile module could be used as a basic unit in a large-scale structure. By aligning several modules in the same axis, a structure in the form of an arc is created. A sequence of several of these arcs then produces a vault-structure. The disadvantage of this approach resides in the fact that the modules remain independent from one another. To obtain large structures, connections would need to be added to the chain mail at the upper and lower points.

Textile principles are therefore of interest both in structural and architectural terms. By twisting the material, internal tensions are created and stiffening becomes apparent. The properties and the aspect created by interlacing elements are not only highly aesthetic, but also embody great structural strength and high potential for spatial quality.

The coherence between structure and space results in high quality architecture. It is important to note that these structures remain limited to single-story buildings. The question remains: How can these structures be roofed over to further their architectural form? The linking elements could also hold a covering membrane together. A great deal of further exploration remains open, though these first research endeavors hold great promise and are an excellent spring board.

Reference

1 Hudert, M. "Timberfabric: Applying Textile Assembly Principles to Wood Construction in Architecture." *Thesis no. 5553,* EPFL, Lausanne, 2007.

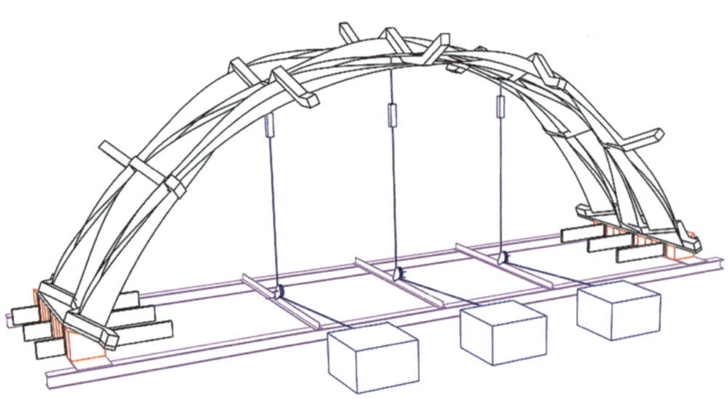

Fig. 3

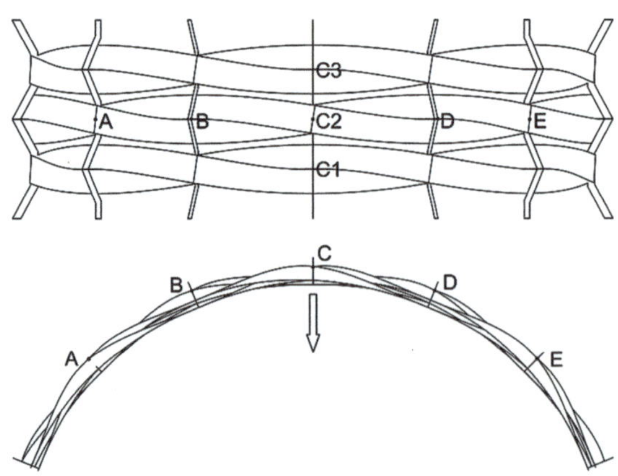

Fig. 4

Fig. 3 Test procedure

Fig. 4 Spatial definition of stress points

Fig. 5 Large module under compression test

Fig. 6 The module on show during the Timber Project exhibition at EPFL

Fig. 5

Fig. 6

5 Customized construction

5.1

Innovative wood–wood connections

Yves Weinand

The technology of the connections is essential in all design processes. In timber construction, half the costs are determined by the volume of material used and the other half by the installation and connection costs. Unlike structural systems of reinforced concrete or steel, timber construction requires a greater focus on integrated planning. For this integrated planning process, the preselected geometry of the connection technology has to be incorporated into the planning process in advance. In timber construction, connection techniques are transmitted by adhesive bonds (chemical forces), by welded connections (physical forces), and by mechanical or integral mechanical connections (mechanical forces).

Regarding the latter category, with integral mechanical connections such as dovetail joints, Christopher Robeller has demonstrated that European dovetail connections can be distinguished from Japanese equivalents by the selected sliding layer. While galvanized wood–wood connections are inserted alongside the dovetail in the European tradition, in the Japanese tradition they are inserted diagonally across the dovetail. In addition to the insertion angle, of interest is the principle of different types of integral connections, which are determined by different geometric boundary conditions.

If the connection has geometric properties, then the following consideration can be added: assembly sequences can be defined that cause specific elements in specific places to be locked at any given moment of the installation phase. Depending on the sequence definition, so-called "lock graphics" arise that describe the locking of an element in the overall system.

Building on this accumulated knowledge, a number of avenues of research were undertaken at IBOIS that examined the relationship between the overall geometry, the assembly, and the connection typology of the individual joints. The investigations focused on the insertion angle, the number of degrees of freedom, tool-related conditions for milling, and

Fig. 1

Fig. 2

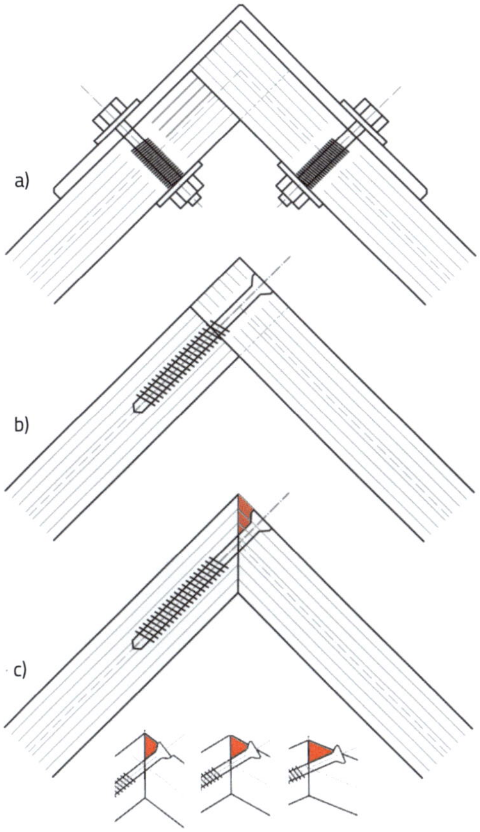

Fig. 3

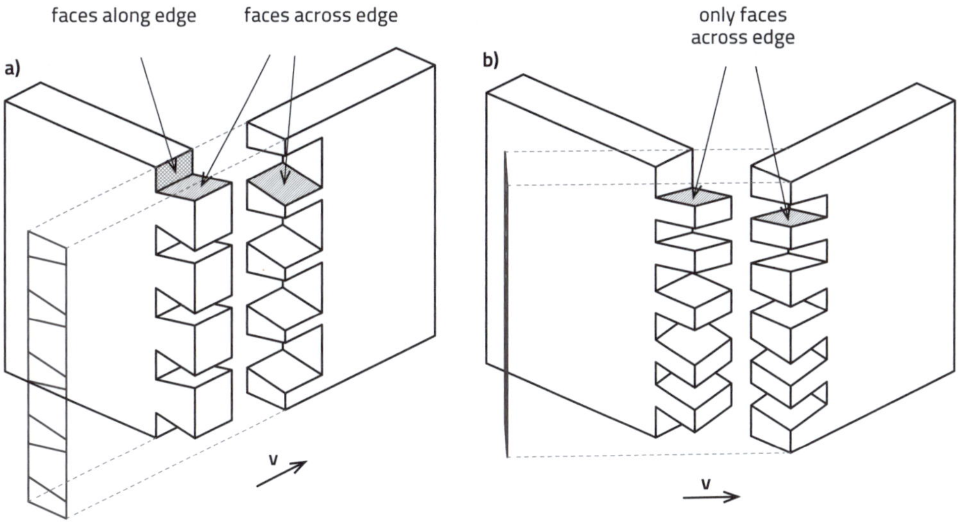

Fig. 4

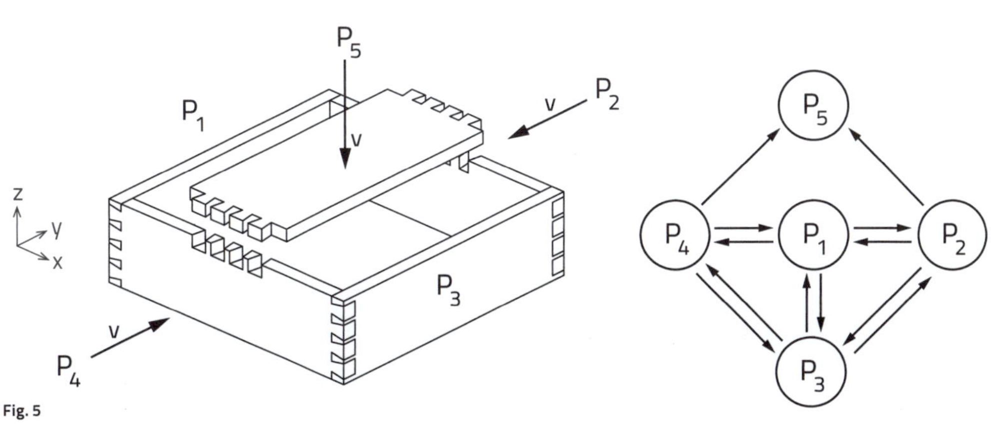

Fig. 5

Fig. 1 Folding structure developed with the origami tool

Fig. 2 Failure mechanism of the shell: opening of bolted joints and connections

Fig. 3 a) Steel corner connections, two orthogonal plates b) Screwed corner connection of two orthogonal plates c) Screwed corner connection of two cut plates with angle variations

Fig. 4 Diagram of a sliding vector
a) European tradition b) Japanese tradition

Fig. 5 Lock diagram

Fig. 6 a

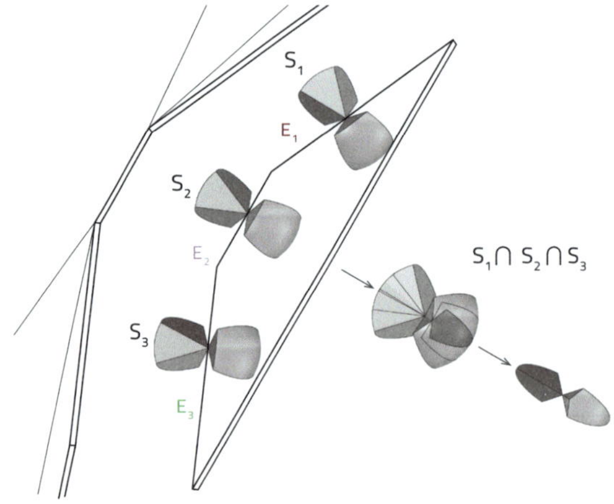

Fig. 6 b

Fig. 6 a, b, c, and d Summary based on research at multiple scales undertaken at IBOIS

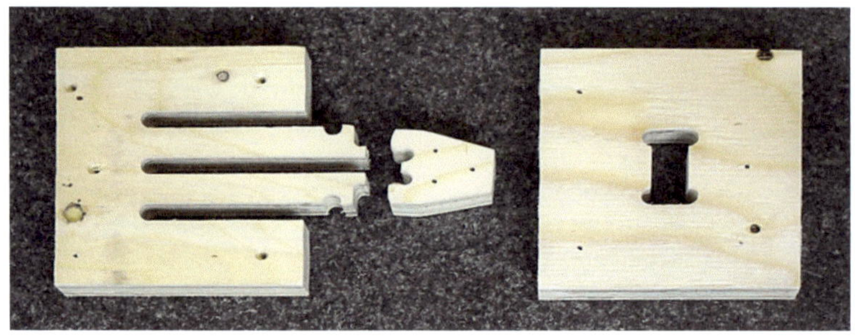

Fig. 6 c

Fig. 6 d

the mechanical properties of the innovative construction. By combining structurally strong veneer plywood panels and efficient wood–wood connections within predefined spaces and where the insertion angle for one or more panels is provided, new sustainable timber structures can be realized. Ideally, these should consist exclusively of wood, which also simplifies the future sorting process for disposal.

In retrospect, a new structural design concept emerged. Usually, a form is first proposed and then discretized or divided, and finally an overall shape is proposed, generally by the architect. The size of each element and the nature of the connection of the elements are then, however, usually determined by the engineer. The subdivision and connection methods are therefore determined by whoever designs the overall form.

The assembly process is ultimately additive in nature. The overall shape is generated by the connections of the individual elements and the mechanical connections. The realization of nonstandard architecture of this kind generally fails due to high costs. However, when the conception of structures is undertaken as described here, the process is reversed. Considerations are not regarded in separate, successive phases—first global and then local. Rather, the interaction of both the detailed scale of the connections and the global scale of the overall form are considered and evaluated simultaneously. Locally defined geometric constraints lead to connection types, which in turn affect the shape as a whole. Civil engineers are interested in the overall form and how the detailed design of elements of the overall form can lead to "integrated" structures, where the relationship between form and structure is tight and inextricably interconnected. This "woven quality" gives the formed structures a special beauty.

Tribute to Eladio Dieste

In numerous constructions, Uruguayan architect and civil engineer Eladio Dieste (1917–2000) demonstrated how form and structure can function efficiently together. For his department store in Montevideo, Dieste envisioned the assembly of a double-curved shell structure using discrete elements—in this instance, bricks. Steel reinforcement was inserted into the joints to absorb the tensile stress. The transformation of the cross-section of an arc from a line to an S-shaped cross-section in the apex of the arc made the double curvature necessary. The S-shaped cross-section provides an increased moment of inertia and thus a higher buckling resistance at the apex of the arch.

The manufacture of the double-curved surfaces also poses a challenge in timber. In the project shown here, Dieste's bricks were replaced by laminated veneer lumber panels. The reinforcing steel was replaced with dovetail connections, which bind the boxes together. A fundamental geometric difference is that the curvature

Fig. 7 a

Fig. 7 b

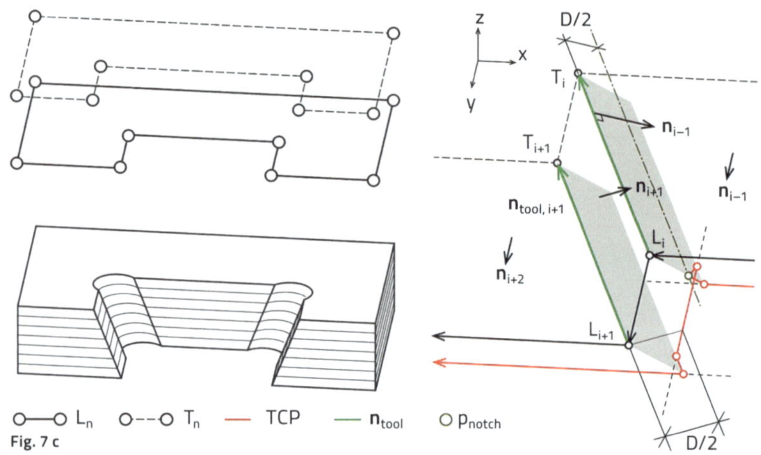

Fig. 7 c

Fig. 8 a

Fig. 8 b

Fig. 8 c

Fig. 7 a–c Milling, geometric description of the machine path that determines the angle of the cutter head to the plate

Fig. 8 The time required for the production of geometrically complex shuttering decreases: a) single-shell element with its pre-milled wood–wood connections; b) shell structure; c) complete molded fiber-reinforced concrete shell

Fig. 9 Development of a concrete form to a timber–concrete shell. The idea of a completely maximized surface structure is formed by two superimposed layers of wood and concrete, which achieve the necessary adhesion solely due to geometric constraints. Since the wood–wood connections along the edges of the plates penetrate into the concrete, these assume the integrating function typically provided by pins.

Fig. 10 Axonometric and section of the double-curved shell proposed by Dieste.

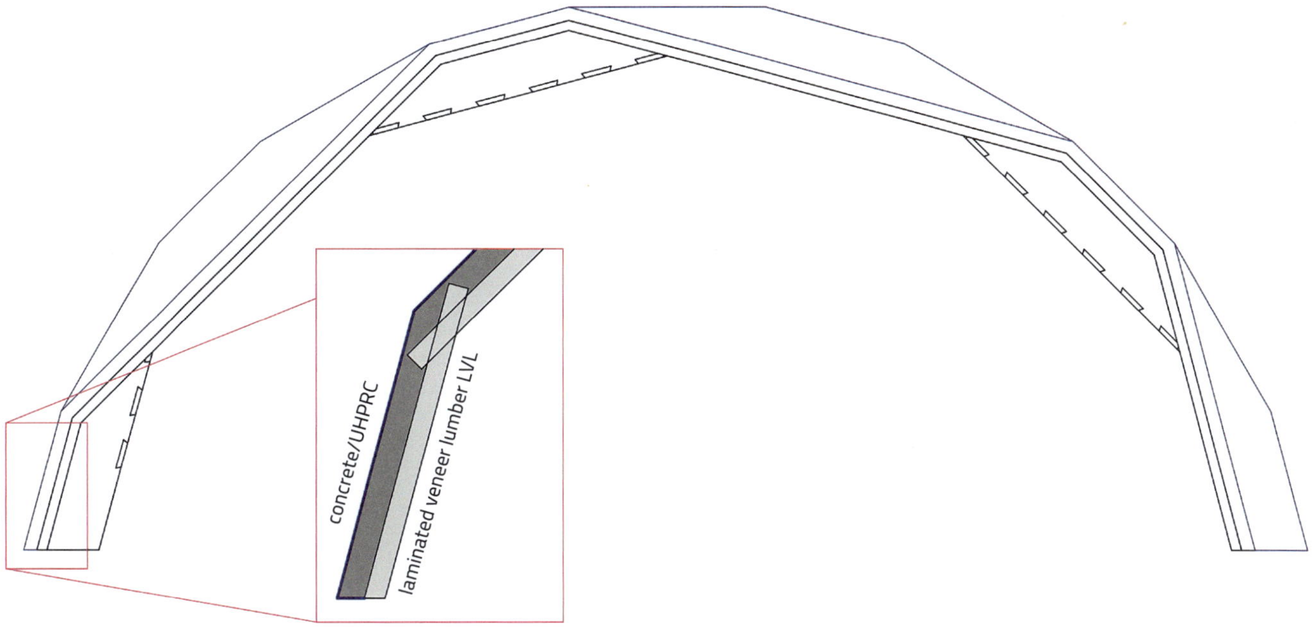

Fig. 9

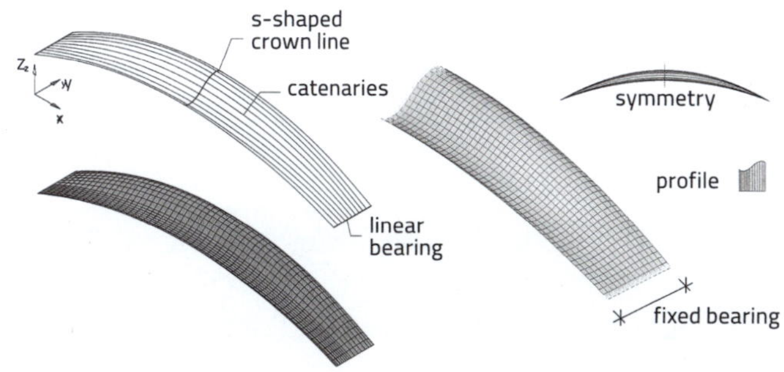

Fig. 10

of the Dieste hangar is achieved by varying the strength of the mortar joints. All the bricks have identical dimensions, but the joints compensate for the variation of the curvature. Here, then, the discretized frame construction must hold the angular position locally between the frames both transversely and longitudinally. Thus, the frame geometry is twisted in the local inclination. In total, twenty-three 6.5-m-wide sheets should be created with a variable span of between 32 m and 52 m and a height attained of 13 m. The scope of the building is approximately 170 m.

The subdivision of the arcs into individual segments was generated by a plug-in and the tool Grasshopper. Subsequently, all the planes were arranged in an orthogonal grid, with all surface ribs positioned vertically. Thus, vertical planes could be created that clearly demarcate the start and end of each arc. Then the solver had to smooth all areas, as the design was composed of flat panels. This planarization amounted to calculating the optimum position of the flat surfaces in the space, so that the various facets could follow the double curvature as closely as possible.

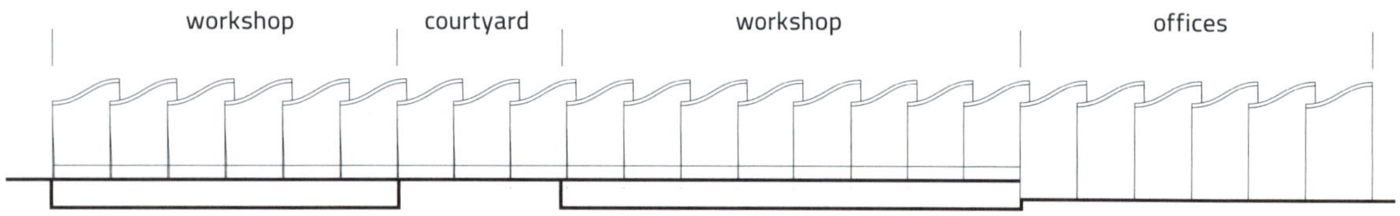

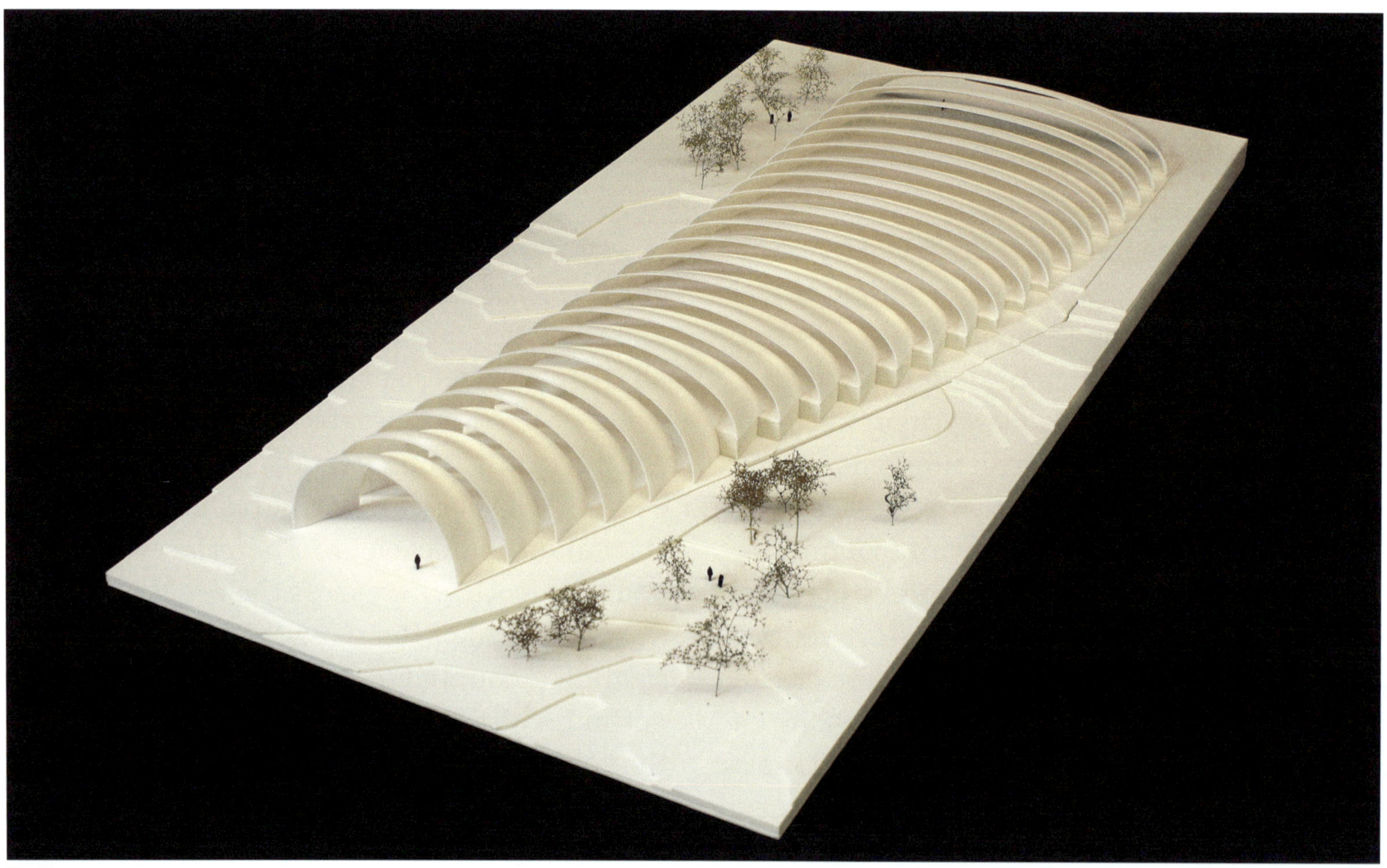

Fig. 11

Fig. 11 Plans and model: architecture collective Valentiny hvp architects and Weinand student bureau,
Floor plan, scale 1:1000,
Longitudinal section, scale 1:1000

Fig. 12 Planarization of the surfaces

Fig. 13 a) Axonometric and parameterized surface of the plug-in system **b)** Milling code **c)** Installation of a prototype at IBOIS

First, a system was developed that, as a plug-in system, required a sequential assembly. A frame does not exist by itself, but rather abuts the next frame, so that at least one lateral surface belongs to two frames (or extends beyond the frame and is integrated into the next). In this manner, a woven system is created. However, a disadvantage was that the modules were not prefabricated, but rather the structure was created by a continuous plug-in system, which could also be classified as a reciprocal assembly method.

From this, the idea arose of combining both requirements. Thus, on the one hand a modular system was developed that simplified the assembly process, and on the other hand an interwoven structural system was developed. The overall geometry is subdivided again, not in rectilinear modules but rather on the basis of a subdivision principle reminiscent of a herringbone pattern. The vertical planes remain. The newly selected subdivision principle affects the subdivision in plan.

The insertion angle is no longer perpendicular to a side surface, but continues diagonally, so that two side surfaces of a frame are simultaneously affected by it. This diagonal insertion angle is reminiscent of the Japanese tradition of dovetail connections presented above, where the insertion angle is selected beyond the orthogonal space. This measure makes it possible to combine the rational principle of serial assembly of modules practiced in

Fig. 12

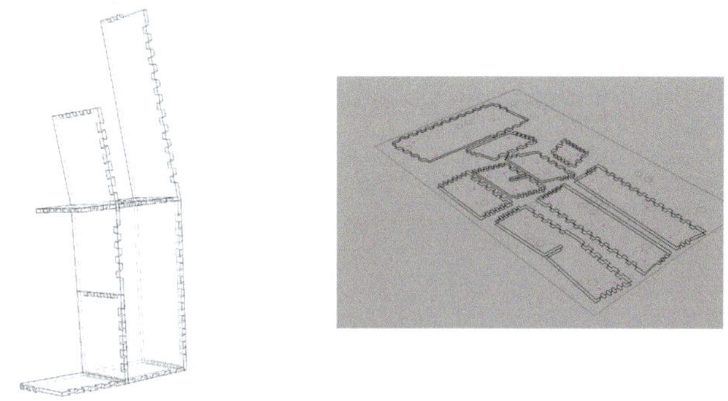

Fig. 13 a

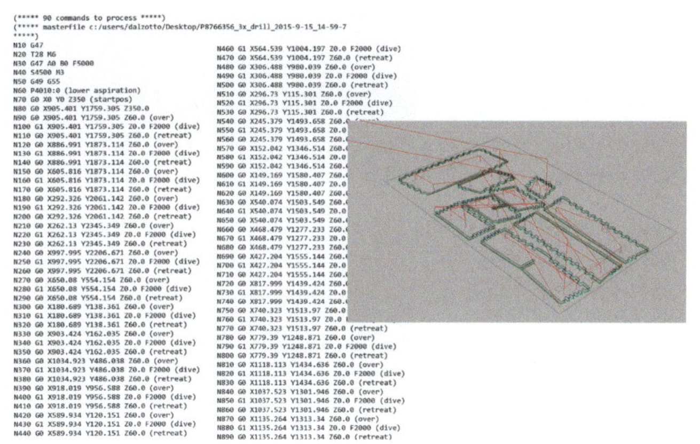

Fig. 13 b

Fig. 13 c

the West with the Eastern principle of interlocking the modules. The interlocking is significant, as both tensile and compressive membrane forces are absorbed into the double-curved shell. In the interlocking here, local displacement or disengagement in relation to the adjacent frame is prevented.

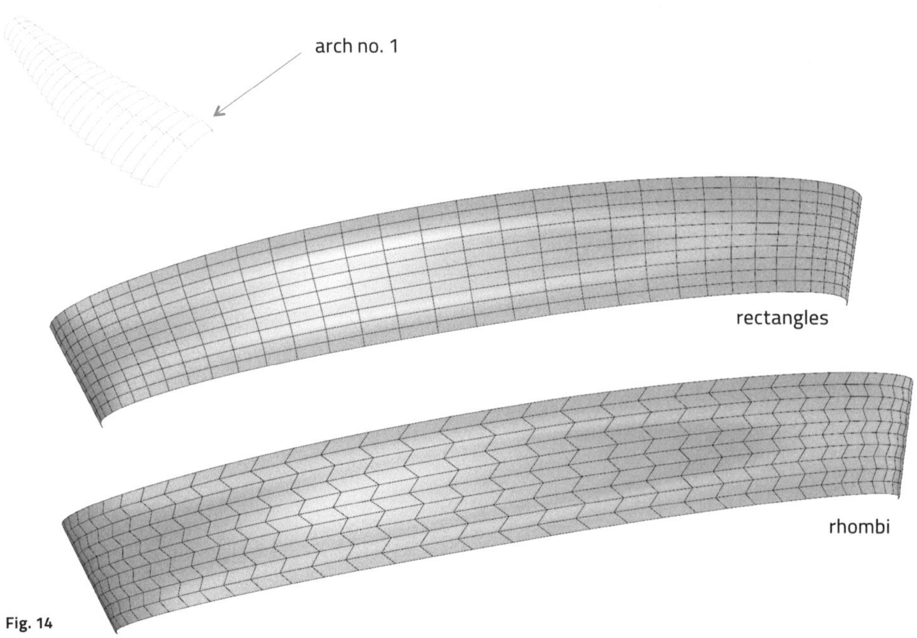

Fig. 14

Fig. 14 Parametric design tool to generate the herringbone pattern

Fig. 15 Parametric design tool generates a pattern similar to the herringbone for the subdivision system of the frames. The diagonal insertion angle relates simultaneously to at least two sides of the same frame.

Fig. 16 Visualization of the herringbone pattern using the example of three primary arches

Fig. 17 Visualization of the interlocked frames. The degree of freedom of the interlocking must be subordinate to the global plug-in sequence.

Fig. 18 Frame prototype images shown at scale 1: 1. Two frames are connected or interlocked here.

Fig. 19 Mechanical test body. Here, the dovetails are tested in shear.

Fig. 20 View of the shell construction from above and below

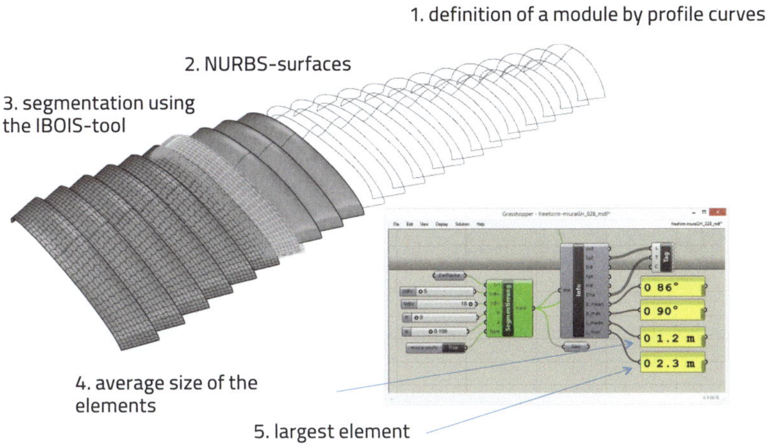

Fig. 15

Fig. 16

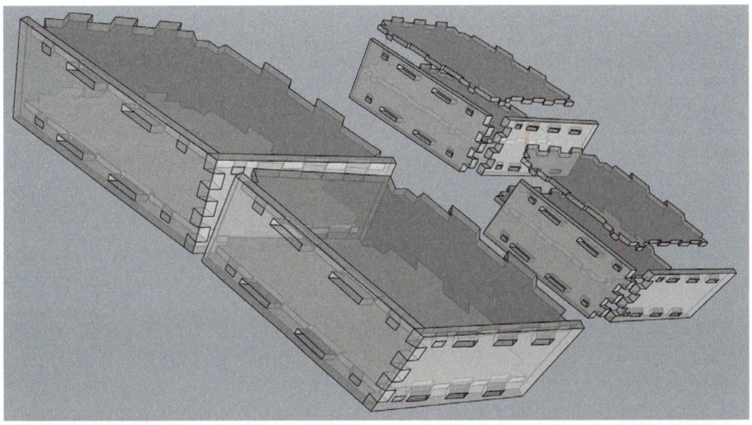

Fig. 17

Fig. 18

Fig. 19

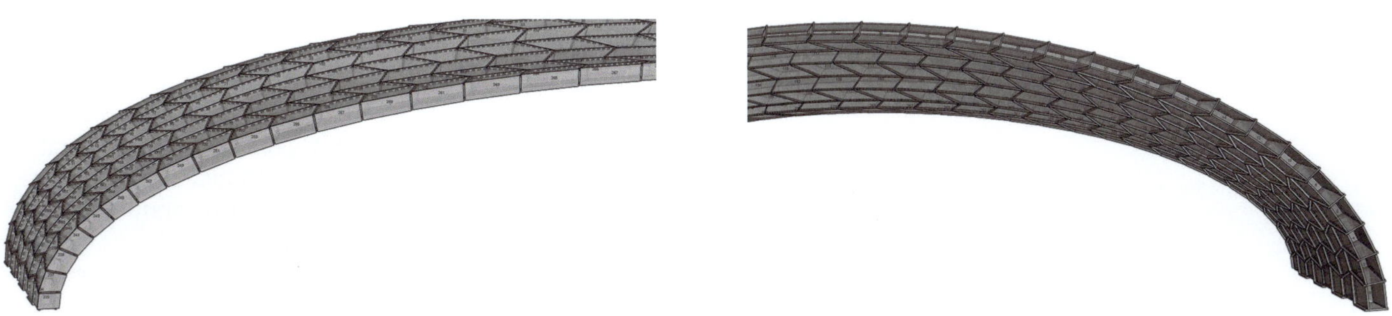

Fig. 20

5.2

Interlocking folded plates—integral mechanical attachment for structural wood panels

Christopher Robeller and Yves Weinand

Automatic joinery has become a common technique for the jointing of beams in timber framing and roofing. It has revived traditional, integral joinery, such as mortise-and-tenon connections. Just recently, the automatic fabrication of traditional cabinetmaking joints has been introduced for the assembly of timber panel shell structures. First prototypes have been assembled with such integrated joints for the alignment and assembly of components, while additional adhesive bonding was used for the load-bearing connection. However, glued joints cannot be assembled on site due to lack of controlled conditions, thus resulting in several design constraints.

In this paper, we propose the use of dovetail joints without adhesive bonding in the case study of a timber folded-plate structure. Through their single-degree-of-freedom (1DOF) geometry, these joints block the relative movement of two parts in all but one direction. This presents the opportunity for an interlocking connection of plates, as well as a challenge for the assembly of folded-plate shells, where multiple, non-parallel edges per plate have to be jointed simultaneously.

Keywords ***integral attachment, timber folded-plate structures, digital fabrication, design for assembly***

1 Introduction

Architectural designs are often inspired by folded shapes such as Japanese origami; however, folding principles like this can rarely be directly applied to building structures. Rather, many folded plates have been cast as concrete thin-shells in the 1960s. These constructions were labor intensive and required elaborate formwork for in-situ casting.

Prefabricated constructions with discrete elements made from fiber-reinforced plastics were researched in the 1960s.[1]

Folded plates built from laminated timber panels were presented by Regina Schineis[19] (Glulam) and Hans Ulrich Buri[2] as cross-laminated timber. These designs combine the elegant and efficient shape of folded plate shells with the advantages of structural timber panels, such as CO_2 storage, and a favorable weight-to-strength ratio. However, a major challenge in the design of a timber folded plate is presented by the joints: Since timber panels cannot be folded, a large number of edgewise joints has to provide two main functions. One of these functions is the load-bearing behavior, where connector features of the joints have to provide sufficient stiffness and rigidity. The second main function of the joints is the assembly of the parts, where locator features of the joints are essential for the precise, fast positioning, and exact alignment of the parts.

Benjamin Hahn[5] examined the structural behavior of the first timber folded-plate shell, which was built from plywood and assembled with screwed miter joints. He concluded that the load-bearing performance could be improved significantly with a greater number of resistant connections.

Inspiration for such improvements may be found in integral mechanical attachment techniques, the oldest known technique for the jointing of parts, where the geometry of the parts themselves blocks their relative movements.[13] Such integrated joints have recently been rediscovered by the timber construction industry. Since 1985, mortise-and-tenon joints have not been used in timberframe and roof constructions.[7] Only very recently, integrated joints have also been proposed for the edge-

Fig. 1

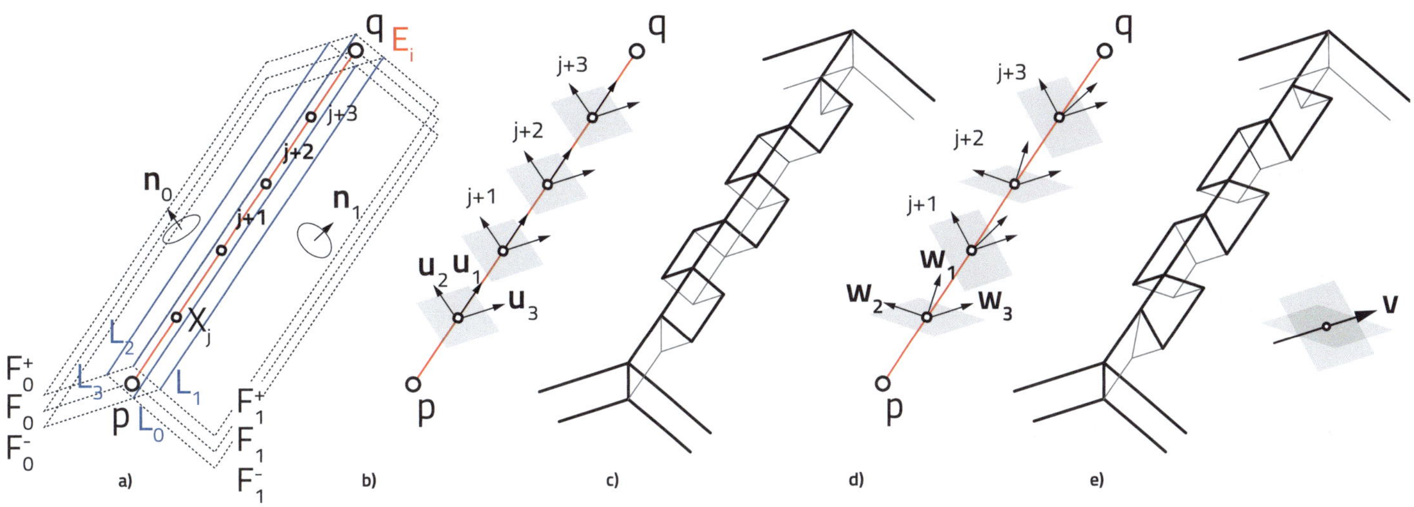

Fig. 2

Fig. 1 Folded thin-shell prototype built from 21-mm LVL panels, assembled with single-degree-of-freedom dovetail joints without adhesive bonding. Components interlock with one another.

Fig. 2 Joint geometry a) Basic parameters b) Intersection planes (gray) normal to $\overrightarrow{pq}$ c) 3DOF joint d) Rotated intersection planes (gray) normal to $\vec{w}_j$ e) 1DOF joint

wise joining of timber panels. In the ICD/ITKE Research Pavilions 2011[12] and 2013[11], finger-joints have been applied to plywood panes and an application of dovetail joints for cross-laminated timber panels (CLT) was presented in the IBOIS Curved Folded Wood Pavilion 2013.[18] In these prototype structures, the integrated joints have played a vital role for the assembly of the components. They have also contributed to the load-bearing connection of the parts, though additional adhesive bonding was needed. With few exceptions,[6] such glued joints cannot be assembled on site, because they require a curing period under controlled conditions, with a specific constant temperature and humidity.[15] Therefore, their application is limited to off-site assembly of larger components, which complicates both transport and handling while still requiring additional connectors for the final assembly.

In this paper, we propose the use of dovetail joints without additional adhesive bonding, in the case study of a timber folded-plate shell **(Fig. 1)**.

Through their single-degree-of-freedom (1DOF) geometry, these joints block the relative movement of two parts in all but one direction. This presents the opportunity for an interlocking connection of plates, as well as a challenge for the assembly of folded-plate shells, where multiple, non-parallel edges per plate have to be joined simultaneously.

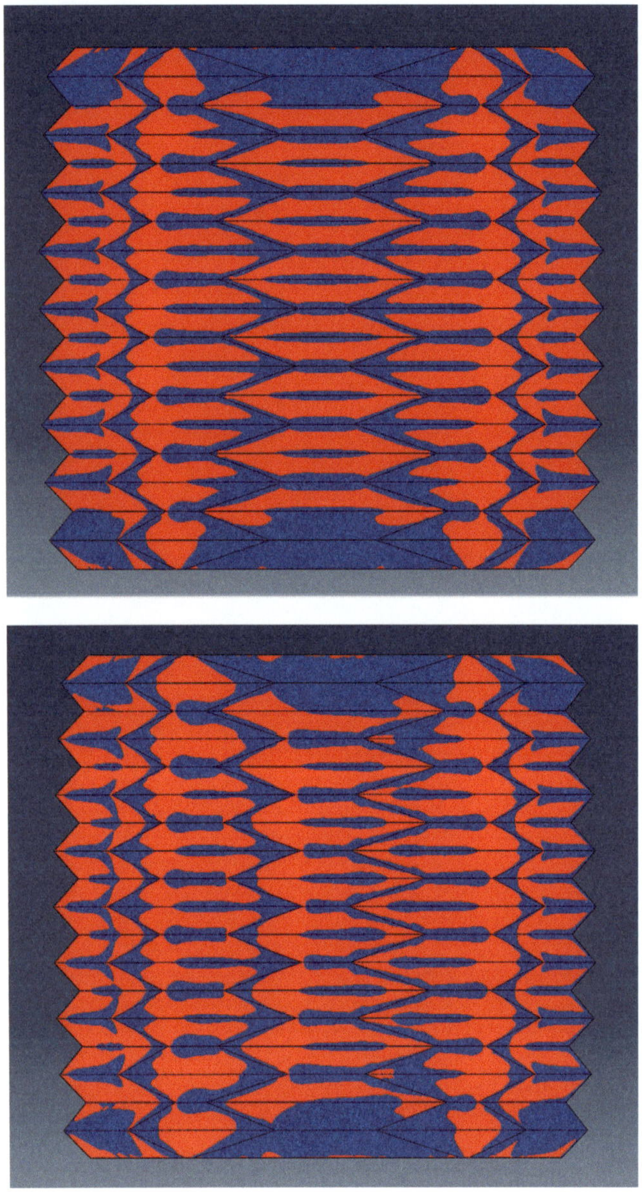

Fig. 3

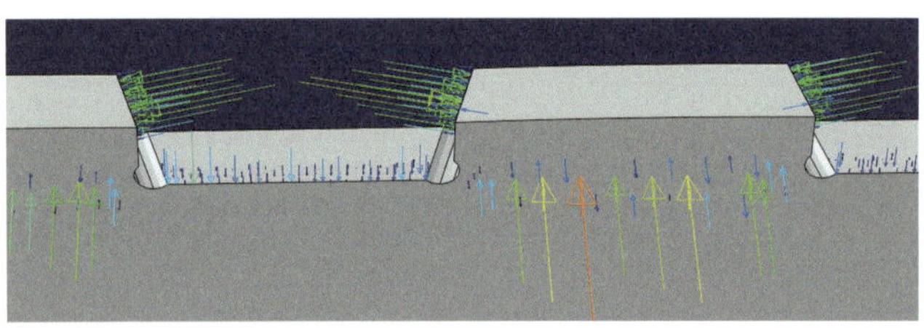

Fig. 4

Fig. 3 FEM analysis (plan view) of a 3x3 m, 21-mm Kerto-Q folded-plate, thin shell assuming fully stiff joints. Distribution of traction (red) and compression (blue) stresses in the y direction5. Top: gravity load case. Bottom: asymmetrical snow load

Fig. 4 FEM simulation of bending on a dovetail joint connecting two 21-mm, Kerto-Q, LVL panels. The bending moment applied is transformed into compression, normal, and shear forces parallel to the inclined contact faces.

1.1 Dovetail joint geometry and mechanical performance

Using polygon mesh processing, we describe an edgewise joint based on its edge E. From the mesh connectivity, we obtain the edge vertices p and q and the adjacent faces F_0 and F_1 with their face normals n_0, n_1. The polygon mesh is used to represent the mid-layer of timber panels with a thickness t and offset F_1 and F_2 at $\pm\frac{t}{2}$ to obtain the lines L (Fig. 2a). From a division of E, we obtain the points X_j for a set of reference frames $\{u_1, u_2, u_3\}$, where $u_1 || \vec{pq}$ and $u_2 || n_0$ (Fig. 2b). A finger-joint geometry is obtained from an intersection of planes located at X_j, normal to u_1, with four lines L.

Without additional connectors, finger-joints are a kinematic pair with three degrees of freedom (3DOF), also called *planar joints*. They can resist shear forces parallel to the edge and in-plane compressive forces. However, depending on the plate geometry, thickness, and most of all rotational stiffness of the connection detail, bending moments are also transferred between the plates. Furthermore, due to the rotation of the plate edge caused by bending, in-plane traction forces appear perpendicular to the edge line and their magnitude increases under asymmetrical loads. Such forces, which occur as a result of out-of-plane loading, cannot only be supported by shear and in-plane compression-resistant joints.

On a dovetail joint (Fig. 2d,e), the intersection planes on the points X_j are normal to a rotated vector w_1. It is obtained from a rotation of the reference frame $\{u_1, u_2, u_3\}$ about u_3 at an alternating angle $\pm\theta_3$. The resulting rotated side faces reduce the dovetail joints, degrees of freedom to one translation $\vec{w_3}$ (1DOF). Simek and Sebera[20] have recommended $\theta_3 = 15°$ for spruce plywood panels. Such prismatic joints can only be assembled or disassembled along one assembly direction $\vec{v} = \vec{w_3}$. In addition to the finger-joints resistant to shear and compressive forces, dovetail joints can, without adhesive bonding, also resist bending moments and traction forces that are not parallel to $\vec{v}$. Due to the inclination of the side faces of the joint, resistance to these forces can be improved significantly. In this manner the inclined faces assume the role that the glue would have in a finger-joint. (Fig. 4)

1.2 Fabrication constraints

One of the main reasons for the resurgence of finger and dovetail joints is the option of having automatic fabrication. However, the mechanical performance of the joints depends on fabrication precision. At the same time, fast machine feed rates are important for time-efficient production. We have fabricated joints such as these with a robot router and a gantry router, achieving higher precision with the gantry machine, which is stiffer and provides a higher repeat accuracy.

The variability of the machine-fabricated joints is enabled by the 5-axis capability of modern routers. Although traditional edgewise joints in cabinetmaking were used for orthogonal assemblies, both the finger

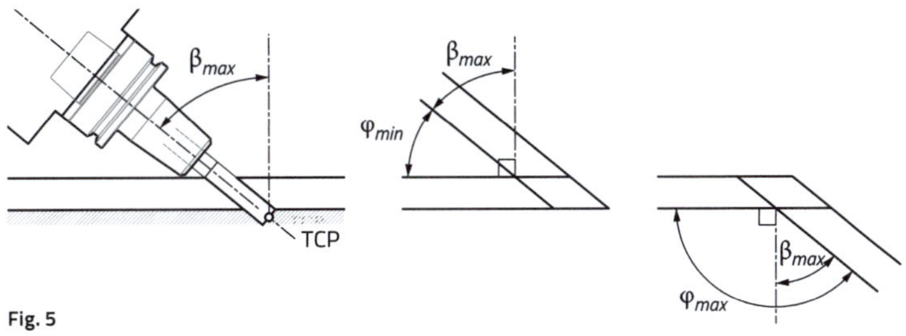

Fig. 5

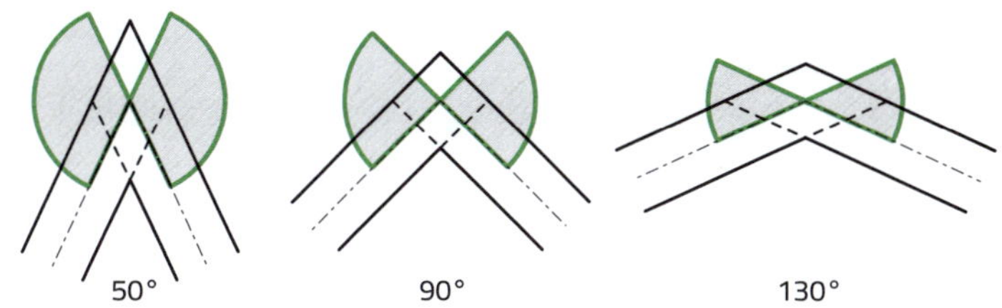

Fig. 8

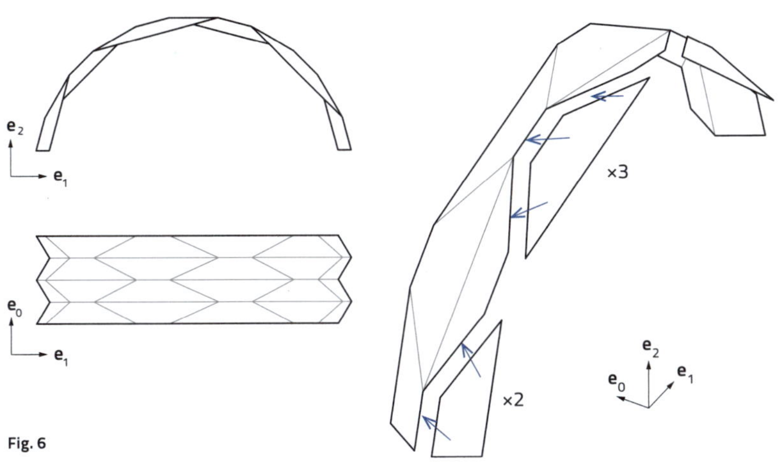

Fig. 6

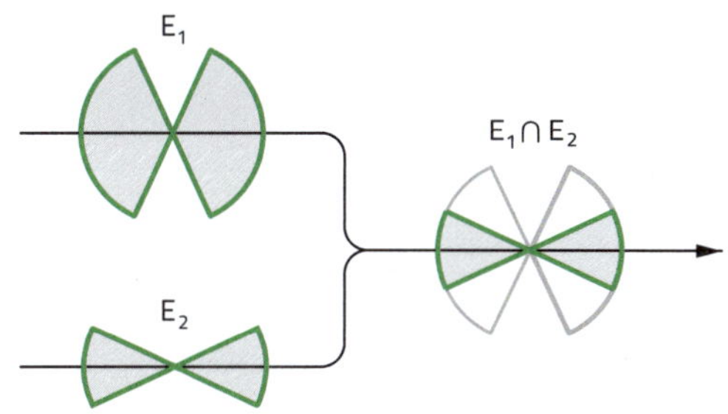

Fig. 9

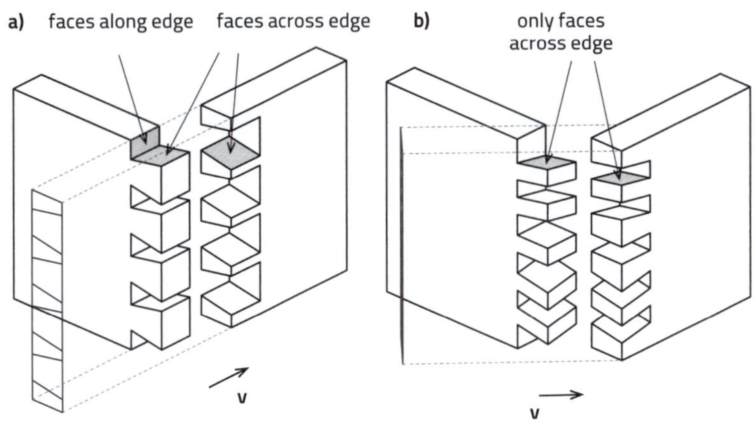

Fig. 7

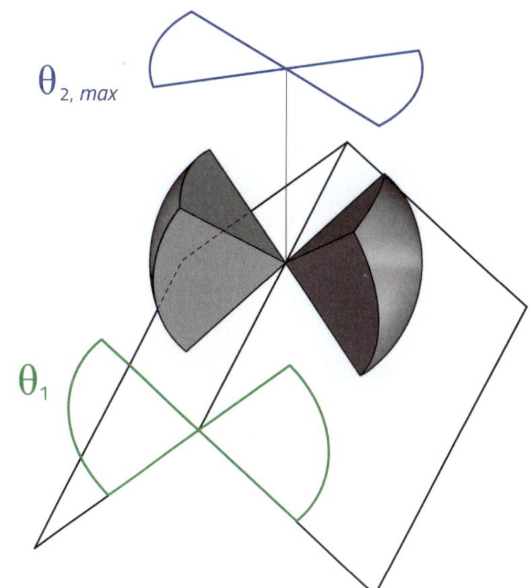

Fig. 10

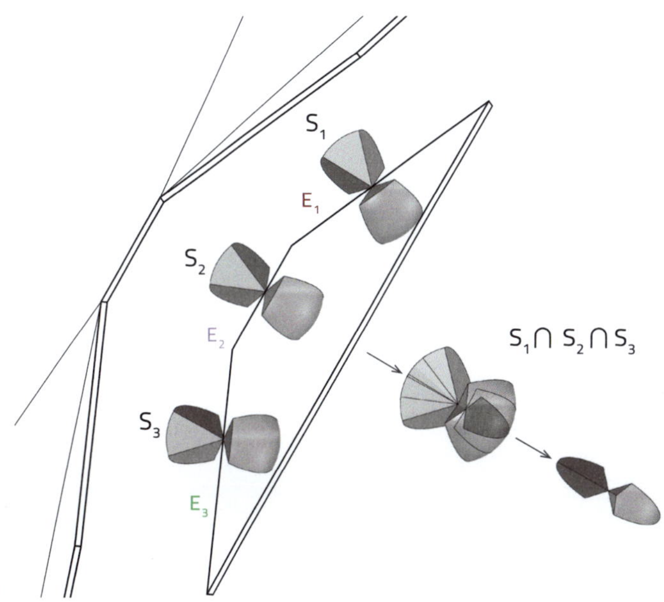

Fig. 11

Fig. 5 Fabrication constraints. Side-cutting techniques used for the automated fabrication of 1DOF edgewise joints with common 5-axis CNC routers. The maximum tool inclination max is a result of the tool and the tool holder geometry. From this we obtain the range of possible dihedral angles between panels.

Fig. 6 The assembly of a folded plate from discrete elements (left) requires the simultaneous assembly of non-parallel edges (right). The insertion direction of our 1DOF joints is rotated, to make the insertion vectors of simultaneously jointed edges parallel.
We chose a hexagon reverse fold pattern, which only requires moderate rotations.

Fig. 7 **a)** Dovetail joint, **b)** Nejiri Arigata Joint

Fig. 8 2D vector subset

Fig. 9 2D simultaneous assembly

Fig. 10 3D vector subset

Fig. 11 3D simultaneous assembly

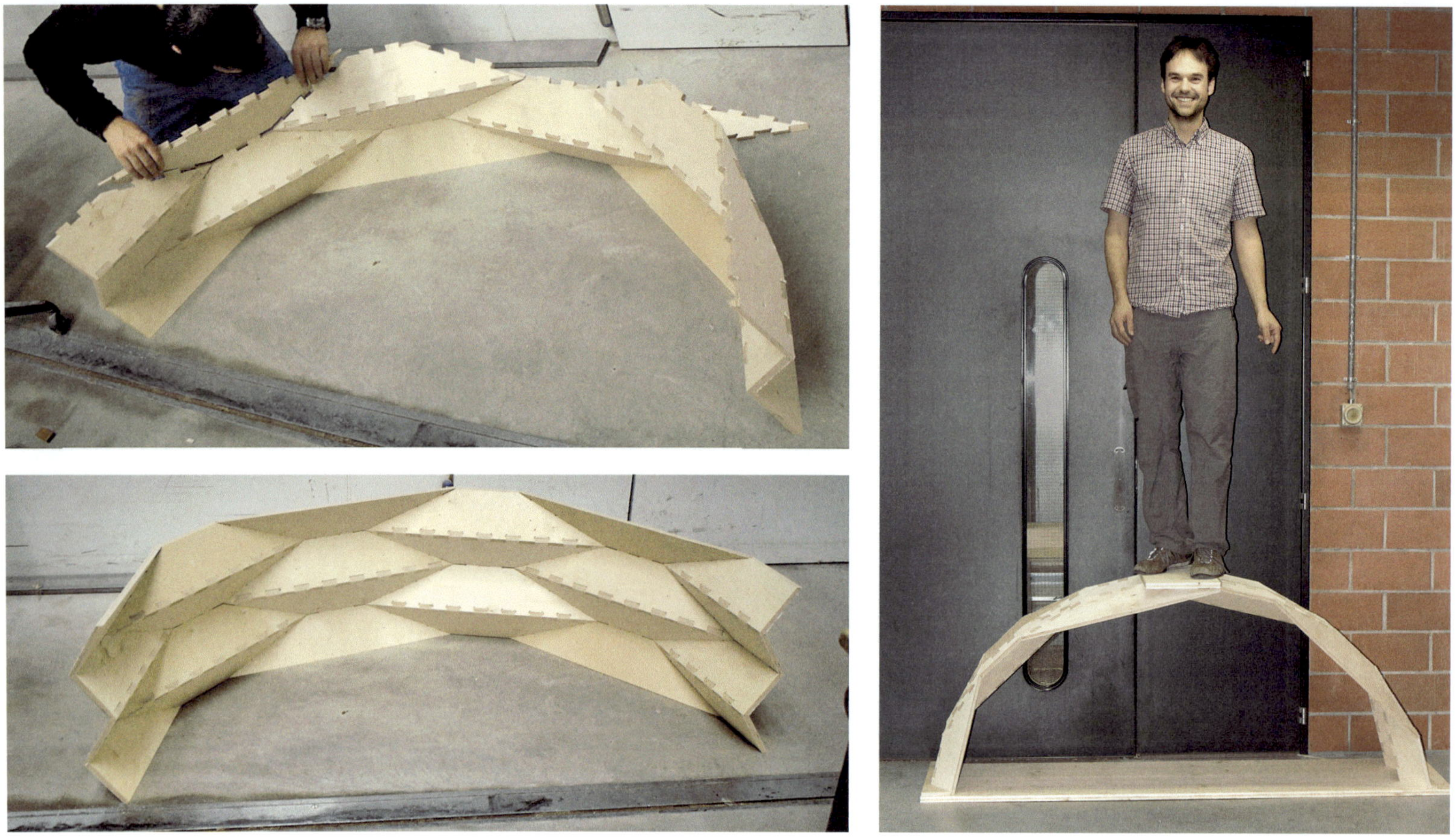

Fig. 12

Fig. 12 Folded-plate arch prototype built from 12-mm birch plywood (9-layer, I-I-I-I-I). Assembled without adhesive bonding or metal fasteners. (Span 1.65 m, self-weight 9.8 kg)

Fig. 13 Series of 3-point flexural tests on a small-scale interlocking arch prototype built from Metsawood 12-mm birch plywood panels.

Fig. 14 Double-curved folded plate: The radius (R=17m) of the transverse curvature is determined by the folded plates' maximum amplitude h^2, which is inversely proportional to the number of segments m of the cross-section polyline (grey). We obtain this polyline from a circular arc divided into segments of equal length.
The interior angle $\gamma = ((m-2) \cdot 180) - m$ of this polyline is proportional to all fold angles φ. The geometry of our prototype was fabrication-constrained to a maximum component length $B \leq 2.5$ m

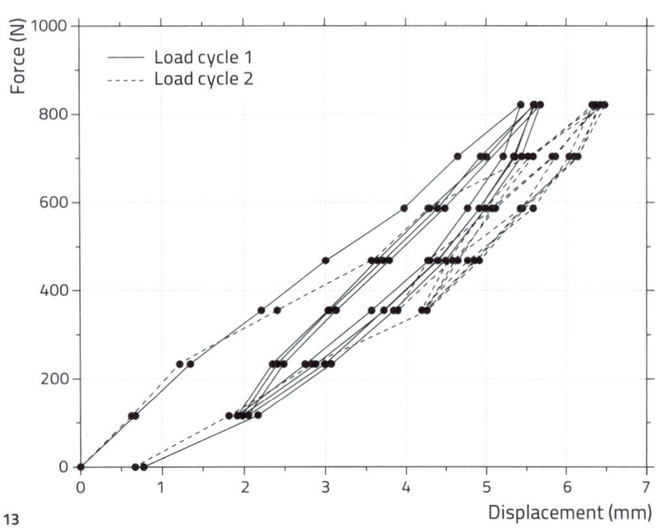

Fig. 13

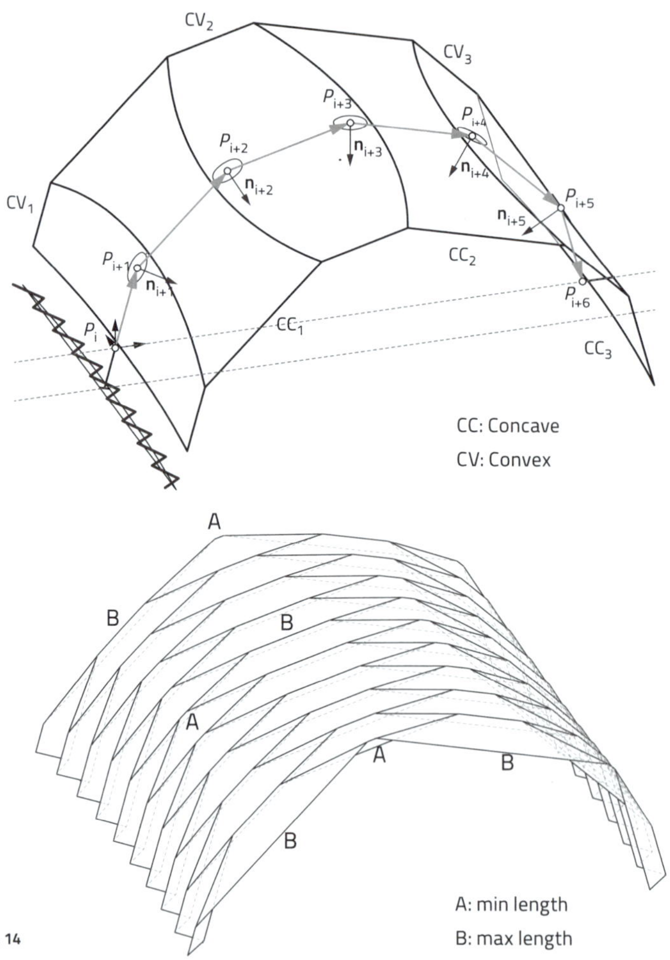

Fig. 14

and dovetail joint can also be applied to non-orthogonal fold angles, which was essential for the reference projects mentioned before. However, there are certain fabrication-related constraints for machine-fabricated dovetail joints. In order to integrate the joint fabrication directly with the panel formatting, we use a side-cutting technique,[8] which is limited to a tool inclination $\beta_{\max}$. We obtain this limit from the specific geometry of the tool, tool-holder, and spindle used for the joint fabrication (Fig. 5).

The parts can be assembled in two ways, as shown in fig. 5, which allows a larger range of dihedral angles φ. From this we obtain the fabrication-constrained most acute fold $\varphi_{\min} = 90° - \beta_{max}$ and most obtuse fold $\varphi_{max} = 90° + \beta_{max}$. With standard cutting tools, this technique allows the jointing of acute folds up to $\varphi = 50°$, which is ideal for folded plate structures. Highly obtuse fold angles $\varphi \geq 140°$, which might be required for smooth segmented plate shells, cannot be fabricated with this method.

1.3 Simultaneous assembly of multiple edges

The assembly of double-corrugated folded plates requires the simultaneous joining of multiple edges per component (Fig. 1), which has implications for both the shell and the joint geometry.

For multiple 1DOF-jointed edges, simultaneous assembly is only possible if the individual assembly directions $\vec{v}$ are parallel. With a normal dovetail joint geometry (Fig. 7a), this is not the case. A simultaneous assembly is only possible for parallel edges, which only allows rectilinear assemblies, such as for drawers or cabinets.

In order to simultaneously join non-parallel edges, the assembly direction v of the joints has to be rotated to make them parallel. This technique is known from Japanese cabinetmaking,[9] where certain joints, like the Nejiri Arigata joint (Fig. 7b), are assembled diagonally along a vector that does not lie on either one of the two planes. European dovetail joints, on the other hand, form a prism with a single tab, using faces both across and along the edge; the Nejiri Arigata joints form a prism using multiple, differently shaped tabs.

This Japanese technique is extended to a vector subset of possible assembly directions. Fig. 8 shows that the rotation about the edge line is constrained to $180° - \varphi_i$. The vector subset is large for acute and small for obtuse fold angles. This is particularly important when joining multiple edges simultaneously, as an intersection has to be found between multiple vector subsets (Fig. 9). If there is an intersection, the parts can be joined simultaneously along any direction within the intersection of the subsets.

Finally, this concept is extended to a three-dimensional rotation (Fig. 10). This is possible through a second rotation θ_2, which is constrained to a maximum value of $\pm\theta_{2;max}$. The limitation results from various other correlated parameters, such as θ_1 and β_{max}. We call the resulting three-dimensional vector subset "rotation window."

With this method, we are able to search for a joining solution for the prototype in fig. 6. We compute rotation windows S_1, S_2, S_3 for the edges E_1, E_2, E_3 and overlay them at their center point. Fig. 11 shows that there is a common vector sub-set $S1 \cap S2 \cap S3$ between these three edges, in which one can choose an assembly direction.

As a result of these limited rotations, the angle between neighboring, simultaneously joined edges cannot be very acute. Folded-plate patterns like the Herringbone, the Diamond, or the Hexagon pattern, which we chose for our prototypes, (Fig. 6) work well for our joining technique. Another essential feature provided by these reverse-folds are the acute fold angles, which easily satisfy the fabrication-constrained range of $\varphi_{min} = 50°$ to $\varphi_{max} = 140°$.

2 Interlocking arch prototype

In an assembly of multiple components (Fig. 12), a step-by-step sequence must be planned for the assembly of parts. The completed structure can only be disassembled piece by piece in the reverse order of assembly. In this way, the elements interlock with one another, like in a burr puzzle.[22]

Each joint consists of two parts, which must be parallel during assembly. We therefore chose a folded-plate geometry with relatively short edges. The manual assembly of long edges may be more difficult but can be simplified with a modified joint geometry. It is important to know the approximate direction of insertion for each part, as this is not easily visible through the joint geometry. Deformations of the arch during the assembly should be minimized. We have assembled this first prototype lying on its side. However, larger assemblies may require temporary punctual supports. Although the in-plane dimensional stability of the Kerto-Q panels is very high, panels may be slightly warped and some force may be necessary during assembly. While we have simply used a rubber hammer, more advanced techniques could be applied.

To understand the mechanical behavior of the built prototype, we have applied a vertical load at the mid-span of the arch and measured the vertical deflection at the same point. The total load of 821 N was applied in two identical load cycles consisting of four loading/unloading sub-cycles. First, a vertical load of 117 N was applied in seven steps, after which the load of the previous four steps was removed. The loading and unloading of the previous four steps was repeated three times, after which the complete load was removed and the residual deflection was measured (Fig. 13).

Under a vertical load equivalent to the arch's dead weight of 9.8 kg (98 N), the deflection measured at mid-span was 2 mm. From this we obtain a span-to-deflection ratio of $L/750$, and the arch's structural efficiency reaches 8.6 when loaded with 821 N (ratio of the maximum load over the dead weight of the arch).

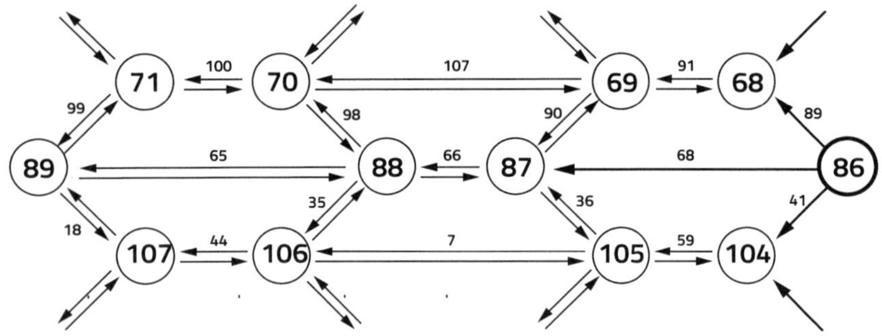

Fig. 15

Fig. 15 Partial connectivity, assembly, and blocking graph of the folded-plate shell prototype (left-to-right assembly). Large numbers represent mesh faces; small numbers represent mesh edges.

Fig. 16 Left-to-right assembly of the interlocking folded-plate shell prototype. Built from Kerto-Q structural-grade LVL panels (7-layer, I-III-I).

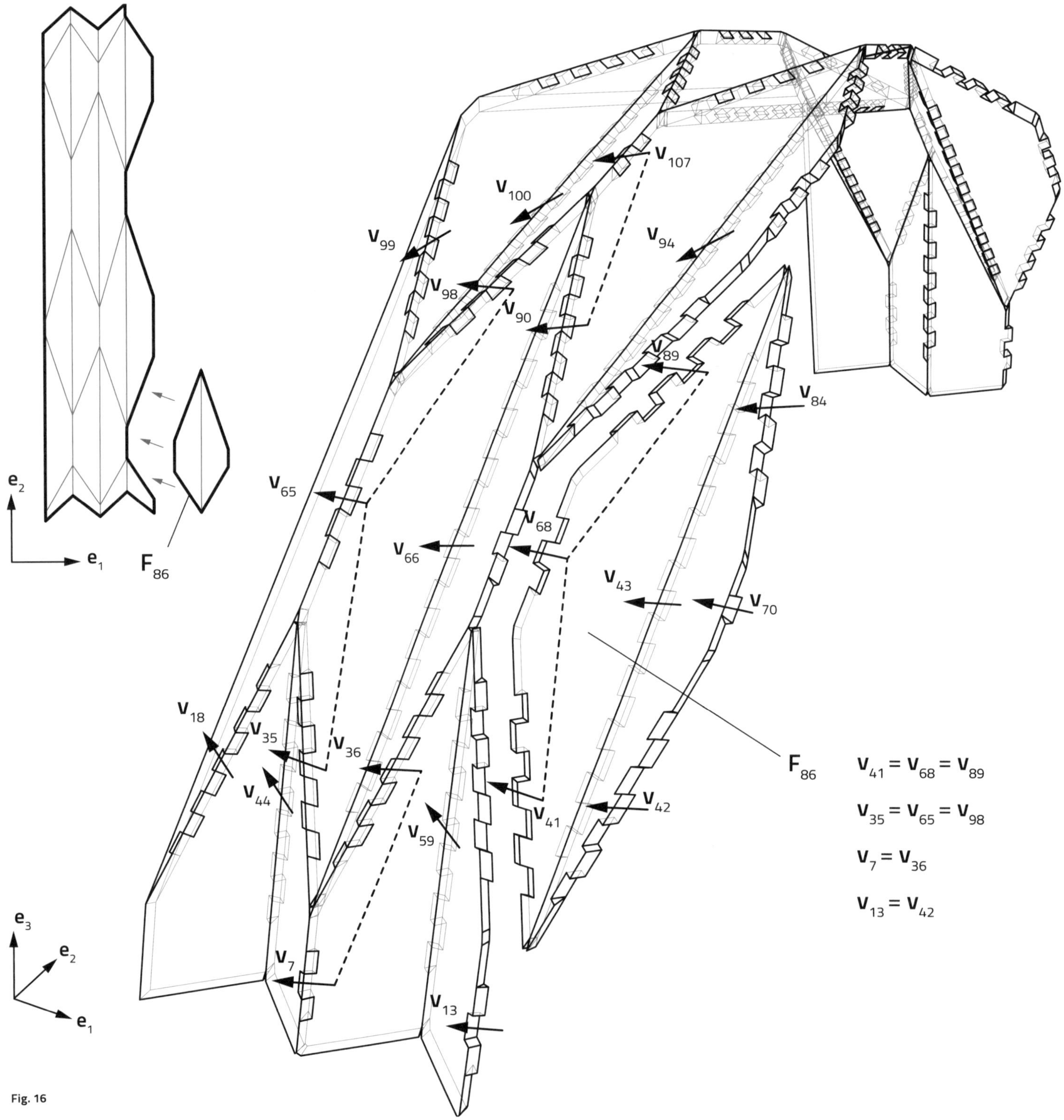

Fig. 16

3 Interlocking shell prototype

3.1 Automatic geometry processing

Using the RhinoPython application programming interface, a computational tool has been developed that enabled the instant generation of the geometry of the individual components, as well as the machine G-Code required for fabrication. The tool processes arbitrary polygon meshes, and generates 1DOF joints for all non-naked edges where the fold angle φ is larger than φ_{min} and smaller than φ_{max} shown in fig. 5 (non-smooth meshes). It also requires the input of edge identifier pairs, identifying those edges which have to be joined simultaneously, and a value for the thickness of the LVL plates. By exploiting this geometric freedom, we have tested our computational tool on the design of a folded-plate shell prototype with an alternating convex-concave transverse curvature. The shell spans over 3 m at a thickness of 21 mm, using Kerto-Q structural grade LVL panels (7-layer, I-III-I). **(Fig. 14)**

Comparing this double-curved folded plate with a straight extrusion (as tested by Hans Ulrich Buri[2]), it can be concluded that the slight double curvature proves to be highly beneficial when it comes to global deflections, for example, those caused by wind loads. Deflections for the double-curved shell geometry in the vertical direction are up to 39 % smaller, and up to 13 % smaller in the lateral direction than those with a straight extrusion.

3.2 Assembly

Due to the different assembly directions of its 239 joints, the 107 components in our prototype interlock with one another in a manner similar to a burr puzzle.[22] Fig. 15 shows a section of a so-called "non-directional blocking graph" (NDBG), developed by Wilson and Latombe.[21]

In a NDBG graph, single arrows indicate that individual parts can be removed from the assembly. Two opposite arrows between parts indicate that the connection is fixed. In order to remove fixed parts, the fixed parts have to be removed first. Our graph illustrates a left-to-right assembly. Part number 86 is being inserted on the right side. It connects to three other plates and blocks all other parts in the graph. In such a configuration, the final part, called the key, remains removable.

Fig. 16 shows the parts from fig. 15 in three dimensions, demonstrating how the component based on mesh face F_{86} is inserted. Its three edgewise joints E_{41}, E_{68}, and E_{89} have to be assembled simultaneously. The three assembly vectors of the edges $\vec{v_{41}}$, $\vec{v_{68}}$, and $\vec{v_{89}}$ have been rotated to be parallel. The same applies for the adjacent edges on the left side of the faces F_{67}, F_{69}, F_{88}, F_{103}, and F_{105} **(Fig. 15)**. Within the rotation window of the edge, we can freely rotate $\vec{v}$ for these edges (the greater the angle between $\vec{v}$ and the main direction of traction e_1, the better).

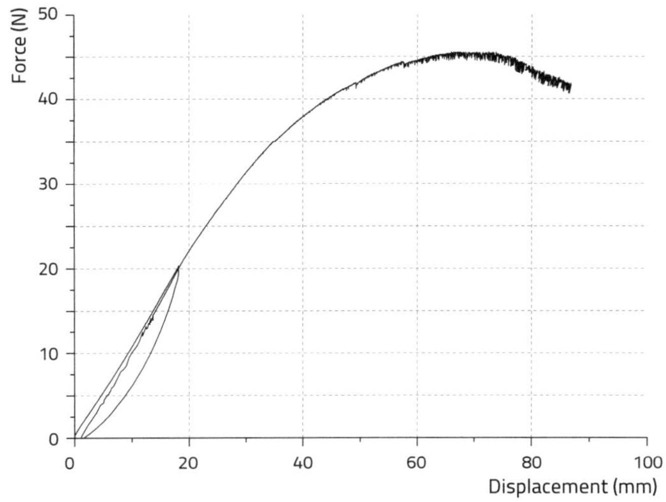

Fig. 17 Load-displacement curve of the shell prototype. A longitudinal line load was introduced along the top of the shell. Vertical displacement was measured at the center point.

3.3 Completed shell prototype and load test

Fig. 17 shows the completed folded-plate prototype, with a span of 3 m and a shell thickness of 21 mm. Boundary conditions that restrain displacements of the supports in every direction, but allow rotations, were applied on both sides. A longitudinal line load was introduced along the top of the shell and vertical displacement was measured at the center point **(Fig. 18)**.

The prototype structure was also modeled in FE analysis software (ABAQUS) and loaded in the same way. The plates were modeled using shell elements, where the mid-surface is used to represent the three-dimensional plate, and transverse shearing strains are ignored. Connections between the plates were considered as being completely rigid in order to obtain minimal displacements of the structure. By comparing displacements of the structure with completely stiff joints with those measured on the prototype, we obtained information about the actual semi-rigidity of the joints. The results obtained from the testing of the large scale prototype showed that the load of 25 kN, which corresponds to the proportional limit of the load-displacement curve, causes a vertical displacement of 23 mm. In the FE model, the load applied in the same manner caused a vertical displacement of only 2.6 mm.

4 Conclusion

A timber folded-plate shell combines the structural advantages of timber panels with the efficiency of folded plates. However, in such discrete element assemblies, a large number of semi-rigid joints must provide sufficient support for the adjacent plates in order to ensure an efficient load-bearing system. This remains a challenge with great potential for improvements.[5]

Integrated edgewise joints present an interesting addition and an alternative to state-of-the-art connectors: Unlike adhesive bonding, such joints can be assembled rapidly on site. Moreover, compared with costly metal plates and fasteners, typically required in large quantities,[14] the fabrication of integrated joints does not increase costs. The replacement or reduction of metal fasteners with an integrated mono-material connection includes advantages such as improved aesthetics, ease of recycling, or a homogeneous thermal conductivity of the parts, which can reduce condensation and decay.[4] Another particular advantage is the ability to connect thin panels. The current technical approval for the Kerto-Q panels does not permit screwed joints on panels with a thickness of less than 60 mm.[3]

Recent experimental projects, introduced in Chapter 1, have already demonstrated initial applications of integrated edgewise joints for timber panels. This paper followed up on these projects, examining the particular advantages, possibilities, and challenges of 1DOF joints for timber folded-plate shells. We have demonstrated how this joint geometry helps in resisting the forces that occur in such structures. In addition to the load-bearing connector features, the joints provide locator features, which allow precise positioning and alignment of the parts through the joint geometry. This improves both accuracy and ease of assembly.

Furthermore, we have presented a solution for the simultaneous assembly of multiple edges per panel, which is essential for the application of 1DOF joints in a folded-plate shell structure. The per-edge "rotation window" introduced in section 1.3 integrates the joint constraints related to assembly and fabrication. It can be processed algorithmically and provides instant feedback on whether or not a set of non-parallel edges can be joined simultaneously. This provides a tool for the exploration of a variety of alternative folded-plate shell geometries.

Fig. 18 Folded-plate shell prototype, built from 21-mm LVL panels. With a self-weight of 192 kg, the prototype with a span of 3 m was tested with a line-load up to 45 kN.

The prototypes presented in this paper already suggest possible patterns and demonstrate the reciprocal relationship between the geometry of the plates and the joints. Two built structures allowed us to test and verify the proposed methods for fabrication and assembly while providing valuable information about the load-bearing capacity of the integrated joints.

For application in a large-scale building structure, further research is required to determine if the integrated joints can replace additional connectors entirely or reduce their number. A possible combination of integrated joints with additional metal fasteners has recently been demonstrated in the LaGa Exhibition Hall.[10] Another possibility would be a combination of the 1DOF joints with integrated elastic interlocks.[17]

Acknowledgments

We would like to express our gratitude to Andrea Stitic and Paul Mayencourt for their support with generating the finite element models and load-testing the prototypes, and to Gabriel Tschanz and Francois Perrin for assisting us with the fabrication and assembly of prototypes. We would also like to thank Jouni Hakkarainen and the Metsa Group for supplying valuable information and material.

References

1 Benjamin, B. S. and Z. S. Makowski. "The Analysis of Folded-Plate Structures in plastics." In: *Proceedings of a conference held in London,* 149–163. London: Pergamon Press, 1965.

2 Buri, H. and Y. Weinand. "BSP Visionen – Faltwerkkonstruktionen aus BSP-Elementen." In: *Grazer Holzbau-Fachtage, 2006.*

3 DIBt. Allgemeine bauaufsichtliche Zulassung Kerto-Q Z-9.1-100, Paragraph 4.2 and Attachment no. 7, Table 5. Deutsches Institut für Bautechnik, 2011.

4 Graubner, W. *Holzverbindungen. Gegenüberstellung von Holzverbindungen Holz in Holz und mit Metallteilen.* Stuttgart, Deutsche Verlags-Anstalt, 1986.

5 Hahn, B. "Analyse und Beschreibung eines räumlichen Tragwerkes aus Massivholzplatten." EPFL, Master's thesis, Lausanne, 2009.

6 HESS. Hess limitless, 2014. www.hess-timber.com/de/produkte/

7 Hundegger. H. Hans Hundegger Maschinenbau GmbH, Hawangen, Germany, 2014. www.hundegger.de/en/machine-building/company/our-history.html.

8 Koch, P. *Wood Machining Processes. Wood Processing.* The Ronald Press Company, 1964.

9 Kondo, T. *Dento kanehozo kumitsugi: Sekkei to seisaku no jissai.* LLP Gijutsushi Shuppankai, Tokyo: Seiunsha, 2007.

10 Krieg, O., T. Schwinn, A. Menges, J.-M. Li, J. Knippers, A. Schmitt, and V. Schwieger. "Computational Integration of Robotic Fabrication, Architectural Geometry and Structural Design for Biomimetic Lightweight Timber Plate Shells." In: *Advances in Architectural Geometry.* 2014. Springer Verlag, 2014.

11 Krieg O. D., Z. Christian, D. Correa, A. Menges, S. Reichert, K. Rinderspacher, and T. Schwinn. "Hygroskin: Meteorosensitive pavilion." In: *Fabricate 2014 Conference Zurich,* 272–279, 2014.

12 La Magna, R. et al. "From Nature to Fabrication: Biomimetic Design Principles for the Production of Complex Spatial Structures." *International Journal of Spatial Structures,* vol. 28, no. 1:27-39, 2013.

13 Messler, R. W. *Integral Mechanical Attachment: A Resurgence of the Oldest Method of Joining.* Butterworth Heinemann, 2006.

14 Neuhaus, H. *DIN EN 1995 (Eurocode 5) – Design of timber structures.* DIN Deutsches Institut fur Normung e. V., 2004.

15 Purbond. National Technical Approval Z-9.1-711 / Single-Component Polyurethane Adhesive for the Manufacture of Engineered Wood Products. Deutsches Institut für Bautechnik, 2011.

16 Robeller, C. "Integral Mechanical Attachment for Timber Folded Plate Structures." PhD Thesis, ENAC, Lausanne, 2015.

17 Robeller, C., Paul Mayencourt, and Yves Weinand. "Snap- t joints: Integrated mechanical attachment of structural timber panels." In: *Proceedings of the 34th International Conference of the Association of Computer-aided Design in Architecture ACADIA, Los Angeles.* Riverside Architectural Press, 2014.

18 Robeller, Christopher, Seyed Sina Nabaei, and Yves Weinand. "Design and Fabrication of Robot-Manufactured Joints for a Curved-Folded Thin-Shell Structure made from Clt." In: *Robotic Fabrication in Architecture, Art and Design 2014,* edited by Wes McGee and Monica Ponce de Leon, 67–81. Springer International Publishing, 2014.

19 Schineis, R. "Gefalteter Klangkoerper Musikprobensaal Thannhausen/Thannhausen Rehearsal Room." In: *10. Internationales Holzbau Forum (IHF),* Garmisch-Partenkirchen, 2004.

20 Sebera, V.,and Milan Simek. "Finite Element Analysis of Dovetail Joint made with the use of CNC Technology. In: *Acta Universitatis Agriculturae et Silviculturae Mendelianae Brunensis,* vol. LVIII. 321–328, 2010.

21 Wilson, R. H. and Jean-Claude Latombe. "Geometric reasoning about mechanical assembly." In: *Artificial Intelligence,* 71(2):371–396, 1994.

22 Wyatt, E. *Puzzles in Wood.* Bruce Publishing Co., 1928.

5.3

Rotational stiffness on the ridges of timber folded-plate structures

Stéphane Roche, Geoffroy Mattoni, and Yves Weinand

Folded-plate structures provide an efficient design using lumber panels of thin laminated veneer. Inspired by Japanese furniture joinery, the multiple tab-and-slot joint was developed for the multi-assembly of timber panels with non-parallel edges without any adhesive or metal joints. As the global analysis of our origami structures reveals that the rotational stiffness at the ridges affects global behavior, we propose an experimental and numerical study of this linear, interlocking connection. The geometry is governed by three angles that orientate the contact faces. Nine combinations of these angles were tested, and the rotational slip was measured with two different bending scenarios: closing or opening the fold formed by two panels. The nonlinear behavior was conjointly reproduced numerically using the finite element method and continuum damage mechanics.

Keywords *semi-rigid, connection, moment-rotation, folded-plate*

1 Introduction

The first generation of folded-plate roofs was constructed in the mid-1950s using plywood.[1] However, technical problems arose from variations in the plywood's structural properties. The aesthetic of the outer layers was paramount, while its characteristic strength could not be fully controlled. Single-fold plate roofs behave like a series of V-section beams. The plywood skin transmits shear like a web and transfers in-plane forces to the roof edges, where the walls and transverse stiffeners bear the vertical and horizontal load components respectively.[1,2] Fifty years after the first attempts at folded-plate roof structures,[1] Jaksch et al.[3] proposed a similar pitched roof made from lightweight, cross-laminated timber (CLT) rather than plywood. More recent attempts have adopted glue and nails, or cold-formed thin steel plate, to create a rigid joint at the ridge. Today, laminated veneer lumber (LVL), CLT, and solid wood panels (SWP) are the most commonly used materials for highly stressed structures. These high-performance panels inspired other architects and researchers to develop novel solutions for reviving folded-plate structures. In 2008, Buri et al. described the application of origami paper folding to timber folded-plate structures at the 10th World Conference on Timber in Japan.[4] The structure's 21-mm plywood panels were assembled with 5-mm, self-drilling screws arranged in staggered rows at the miter joints of the panels. Both sides were beveled at a 60° angle, to form a 120° corner. Because the relative position of the parts had to be set before screwing, pre-mounting with jigs was required. Failure during a loading test occurred through tearing at the ridges and valleys, a direct result of the weakness of the connection.[5] Although origami folded-plate structures are promising complex timber structures, the connections of their thin panels have to be improved. Robeller et al. described the first approach to the problem in 2014.[6] A double-curved, folded-plate prototype was constructed using a five-axis CNC machine. The cut LVL plates had integrated jointing at their edges. Based on linear connections used in furniture (e.g., dovetail or Japanese Nejiri Arigata joints), these multiple tab-and-

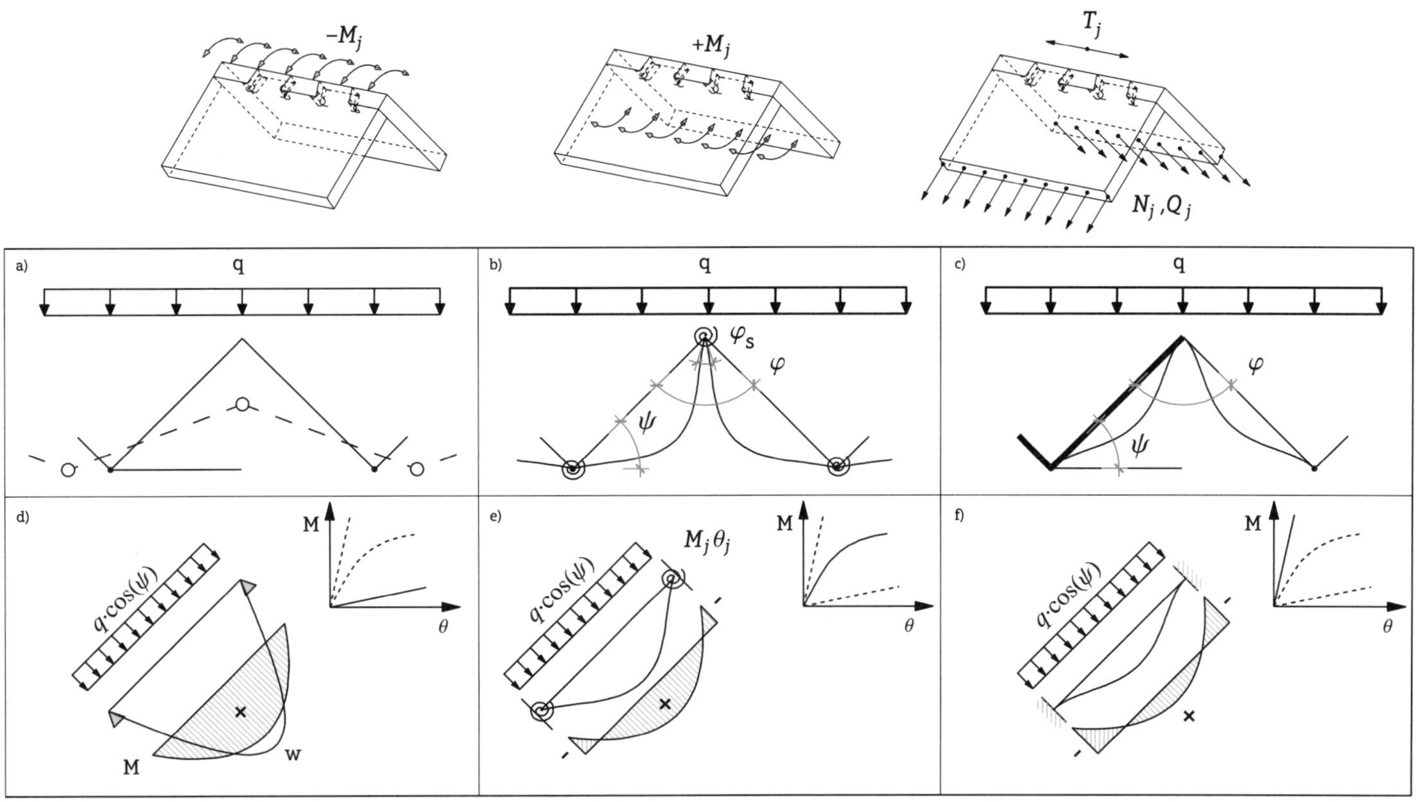

Fig. 1 a) Origami double-curved, folded-plate prototype analyzed by Robeller et al.[6]
b) Bending moments on the ridge
c) Forces along and perpendicular to the ridge
d) Hinged joint at the ridge
e) Semi-rigid joint at the ridge
f) Rigid joint at the ridge

slot joints (MTSJ) with no adhesive bonding allow the simultaneous assembly of up to three ridges or valleys. The panels can then be directly interlocked in situ at their final location. All these previous studies noted a lack in the connection stiffness model, a deficit that could affect the global behavior. Literature on timber connections rarely addresses the structural behavior of panel connections. Most research on connections during the last thirty years has examined joint stiffness within the context of frame structures. Among many others, H.J. Larsen, A.J.M. Lejiten, A. Kevarinmaki, and D.B. Engstrom, Scandinavian members of the Timber Joints Working Group of the European Cooperation in Science and Technology (COST action C1, semi-rigid behavior of civil engineering structural connections), have published specific research on the rotational stiffness of timber joints.[7] Their work showed that the inclusion of semi-rigidity in the structural analysis gave more realistic frame stress and deflection results, thus demonstrating its significant effect on stability. Before the application of limit-state codes, such as Eurocode (EC), to the design of structures, joints were commonly considered as either hinged or rigid. However, the fastened connections usual in timber or steel structures actually behave like semi-rigid connections.[7, 8] These works led EC5[9] to take fastener stiffness into consideration in the design of timber structures. The new rules are nonetheless unsuitable for traditional carpentry joints. Research programs on the moment-rotation behavior of these types of connections include a project by the Czech Ministry of Culture, which is dedicated to investigating the timber joints in historical structures ("Design and Assessment of Timber Joints of Historical Structures," which was launched in February 2012). A study of the bending stiffness of a dovetail joint for plywood panels has also been analyzed. Again, friction is implemented in the three-dimensional, solid, finite-element model, but the studied orthotropic material (used for furniture) was only considered in its elastic stage, as the final displacement was limited to 2 mm.[10] Robeller et al.[6] presented a linear interlocking connection of plates for the in situ assembly of folded plate shells. They introduced a geometric solution for simultaneously joining plates with multiple, non-parallel edges and calculated the mechanical performance of the overall structure by a finite element method (FEM) analysis of perfectly rigid joints. In conclusion, a local simulation of bending on a dovetail joint connecting two panels was undertaken. The bending moment was transformed into compression, normal, and shear forces parallel to the locking faces. Conjointly, Roche et al.[11] conducted numerical and experimental work to test the assumptions of semi-rigid behavior of such a connection. Three-layered box beams assembled at the inter-layers using dovetail joints with differing tab lengths and tab angles were loaded to failure point in three-point bending tests. The connection showed promising stiffness results. For instance, a 110-mm tab-length dovetail joint was stiffer than 110-mm spaced screws. The previous results confirmed the good strength/stiffness ratio of the MTSJ in shear. The shear due to the inter-layer slip is transferred by compression on the locking faces. The moment-resisting performance discussed in Robeller[6] should be also confirmed.

This paper presents a study of the rotational stiffness of a MTSJ as a structural connection in architecture. A dedicated folding machine was developed to test the Japanese pattern of the MTSJ under bending moments. A parametric experimental study explored the rotational stiffness through different combinations of the three angles governing the joint geometry. Nonlinear behavior is reproduced by finite element analysis using an adaptation of the continuum damage model of Sandhaas.[12]

2 The multiple tab-and-slot joint as a structural panel connection

2.1 Forces at ridges and valleys

When a folded-plate structure is uniformly and vertically loaded, slab and plate action induce transverse and longitudinal action in the folds.[3] The ridges typically experience the bending moment, M_j **(Fig. 1b)**, forces perpendicular to the ridge in the panel plane, N_j, and beyond the panel plane, Q_j **(Fig. 1c)**, and forces along the ridge, T_j **(Fig. 1c)**. The study is limited to behavior under the bending moment, and at this stage ignores the effect of transverse forces. As the model of this joint is assumed to be semi-rigid **(Fig. 1e)** during bending of the plates, an additional rotation is induced at the ridges. This is contrary to the rigid model, where the initial angle, φ, between the panels remains unchanged after deformation **(Fig. 1f)**. The MTSJ offers a moment-resisting connection that cannot be provided by a hinge **(Fig. 1d)**. Here we seek to understand the way that geometry, particularly the angles directing the normal to the locking face, affects this semi-rigidity.

2.2 Description of the MTSJ

A detailed description of this "integrated mechanical attachment" is given in Robeller;[6] thus, in the next section a brief introduction to the geometric parameters of the joint is included. The ridge (or valley) connection employed here is a one-degree-of-freedom connection (for the purpose of assembly) that has tabs inserted into slots **(Fig. 2e)**. "Multiple" refers to the interlocked tabs and slots, which are repeated along the common edge of two connected panels. The geometry of the connection defines the relative positioning of the panels and allows a certain degree of load-transfer between them. The assembly of the two panels is directed along the vector of insertion. The locking faces of each part belong to the same contact plane after insertion. The insertion vector and the vector normal to the locking face are obtained by the sequence of rotations **(equ. 1)** following the convention of the Bryant angles **(Fig. 2a–d)**.

$$\begin{gathered}\{P_i, F=(\boldsymbol{u}_1, \boldsymbol{u}_2, \boldsymbol{u}_3)\}\\ R(\boldsymbol{u}_1, \theta_1)\downarrow\\ \{P'_i, F'=(\boldsymbol{u}'_1=\boldsymbol{u}'_1, \boldsymbol{u}'_2, \boldsymbol{u}'_3)\}\\ R(\boldsymbol{u}'_2, \theta_2)\downarrow\\ \{P''_i, F''=(\boldsymbol{u}''_1, \boldsymbol{u}''_2=\boldsymbol{u}'_2, \boldsymbol{u}''_3)\}\\ R(\boldsymbol{u}''_3, \pm\theta_{3;i})\downarrow\\ \{P'''_i, F'''=(\boldsymbol{u}'''_{1;i}, \boldsymbol{u}'''_{2;i}, \boldsymbol{u}'''_{3;i}=\boldsymbol{u}''_3)\}\end{gathered} \qquad \text{Equ. 1}$$

If $\boldsymbol{n}_0$ and $\boldsymbol{n}_1$ are the normals of the two panels, the frame of the joint $F=(\boldsymbol{u}_1, \boldsymbol{u}_2, \boldsymbol{u}_3)$ can be calculated as $\boldsymbol{u}_1=\boldsymbol{n}_0\times\boldsymbol{n}_1$, $\boldsymbol{u}_2=\boldsymbol{n}_0$, and $\boldsymbol{u}_3=\boldsymbol{u}_1\times\boldsymbol{u}_2$. Subsequently, the line segment representing the intersection of the two panel, mid-planes can be uniformly divided into N points $(X_i)_{(i=1\dots N)}$; the distance between points will be the tab length L_j of the joint. The plane P_i will be the plane of normal $\boldsymbol{u}_1$, containing the point X_i **(Fig. 2a)**. Thus, by three successive rotations of the plane, P_i, and its attached frame, F **(Fig. 2a–d)**, the final joint will be obtained, defining the planes, P'''_i for both the tab and slot as the locking faces and the vectors, $\boldsymbol{u}'''_{1;I}$, and $\boldsymbol{u}''_3=\boldsymbol{u}'''_{3;i}$ as their normals and insertion vector respectively **(Fig. 2d)**. This elemental sequence of rotations is represented by the Bryant angles, θ_1, θ_2, and θ_3. In this study, the principle of the nejiri arigata joint described by Robeller [6, 13] is adopted, where $+\theta_3$ is applicable from P_1 to $P_{N/2}$, and $-\theta_3$, from $P_{N/2}$ to P_N.

This geometry allows a multi-edge assembly of one panel with three adjoining plates. Four plates k ($k=0$; 1; 2; 3) are taken from the origami folded-plate structure **(Fig. 3a and b)**. Plate 0 (i.e., $k=0$) shares edges 01, 02, and 03 with plates 1, 2, and 3 respectively. For each edge, the eligible subset of insertion vectors is defined as E^{01}, E^{02}, and E^{03} **(Fig. 3c)**. The fabrication constraint, β_{max}, referring to the maximal tool inclination, sets the limits of the dihedral angle φ^{0k} and the angles θ_1^{0k}, θ_2^{0k}, and θ_3^{0k} that govern the insertion vector subsets and the tab angle respectively.[6,13,14] For example, for a 12-mm-diameter milling tool with a cutting length of 28 mm, β_{max} is 30°, which prevents spindle collision when machining 21 mm panels. If the dihedral angle, φ, is set to 120° **(Fig. 2f)** or if θ_3 is set to 30° **(Fig. 2g)**, then the maximum tool inclination is already reached and θ_1 and θ_2 cannot be different from 0°. Finally, θ_1, θ_2, and θ_3 can only all have non-zero values if $\beta_{max} < 30°$

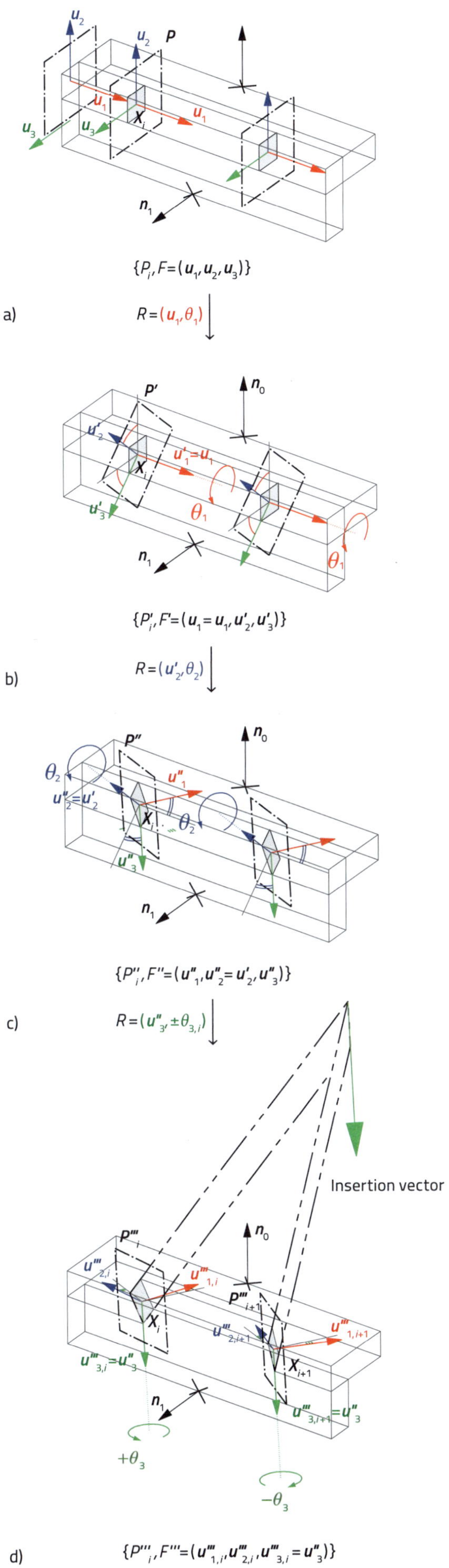

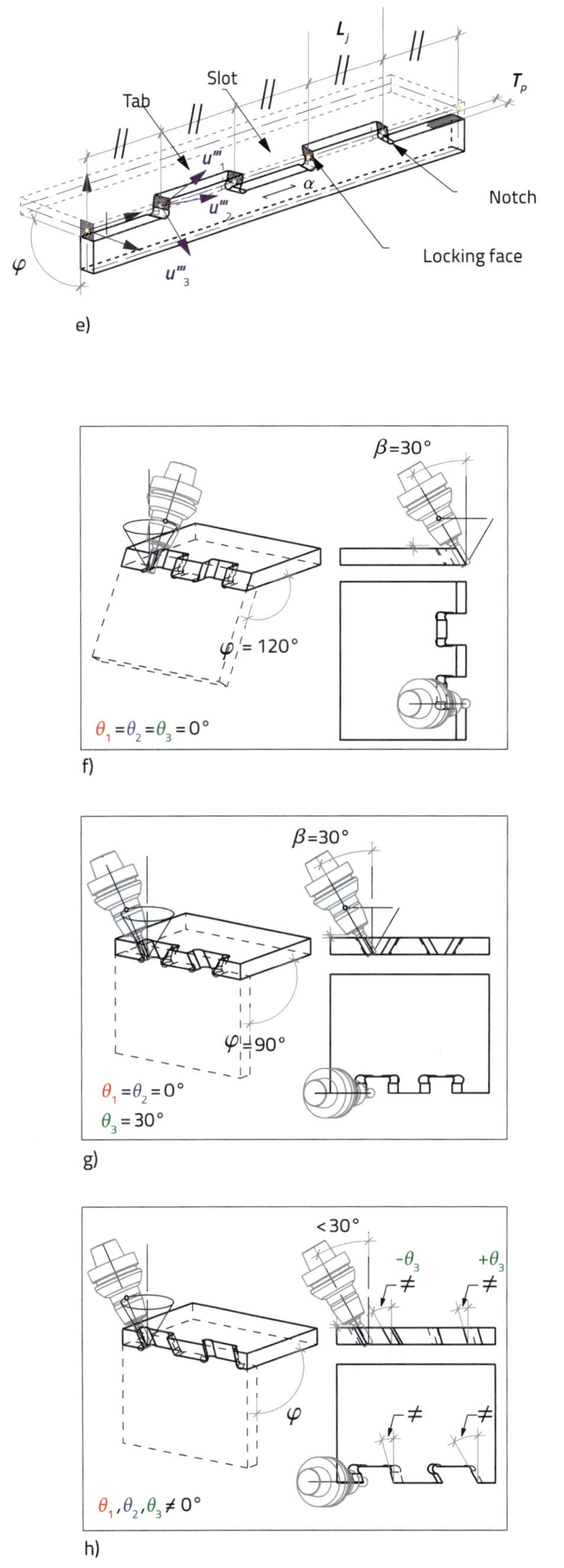

Fig. 2 a–d) Rotation sequence for designing the locking face
e) Terminology of the MTSJ
f) Fabrication constraint due to the dihedral angle
g) Fabrication constraint due to the tab angle
h) Constrained combination of 1, 2, and 3

and $60° < \varphi < 120°$. Geoffroy Mattoni showed that the interdependency between the Bryant angles, dihedral angle, and tool inclination limits the possible combinations.[14] By intersecting the three constrained local vector subsets, a unique vector subset is finally obtained, from which the insertion vector has to be selected.

Given our intention to analyze the rotational stiffness of the joint, angle θ_3 will be taken to be at least 10° where the locking faces are clearly not parallel and offer a certain resistance to rotation about the ridge axis. Detailed analysis determining the parameters for our sample is given in Mattoni.[14]

3 Experimental study

3.1 Description of the samples

Eleven series of two, 21-mm LVL Kerto-Q panels with spruce ply (0-90-0-0-0-90-0) were assembled with eleven particular sequences of joint elements. The geometric properties of the specimens are summarized in Table 1 and fig. 5. No glue was utilized during the assembly of samples (a) to (i), which correspond to nine characteristic combinations of Bryant angles.[14] All had the same tab length of 50 mm. The two panels F0 and F1 were 200 mm long (along the ridge) and 150 mm wide and had two and one full tab respectively. One series (j) was screwed using three Würth ASSY screws (4 mm × 70 mm) with 50 mm spacing. The last series (k) was bonded by PUR gluing. The last two series had butt joints.

3.2 Method

Four specimens of each sample were tested using a dedicated folding-test machine (Fig. 5). To reflect the asymmetry of the joint along the bisector plane of the connected panels, the tests of pushing F0 and F1 were each repeated twice. A 20 kN cylinder pulled two cables to drive the rotation of two moment-transmitting pulleys. Subsequently, the plate supporting two 5 kN cell loads acted as a lever arm on the vertical panel F0 (or F1) of the sample. The horizontal F1 (or F0) panel was rigidly clamped onto the base plate. The location of the rotation center of the rig could be set to coincide with the rotation center of the joint. Rotation and loads were recorded by averaging the values of two inclinometers and adding data from the cell loads respectively. This method was applied in both closing (S01) and opening (S02) tests. A total of 88 specimens were tested.

4 Numerical model

4.1 Material model

The LVL Kerto-Q material is multi-layered due to the circular notch (Fig. 6), which is required for the digital manufacture of the joint, reducing the contact surface to a few layers. As through-thickness stress cannot be ignored, the model uses linear, hexahedral elements with reduced integration (C3D8R). Each of the seven layers has its own orientation (0-90-0-0-0-90-0). The Coulomb friction coefficient is set to 0.3.

4.2 Elastic behavior

Each northern spruce veneer is typically considered orthotropic and behaves elastically in its local orientation frame (Fig. 7). Interfaces are considered to be rigid, thus inter-laminar failure is not addressed. Elastic properties are summarized in Table 2.[15,16] Reliable values for spruce veneer are relatively rare in the published literature.

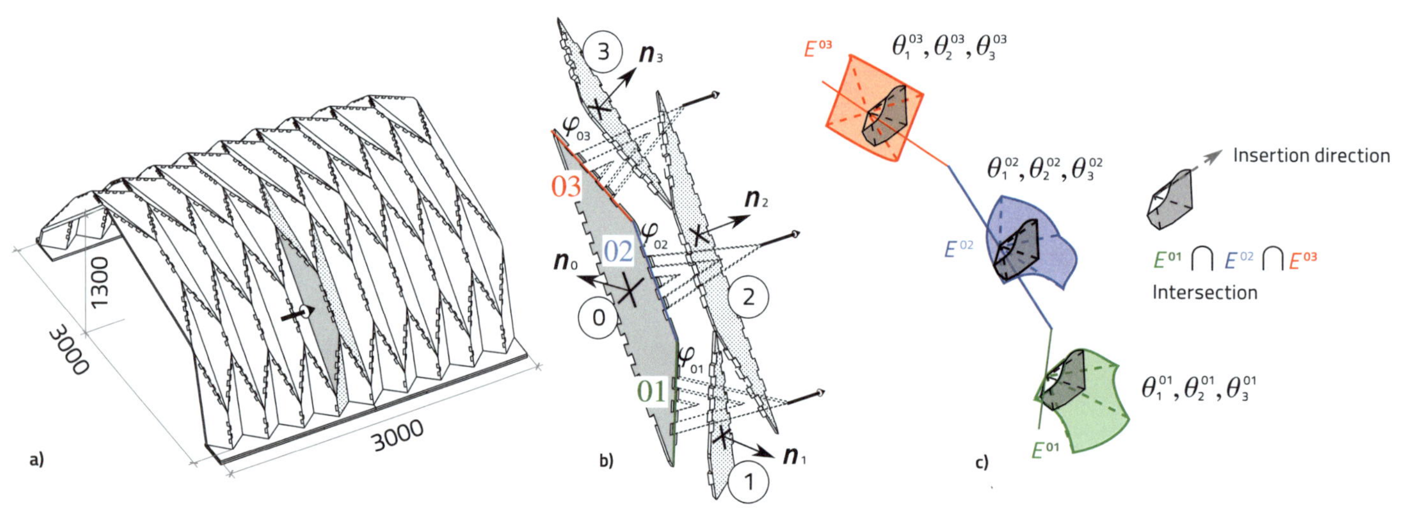

Fig. 3 Assembly constraints a) Plates in situ in the global geometry b) Three-edge simultaneous assembly c) Selection of the insertion vector into the intersection space

Parameters	Symbol	Unit	Value
Insertion angle (1)	θ_1	°deg	→
Insertion angle (2)	θ_2	°deg	
Tab angle	θ_2	°deg	
Dihedral angle	φ	°deg	90
Tab length	L_t	mm	50
Panel thickness	T_p	mm	21
Panel density	ρ_k	kg/m³	480
Angle edge to grain	α	°deg	0
12-mm notch	r	t or b	b
Tool angle T241	β	°deg	30

Table 1 Sample parameters

Property	Symbol	Value	Unit
Elastic modulus 11	E_{11}	6200	Mpa
Elastic modulus 22	E_{22}	210	Mpa
Elastic modulus 33	E_{33}	210	Mpa
Shear modulus 12	G_{12}	350	Mpa
Shear modulus 13	G_{13}	200	Mpa
Shear modulus 23	G_{23}	40	Mpa
Poisson's ratio 12	V_{12}	0.61	
Poisson's ratio 13	V_{13}	0.60	
Poisson's ratio 23	V_{23}	0.50	

Table 2 Elastic properties of spruce

Fig. 4 Combination of Bryant

Fig. 5 Bending tests: **a)** closing mode **b)** opening mode **c)** sample geometries

Fig. 6 Reduced contact on the locking faces

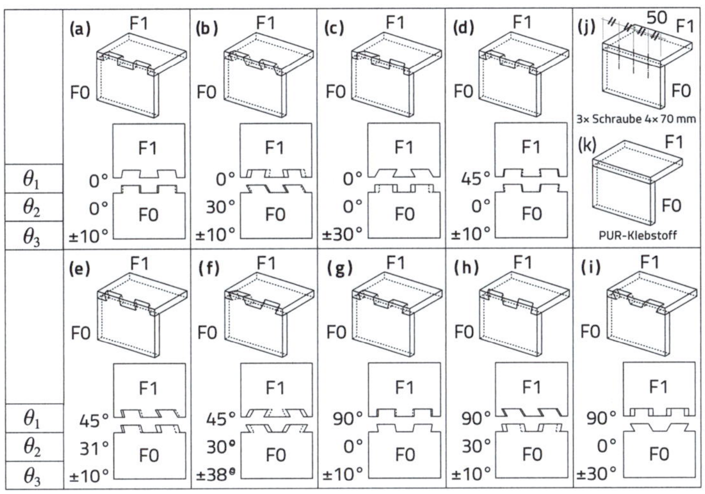

Fig. 4

Fig. 6

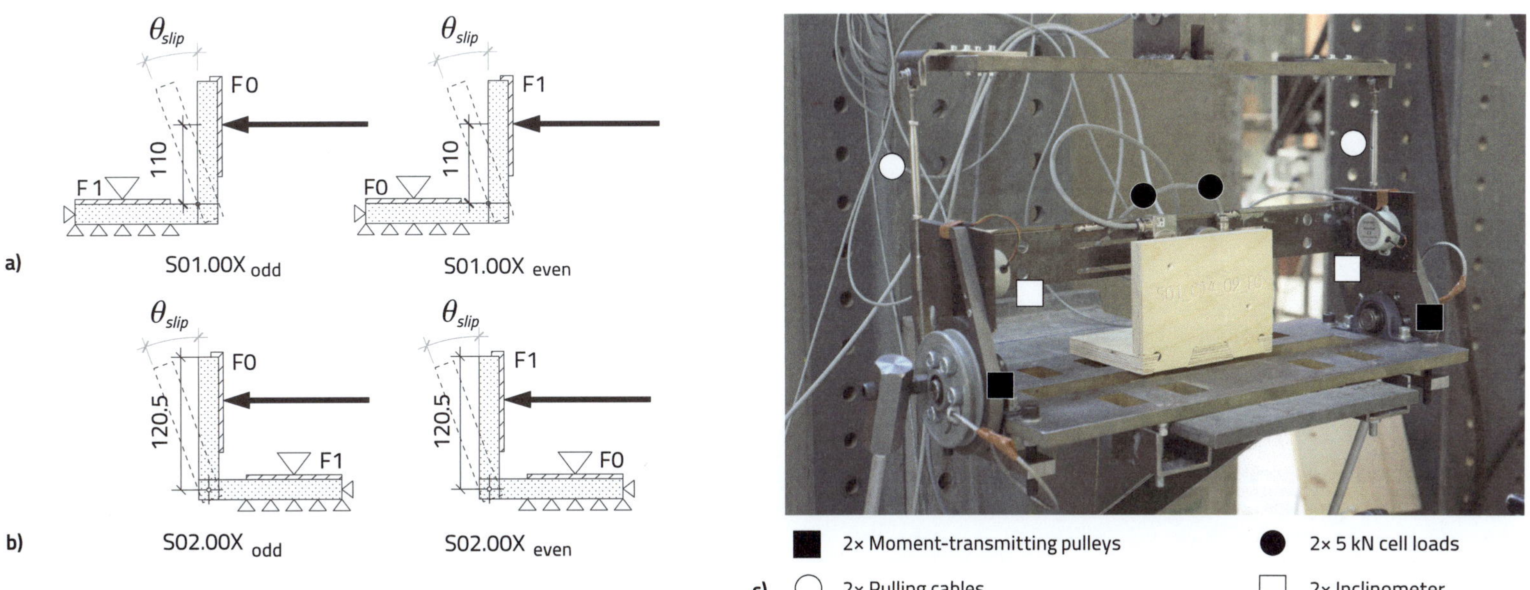

Fig. 5

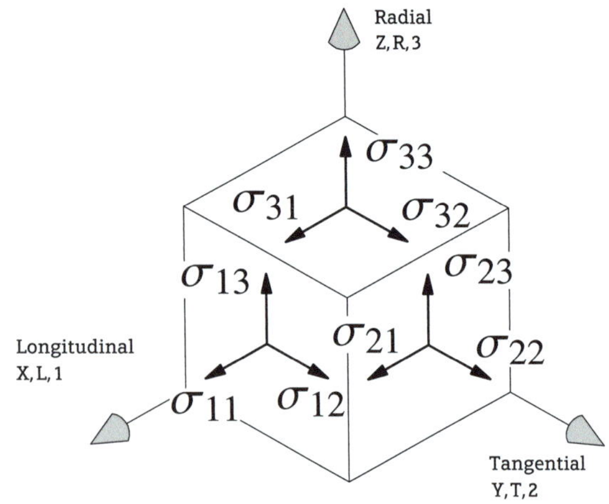

$$\underline{\varepsilon} = (\varepsilon_{11}, \varepsilon_{22}, \varepsilon_{33}, 2\varepsilon_{12}, 2\varepsilon_{13} 2\varepsilon_{23})^T$$

$$\underline{\sigma} = (\sigma_{11}, \sigma_{22}, \sigma_{33}, \sigma_{12}, \sigma_{13}, \sigma_{23})^T$$

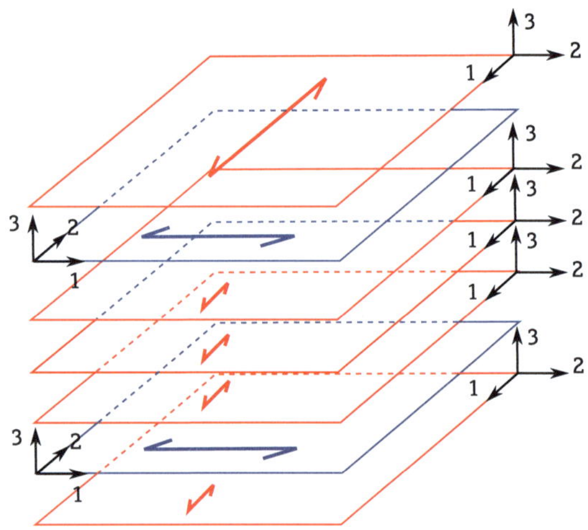

Fig. 7 Material frame and orientation

Property	**Symbol**	**Value**	**Unit**
Tensile strength // to grain f^{t}_{11}	$f_{t;0}$	78	Mpa
Compressive strength // to grain f^{c}_{11}	$f_{c;0}$	50	Mpa
Tensile strength ⊥ to grain f^{t}_{22}	$f_{t;90}$	2.2	Mpa
Compressive strength ⊥ to grain f^{c}_{22}	$f_{c;90}$	10	Mpa
Tensile strength ⊥ to grain f^{t}_{33}	$f_{t;90}$	2.2	Mpa
Compressive strength ⊥ to grain f^{c}_{33}	$f_{c;90}$	10	Mpa
Longitudinal shear strength f_{12}	f_v	6.9	Mpa
Longitudinal shear strength f_{13}	f_v	6.9	Mpa
Rolling shear strength f_{23}	f_{roll}	2.5	Mpa
Fracture energy tension // to grain	$G_{f;0}$	1.6	N/mm
Fracture energy tension ⊥ to grain	$G_{f;90}$	0.5	N/mm
Fracture energy longitudinal shear	$G_{f;v}$	1.2	N/mm
Fracture energy rolling shear	$G_{f;roll}$	0.6	N/mm

Table 3 Mechanical properties of spruce

4.3 Damage approach

To approach nonlinear behavior of the joint, we use a wooden model based on continuum damage mechanics (CDM) (generously provided by C. Sandhaas and described in detail in references 12 and 15). A main feature of the model is its penalization of the elastic moduli. The compliance matrix Cdam, derived for the elastic compliance matrix, defines the modified behavior of damaged elements **(equ. 2–5)**.

$$\varepsilon = C^{dam}\sigma$$ **Equ. 2**

$$\begin{bmatrix} \varepsilon_{11} \\ \varepsilon_{22} \\ \varepsilon_{33} \\ 2\varepsilon_{12} \\ 2\varepsilon_{13} \\ 2\varepsilon_{23} \end{bmatrix} = \begin{bmatrix} C_1^{dam} & 0 \\ 0 & C_2^{dam} \end{bmatrix} \begin{bmatrix} \sigma_{11} \\ \sigma_{22} \\ \sigma_{33} \\ \sigma_{12} \\ \sigma_{13} \\ \sigma_{23} \end{bmatrix}$$ **Equ. 3**

$$C_1^{dam} = \begin{bmatrix} \frac{1}{(1-d_{11})E_{11}} & \frac{-\nu_{12}}{E_{22}} & \frac{-\nu_{31}}{E_{33}} \\ \frac{-\nu_{12}}{E_{11}} & \frac{1}{(1-d_{22})E_{22}} & \frac{-\nu_{32}}{E_{33}} \\ \frac{-\nu_{13}}{E_{11}} & \frac{-\nu_{23}}{E_{22}} & \frac{1}{(1-d_{33})E_{33}} \end{bmatrix}$$ **Equ. 4**

$$C_2^{dam} = \begin{bmatrix} \frac{1}{(1-d_{12})G_{12}} & 0 & 0 \\ 0 & \frac{1}{(1-d_{13})G_{13}} & 0 \\ 0 & 0 & \frac{1}{(1-d_{23})E_{23}} \end{bmatrix}$$ **Equ. 5**

d_{ij} are the damage variables, which depend on the strain state. Here, we recall the failure modes F_m[15] that control the onset of damage.

4.3.1 Brittle behavior

- In tension:

$$F_{t,1} = \frac{\sigma_{11}}{f_{t,0}} \leq 1$$ **Equ. 6**

- In tension and shear:

$$F_{t,2} = \frac{(\sigma_{22})^2}{(f_{t,90})^2} + \frac{(\sigma_{12})^2}{(f_v)^2} + \frac{(\sigma_{23})^2}{(f_{roll})^2} \leq 1$$ **Equ. 7**

$$F_{t,3} = \frac{(\sigma_{33})^2}{(f_{t,90})^2} + \frac{(\sigma_{13})^2}{(f_v)^2} + \frac{(\sigma_{23})^2}{(f_{roll})^2} \leq 1$$ **Equ. 8**

- In shear under compression:

$$F_{v,2} = \frac{(\sigma_{12})^2}{(f_v)^2} + \frac{(\sigma_{23})^2}{(f_{roll})^2} \leq 1 \qquad \text{Equ. 9}$$

$$F_{v,3} = \frac{(\sigma_{13})^2}{(f_v)^2} + \frac{(\sigma_{23})^2}{(f_{roll})^2} \leq 1 \qquad \text{Equ. 10}$$

4.3.2 Ductile behavior

- In compression:

$$F_{c,1} = \frac{-\sigma_{11}}{f_{c,0}} \leq 1 \qquad \text{Equ. 11}$$

$$F_{c,2} = \frac{-\sigma_{22}}{f_{c,90}} \leq 1 \qquad \text{Equ. 12}$$

$$F_{c,3} = \frac{-\sigma_{33}}{f_{c,90}} \leq 1 \qquad \text{Equ. 13}$$

Damage in tension or due to shear will be brittle ($F_{t,1}$, $F_{t,2}$, $F_{t,3}$, $F_{v,2}$, $F_{v,3}$), whereas modes in compression will be ductile ($F_{c,1}$, $F_{c,2}$, $F_{c,3}$). While for $F_m \leq 1$ or $(F_m - \kappa_m) \leq 0$, the material is in the elastic range, and the damage parameter d_m $(\kappa_m \leq 1) = 0$. Damage for mode m initiates when $F_m \geq 1$; thus the history parameter, κ_m, and consequently the damage parameter, d_m, increase. When d_m $(\kappa_m \rightarrow \infty)$ approaches 1, the material is fully damaged; it is calculated according to the failure modes. Fig. 9 indicates that for ductile failure the stress will remain constant after failure initiation, and d_m is expressed as:

$$d_m = 1 - \frac{1}{\chi_m} \qquad \text{Equ. 14}$$

However, for brittle failure, stress diminishes with the strain increment **(Fig. 8)**. This decrease is defined by the failure energy (the area below the curve). In this case, the damage parameter, d_m, can be calculated as:

$$d_m = 1 - \frac{f_{\max}^2 - \frac{2g_f E}{\sqrt{\chi_m}}}{f_{\max}^2 - 2g_f E} \qquad \text{Equ. 15}$$

where the fracture energy is

$$g_f = \frac{G_f}{l_e} G_f \qquad \text{Equ. 16}$$

and l_e is the characteristic length of the element in the numerical model (CELENT parameter in ABAQUS). Replacing G_f with g_f in equation 15 minimizes the mesh dependency.[15] κ_m is the track parameter of the loading history, defined as $\kappa_m^t = _{max}$ $(1; F_m; \kappa_m^{t-1})$. The mechanical properties are given in Table 3.[15,16]

4.4 User-defined field (USDFLD) in ABAQUS

To modify the compliance matrix, we implement our CDM by an USDFLD subroutine, whereas Sandhaas developed a user material subroutine (UMAT). The elastic properties of the material were described in ABAQUES as "field dependent." Six fields are defined to represent the damage: $E_{ii} = (1 - f_i)\ E_{ii}^0$ for $1 \leq i \leq 3$, $G_{ij} = (1 - f_{i+j+1})\ G_{ij}^0$ for $1 \leq i \leq j \leq\ 3$, $V_{ij} = (1 - f_i)\ V_{ij}^0$ for $1 \leq i < j \leq 3$, where $f_1 = d_{11}$, $f_2 = d_{22}$, $f_3 = d_{33}$, $f_4 = d_{12}$, $f_5 = d_{13}$ and $f_6 = d_{23}$.
The subroutine fixes the value of the field according to strain and stress verifications. Fig. 10 presents the simplified algorithm of the subroutine.

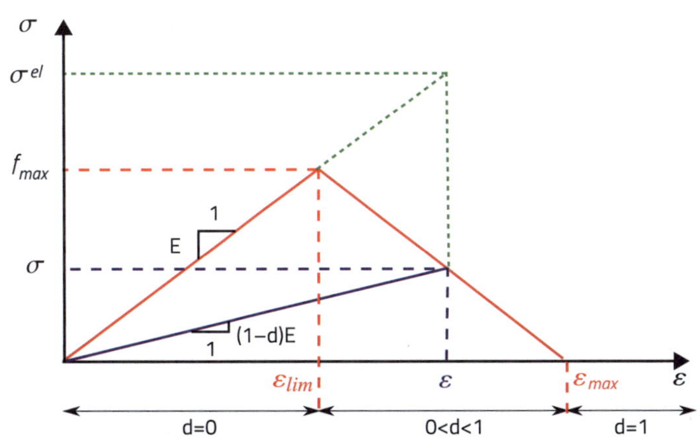

Fig. 8 Brittle behavior

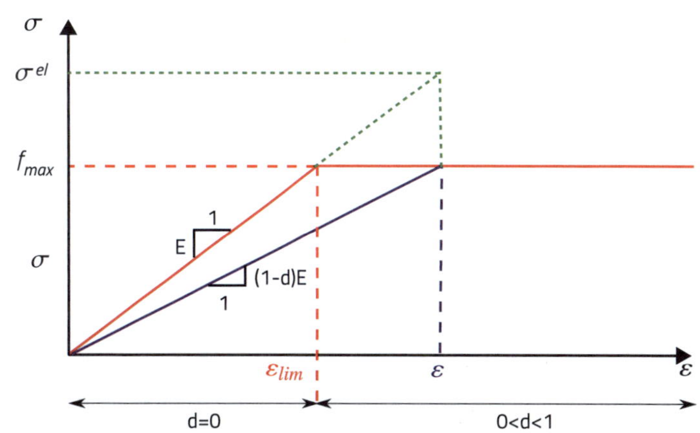

Fig. 9 Ductile behavior

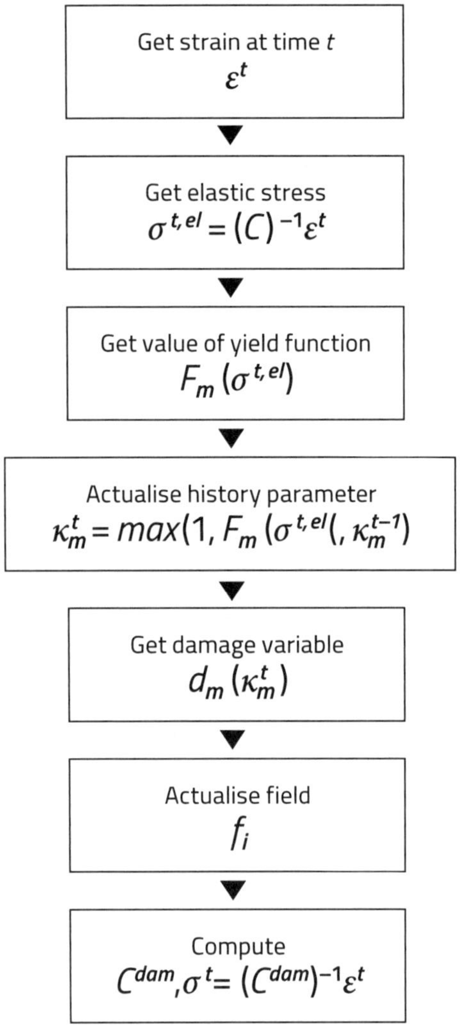

Fig. 10

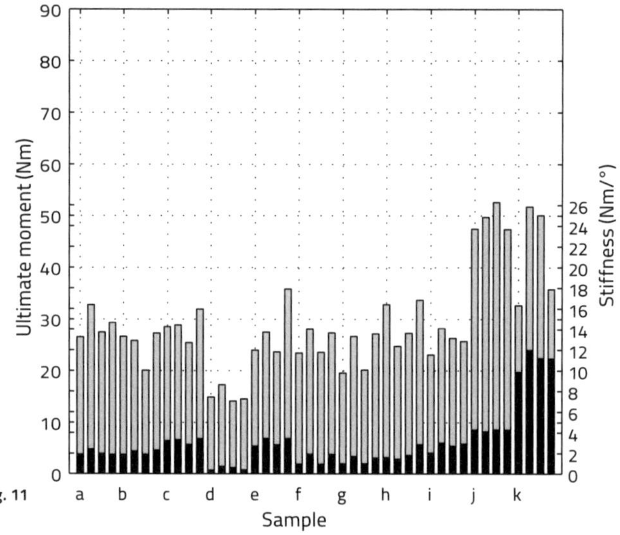

Fig. 11

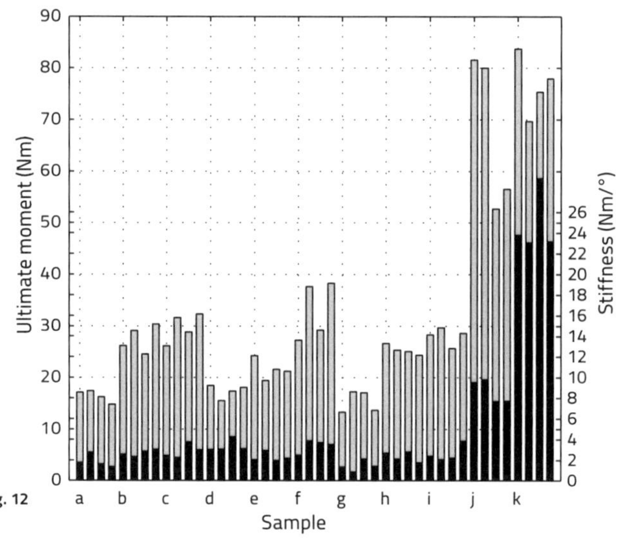

Fig. 12

Fig. 10 Algorithm of the USDFLD subroutine

Fig. 11 S01—Ultimate moment ░ and stiffness ■

Fig. 12 S02—Ultimate moment ░ and stiffness ■

5 Results and discussion

5.1 Experimental results

5.1.1 Replicate comparison

The sequence of rotations that defined the final geometry of each sample gave an asymmetry with respect to the bisector plane of the two connected panels. For each sample, four replicates were tested by pushing the F0 panel for the two first samples and F1 for the last two. Figs. 11 (S01, closing), and 12 (S02, opening) present the ultimate moment and stiffness of each replicate. Despite the asymmetry of the joint, we did not observe a large difference between pushing on F0 or F1, both for S01 and S02. Therefore, the rest of the analysis considers the average of the four replicate values.

5.1.2 Nonlinear moment-rotation curves

Figs. 13 (S01, closing) and 14 (S02, opening) show the average moment-rotation full curves. The sample names in the legend correspond to those in Tables 4 and 5.

After failure, a residual moment remains for all the samples. It never drops below 50% of the ultimate moment except in the case of sample (c), which it does in both directions (S01 and S02). This sample later showed the best compromise of stiffness in closing and opening (0°, 0°, 30°). Samples with the most rigid connections generally showed quicker softening after damage. The hardening differed between closing (S01) and opening (S02), with only sample (h) hardening in S01 and all but (f) and (d) hardening in S02. The MTSJ is globally fairly ductile, although its stiffness in bending is far weaker

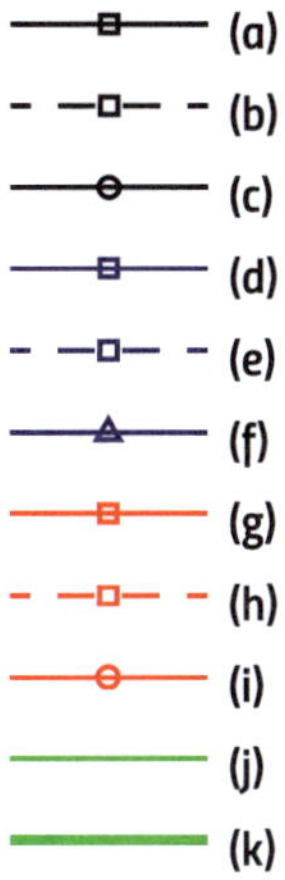

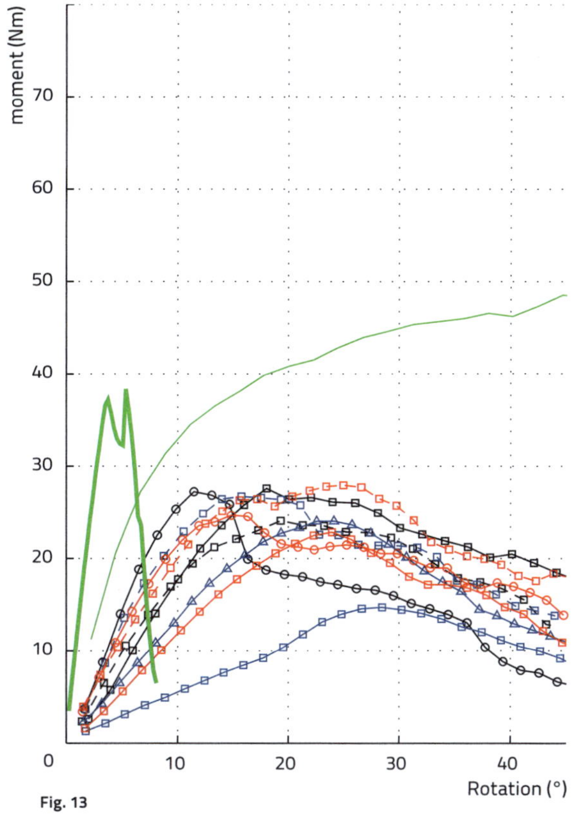

Fig. 13

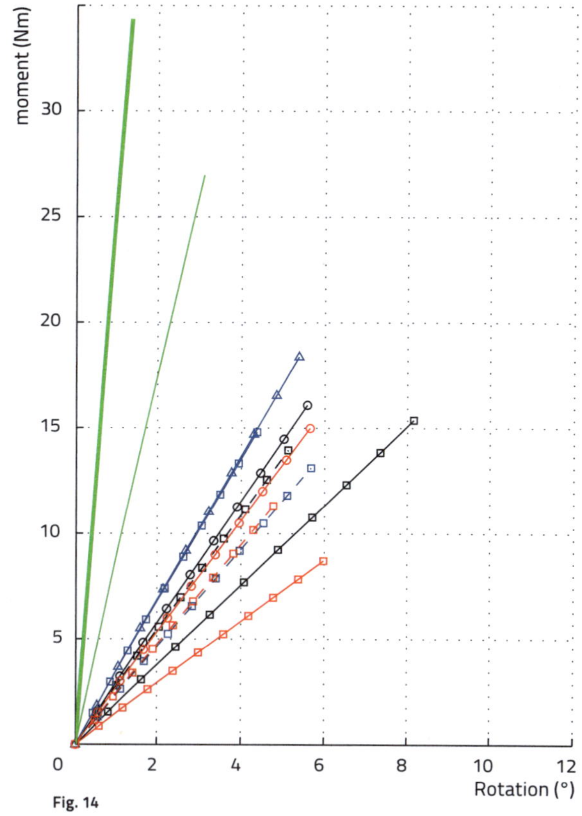

Fig. 14

Fig. 13 S01—Moment-rotation average nonlinear curves for bending while closing

Fig. 14 S02—Moment-rotation average nonlinear curves for bending while opening

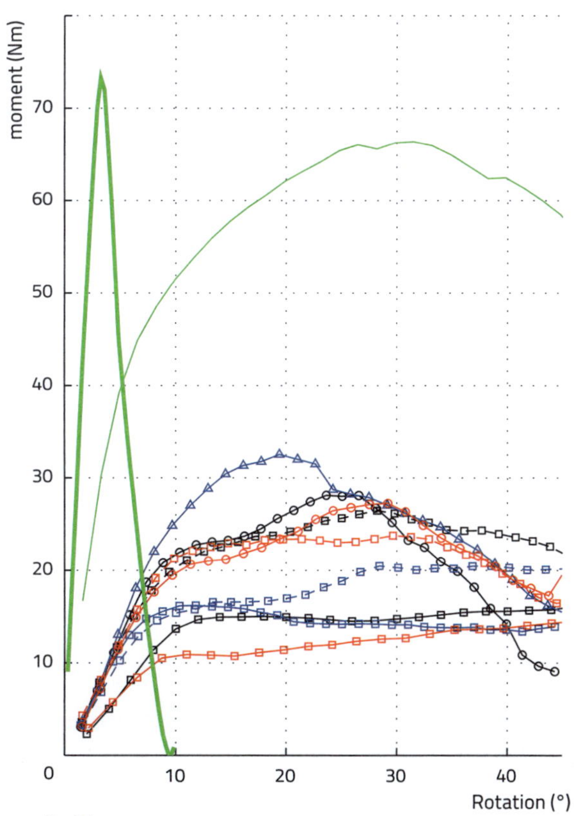

Fig. 15

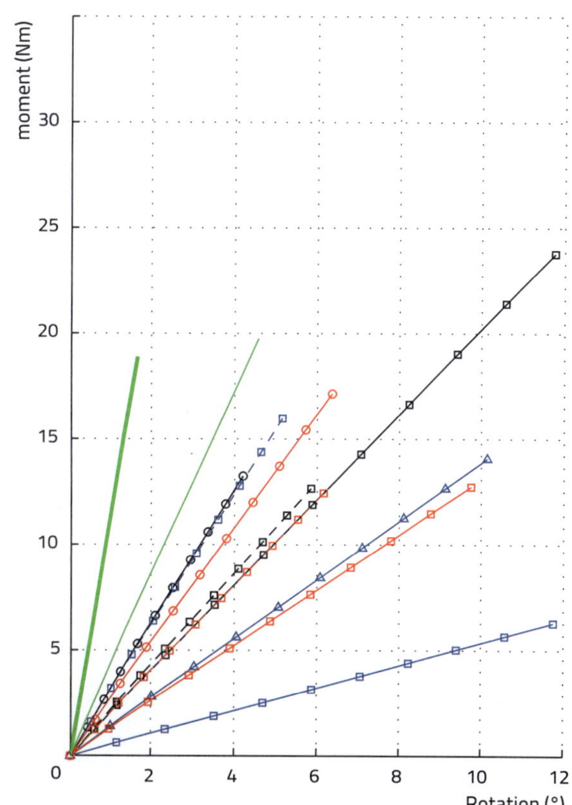

Fig. 16

Fig. 15 S01—Moment-rotation average linear curves (between 10 % and 40 % of m_{max}) for bending while closing

Fig. 16 S02—Moment-rotation average linear curves (between 10 % and 40 % of m_{max}) for bending while opening

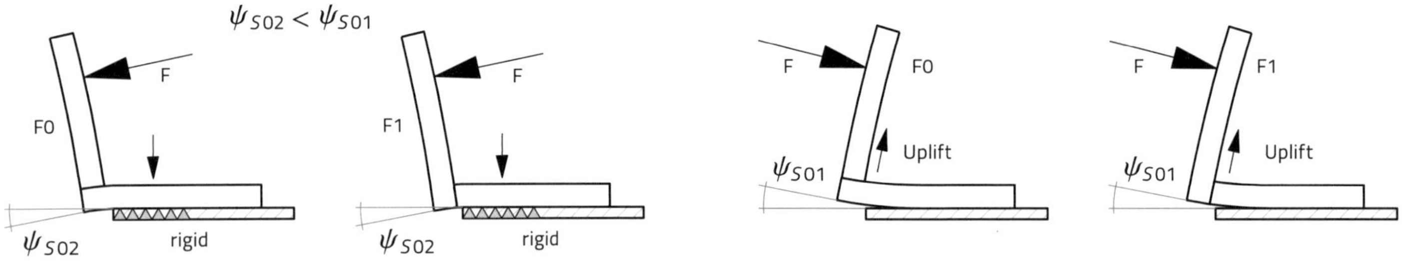

Fig. 17 High-stiffness effect on glued and screwed joints

Sample	θ_1–θ_2–θ_3	Moment max mean Nm/°	Moment max std	Stiffness mean Nm/°	Stiffness std
a	00-00-10	29	2.8	2.0	0.2
b	00-30-10	25	3.3	2.2	0.3
c	00-00-30	29	2.7	3.1	0.2
d	45-00-10	15	1.4	0.5	0.1
e	45-31-10	28	5.7	3.1	0.4
f	45-00-38	26	2.4	1.4	0.6
g	90-00-10	23	4.1	1.3	0.3
h	90-30-10	30	4.3	2.0	0.5
i	90-00-30	26	2.1	2.7	0.4
MTSJ		25.5	4.4	2.04	0.9
j	screwed	49	2.5	4.3	0.1
k	glued	43	9.7	11.3	1.0

Table 4 S01—Mean and standard deviation

Sample	θ_1–θ_2–θ_3	Moment max mean Nm/°	Moment max std	Stiffness mean Nm/°	Stiffness std
a	00-00-10	16	1.2	1.7	0.6
b	00-30-10	28	2.7	2.8	0.4
c	00-00-30	30	2.9	2.8	0.8
d	45-00-10	17	1.3	3.2	0.7
e	45-31-10	22	2.0	2.4	0.4
f	45-00-38	33	5.7	3.0	0.6
g	90-00-10	15	2.1	1.6	0.7
h	90-30-10	25	0.9	2.4	0.5
i	90-00-30	28	1.7	2.6	0.7
MTSJ		23.8	6.41	2.5	0.6
j	screwed	68	15.2	8.9	1.1
k	glued	77	5.8	25.3	2.9

Table 5 S02—Mean and standard deviation

Row	Samples	θ_1	θ_2	θ_2	Mean S01 N m/°	Dev. %	Mean S02 N m/°	Dev. %
1	a,b,c	0			2.44	20	2.44	−2
2	d,e,f	45			1.67	−18	2.86	14
3	g,h,i	90			2.01	−2	2.21	−12
4	a,c,d,f,g,i		0		1.85	−9	2.50	0
5	b,e,h		30		2.42	19	2.52	1
6	a,b,d,e,g,h			10	1.86	−9	2.35	−6
7	c,f,i			30	2.41	18	2.82	13
8	a,c	0	0		2.59	27	2.26	−10
9	b	0	30		2.16	6	2.82	13
10	a,b	0		10	2.09	2	2.25	−10
11	c	0		30	3.15	54	2.83	13
12	d,f	45	0		0.97	−53	3.11	24
13	e	45	30		3.09	51	2.36	−6
14	d,e	45		10	1.82	−11	2.78	11
15	f	45		30	1.39	−32	3.02	21
16	g,i	90	0		2.00	−2	2.12	−15
17	h	90	30		2.02	−1	2.39	−5
18	g,h	90		10	2.00	−2	2.12	−15
19	i	90		30	2.69	32	2.62	5
20	a,d,g		0	10	1.29	−37	2.17	−13
21	c,f,i		0	30	2.41	18	2.82	13
22	b,e,h		30	10	2.42	19	2.52	1

Table 6 Comparative analysis of the combination of Bryant angles

than that of the screwed and glued connections. The fastened and bonded samples showed different experimental results for closing and opening.These types of connections are far more rigid than the MTSJs; they were stronger than the panel strength and stiffness during deformation. Bending occurred in panels F0 and F1, and fig. 17 explains this effect on the stiff joint.

In S02, the reference panel was pressed onto the support with increasing load, reducing local bending near the joint. In S01, the reference panel was lifted by the loading, and was free to bend. Under a given load, the sample deflected more in S01 than in S02. In both cases, the glue was subjected to tensile stress, which was unfavorable for adhesive bonding in that direction. As the glued and screwed samples were stiffer in S02, their moments in S02 were higher than in S01 for a given rotation. The highly brittle failure of the glued connection is clearly visible in figs. 13 and 14. The less rigid screwed joint behaved similarly to the glued joint, but with smaller differences between S01 and S02. Both connections were tested as elements of comparison. After testing, the samples connected with MTSJs showed no bending within the panels. The relative weakness of the connection resulted in no high-stiffness effect. The rest of the analysis will only focus on the MTSJs.

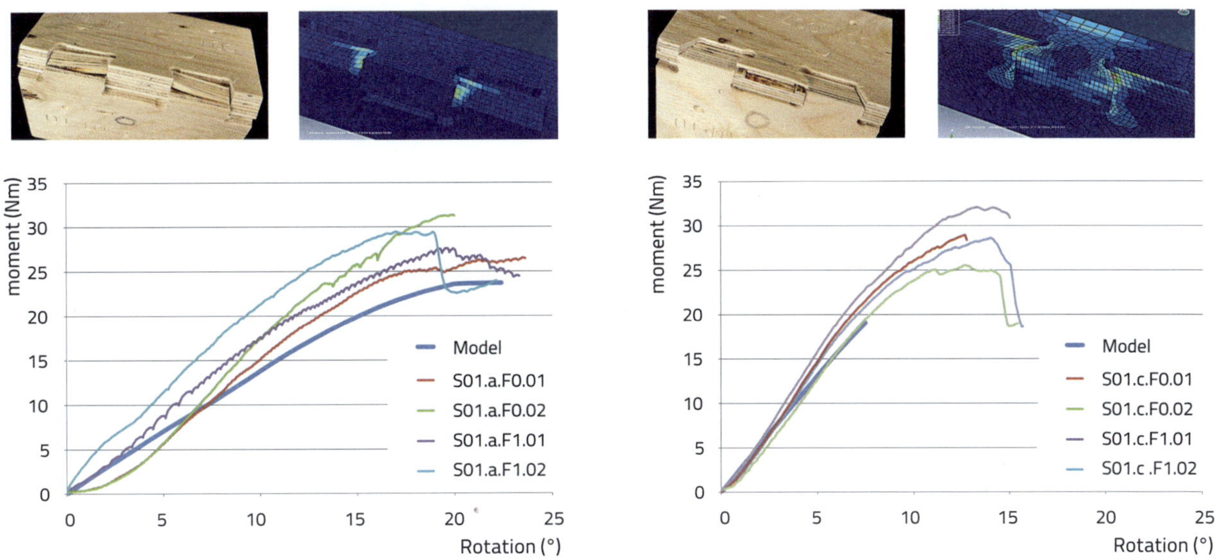

Fig. 18 S01—Numerical curves versus experimental results, a) Sample a (0°; 0°;10°), b) Sample c (0°; 0°; 30°)

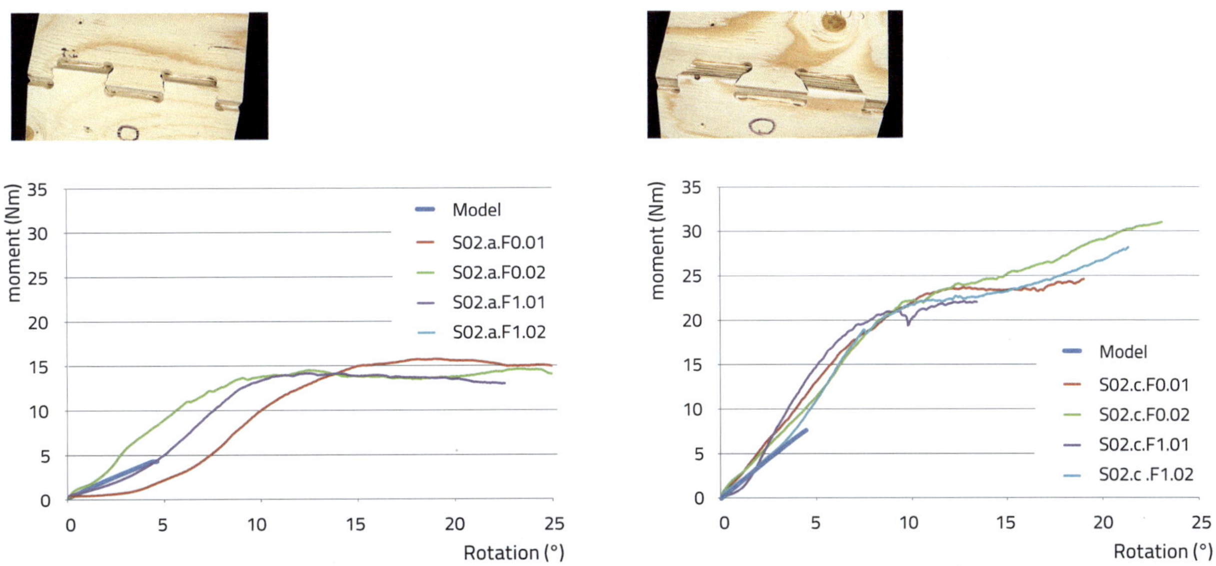

Fig. 19 S02—Numerical curves versus experimental results, a) Sample a (0°, 0°, ±10°), b) Sample c (0°, 0°, ± 30°)

Fig. 20 MTSJ: open slot (left), closed slot (right).

5.1.3 Linear moment-rotation curves

Figs. 15 and 16 were obtained by taking an average of the linear fits of four replicates of each sample. The limits of the linear range were chosen between 10% and 40% of the ultimate moment. The curves are displayed in this range. The sample names in the legend correspond to those in Tables 4 and 5.

5.2 Global analysis

The total average stiffness in closing (2.04 Nm/°) was 18.5 % lower than the stiffness in opening (2.5 Nm/°), whereas the ultimate moments were not significantly different (25.5 Nm ⟷ 23.8 Nm, less than 8 %). The range of stiffness on S02 was also narrower than that of S01. This is confirmed by the standard deviations listed in Tables 4 and 5. In the tables, "MTSJ" refers to the group of nine MTSJ samples (36 samples per loading case).

Sample	θ_1–θ_2–θ_3	Model stiffness mean Nm/°	Exp. Stiffness mean Nm/°	std
a	00–00–10	1.4	2.0	0.2
b	00–30–10	1.7	2.2	0.3
c	00–00–30	2.6	3.1	0.2
d	45–00–10	0.6	0.5	0.1
e	45–31–10	1.4	3.1	0.4
f	45–00–38	1.1	1.4	0.6
g	90–00–10	0.7	1.3	0.3
h	90–30–10	1.0	2.0	0.5
i	90–00–30	1.8	2.7	0.4

Table 7 S01—Model stiffness versus experimental stiffness

5.3 Comparative analysis of stiffness

To evaluate the effects of the Bryant angles and their partial combinations on the semi-rigidity of the MTSJ, mean values of stiffness are listed in Table 6. For each group of samples that has the same values for θ_1, θ_2 or θ_3, or any combination thereof, the mean value of stiffness is compared with the mean of all samples (see Tables 4 and 5) for both S01 and S02. The deviations from the global means are listed in the columns headed "dev.%," and they are considered to be meaningful when greater than 10%. The full combinations ($\theta_1 \theta_2 \theta_3$) are listed in Tables 4 and 5, but trends cannot be clearly extracted from them. Here, we analyze trends revealed in the experimental results. Each sample was tested in four replicates. For clarity, the combinations are written as (θ_1 θ_2 _), (θ_1 _ θ_3) and (_ θ_2 θ_3) in the text. Any value greater than or equal to 30° is noted as 30° in the brackets. $\theta_1 = 0°$ gives the best stiffness in closing (S01), whereas $\theta_1 = 45°$ reduces the stiffness in opening (S02). Connections with $\theta_1 = 90°$ are possibly weaker, particularly in S02. The effect of θ_3 is obvious, as increasing the value to 30° or more increases the rigidity in both loading directions. For θ_2, which seems to improve the stiffness in S01, the analysis must be completed in combination with θ_1. (00 00 _) and (45 00 _) provide opposite results, with stiffness increasing in S01 and decreasing in S02 for the former, and vice versa for latter. In contrast, (00 30 _) was stiffer in S02 and (45 30 _) was stiffer in S01. Combining $\theta 2$ with $\theta 1 = 90°$ did not greatly alter the results relative to the total mean. The combinations (_ θ_2 θ_3) confirmed the positive effect of θ_3 on the stiffness, except in the case of sample (f) ($\theta_3 = 38°$), which showed a rather low stiffness in closing. The tabs were stretched in S01 for this configuration. Above a certain value (>30°), the normal of the locking faces approaches the orientation where the tension perpendicular to the grain starts reducing the strength and stiffness. The sensitivity to large θ_3 angles could be explored in future works. (_ 00 30) and (_ 30 10) both have a similar effect on S01, and again highlight the beneficial effect of $\theta_3 = 30°$. However, $\theta_2 \geq 30°$ did not have any major effect during opening (S02). Finally, the best compromise of the nine samples is $\theta_1 = 0°$, $\theta_2 = 0°$, $\theta_3 = 30°$, which takes advantage of the positive effects described above. According to the trends shown in this study, a joint definition with $\theta_1 = 0°$, $10° \leq \theta_2 \leq 30°$, $15° \leq \theta_3 \leq 30°$ could behave efficiently. As described in section 2.2, the available combinations are limited by the tool inclination and the multi-edge assembly constraints. The range of possible tool inclinations could be increased by altering the shank length and the tool-holder geometry.

5.4 Numerical model versus experiments

A numerical model was built to reproduce the experimental set-up as faithfully as possible. For analyzing the rotational stiffness, the elastic range would have been sufficient. As the ultimate moment of the connection is monitored experimentally, comparison with a numerical value would be useful, and a damage model was implemented to achieve this. Some preliminary results are presented as follows:

In both S01 and S02, numerical strain and damaged element were well observed experimentally. Consider, for example, sample (a) in fig. 18: the numerical model matched the experimental curves well under closing. However, at this stage, our model aborted before reaching the ultimate strength when the samples were tested while opening (Fig. 19). The experiment halted before reaching the yield point, thus preventing the observation of the ultimate moment. The continuum damage model worked well, but convergence problems were encountered during the contact step, so that only the stiffness values of the model in S01 were listed in Table 7 and compared to experimental values. They were generally significantly lower than the averages of the experimental results. This was probably due to the deviation from actual values of both material properties and the friction coefficient (set to 0.3 in our case). The development of the numerical model remains ongoing, and pending issues have to be solved by:

- Re-defining the contact step and the Coulomb friction coefficient
- Refining the elastic and mechanical properties by material testing on Kerto-Q panel and spruce veneer
- Identifying dislocated elements with large rotation and mesh modification
- Controlling the time increment
- Conducting mesh optimization
- Migrating from ABAQUS/Standard to ABAQUS/Explicit.

6 Conclusion

Producing strong edge-to-edge connections in thin panels is challenging. Usual metal fasteners such as screws may not comply with the required edge distance. The MTSJ presented in this paper is an innovative solution. It allows the in-space, multi-edge interlocking connection of panels. The joint is milled during cutting. Its shear performance has already been assessed mechanically[11], and the semi-rigidity of the MTSJ has demonstrated properties competitive with wood screws. The bending stiffness on the ridges of the folded-plate structures is considered here. The first tests on the rotational stiffness of MTSJ with nejiri arigata design revealed ductile behavior with relatively low stiffness compared with screwed and glued joints. The joints were tested in their minimum configuration with one and two tabs per panel. The effect of the tab angle θ_3 is significant, and its combination with the other angles is promising for further analysis. The sample with $\theta_1 = 0°$, $\theta_2 = 0°$, $\theta_3 = 30°$ behaved well, but joints defined as $\theta_1 = 0°$, $10° \leq \theta_2 \leq 30°$, $15° \leq \theta_3 \leq 30°$could be even more efficient. A numerical model based on continuum damage mechanisms showed promising results, although it also showed limitations that must be addressed. These preliminary results will facilitate the development of stiffer joints, and further work is already under way. Fig. 20 presents the evolution of the MTSJ with closed slots; the sample (a) ($\theta_1 = 0°$, $\theta_2 = 0°$, $\theta_3 = 30°$) is shown in a joined position for comparison. Bending experiments have already demonstrated its stiffness to be equivalent to that of a screwed joint. The connections have already been used in a double-layer, curved shell.

References

1 Wardle, T. "Old Growth: Our Timber Engineering Heritage. Design Notes for Plywood Folded-Plate Roofs." In: *New Zealand Timber Design Journal,* Issue 1, Vol. 19, 548–551, 2001.

2 McDowall, M., M. Leigh, and M. Punton. "Exotic Structural Forms—Folded plates." In: *Structural Plywood & LVL Design Manual,* Chapter 12, Engineered Wood Products Association of Australasia, ,154–158, 2007.

3 Jaksch, S., A. Fadai, and W. Winter. "Folded CLT structures—Developments in Design and Assembly Strategies." In: *World Conference on Timber Engineering,* Vol.1, 18–22, 2012.

4 Buri, H., and Y. Weinand. "Origami—Folded Plate Structures, Architecture." 10th World Conference on Timber Engineering, Miyazaki, Japan, 2008.

5 Buri, H. "Origami—Folded Plate Structures."Thesis, Swiss Federal Institute of Technology Lausanne, 2010.

6 Robeller, C., A. Stitic, P. Mayencourt, and Y. Weinand. "Interlocking Folded Plate—Integrated Mechanical Attachment for Structural Wood Panel." In: *Advances in Architectural Geometry,* University College London, London: Springer, September 18, 2014.

7 Haller, P., Cost C., eds. *Semi-rigid Timber Joints, Structural Behaviour, Modelling and New Technologies.* European Commission, 1999.

8 Jaspart, J.P. "General Report: Session on Connections." In: *Journal of Constructional Steel Research,* Vol. 55, 68–89, 2000.

9 EN 1995-1-1, *Eurocode 5: Design of Timber Structures, Part 1-1: General-Common Rules and Rules for Buildings.* European Committee for Standardization, Brussels, 2004.

10 Sebera, V., and M. Simek. "Finite Element Analysis of Dovetail Joint Made with the Use of CNC Technology." In: *Acta Universitatis Agriculturae Et Silviculturae Mendelianae Brunensis,* Vol. LVIII, Nr. 3, 321–328, 2010.

11 Roche, S., C. Robeller, L. Humbert, and Yves Weinand. "On the Semi-Rigidity of Dovetail Joints for the Joinery of LVL Panel." In: *European Journal of Wood and Wood Products,* 2015.

12 Sandhaas, C. "3D Material Model for Wood, Based on Continuum Damage Mechanics." In: *Stevinrapport 6-11-4, Stevin II Laboratory,* University of Technology Delft, 2011.

13 Robeller C. "Integral Mechanical Attachment for Timber Folded Plate Structures." Thesis, Swiss Federal Institute for Technology Lausanne (EPFL), 2015.

14 Mattoni, G. "Folded Plate Structure, Design and Analysis of Woodworking Joints for Structural Timber Panels." M.A. Thesis, ENPC and EPFL, 2014.

15 Sandhaas, C. "Mechanical Behaviour of Timber Joints With Slotted-In Steel Plates." Thesis, University of Technology Delft, 2012.

16 Ivanov, I., T. Sadowski, M. Filipiak, and M. Knec. "Experimental and Numerical Investigation of Plywood Progressive Failure in CT Tests." In: *Budownictwo i Architectura,* 79–94, 2008.

Reprinted from *International Journal of Space Structures,* vol. 30, No. 2, 2015.

5.4

"Digital fabrication leads to a new building culture"

Interview between Matthias Kohler (chair of Architecture and Digital Fabrication, ETH, Zürich) and Yves Weinand February 17, 2015

Yves Weinand: When I look at your work, it's clear that you have a strong interest in the process and in the exemplification of the process. Of course you also consider constraints, such as a flat panel or a sheet of paper, for instance.
As a general goal, we are looking for a structural form. We are looking for a connection between structure and form. I don't know exactly how to describe it, so I tend to speak about a "close relationship" linking structure and form.
We realize that when we integrate local observations, the connections also become stronger. In general, we think in terms of global form-finding tools. But I believe that there is a link between the overall geometric situation (geometric parameters) and the local situation. We have different angles and different approaches. Could you first develop the process orientation of your own work and later concentrate on how an architectural synthesis might be achieved?

Matthias Kohler: That's interesting, as I think this rejuvenated relationship between global form and local performances—which in our case particularly relates to the process of making—is a debate that we need to have in future architectural discussions. Our research is centered on the question of how digital fabrication principles might lead architecture to a new contemporary building culture. By using digital design methods that seamlessly inform new construction techniques, the design of details becomes an essential part of an architectural design. You addressed our interest in process—it is true, we start from taking a material and exploring its properties, associating it with specific fabrication methods and construction systems in order to see how these relate back to the material's inherent capacity, and ultimately trying to gain an understanding of the limits of what we can achieve with it. By the careful analysis of digitally controlled processes, we seek expanded design opportunities, which are different from what is achievable with traditional building techniques. This highly constructive mode of thinking about design with digital technologies is at the core of our research. Since our objective is not only an outcome, but also the precise control of each progressive step, it is often the case that the local constraint, i.e. the detail, is significant in the search for entirely new processes of creating physical form. So through digital fabrication we establish a direct connection between material and its processing. Precisely this design of processes—of how you make things—creates new opportunities for architectural design. However, let's not be naïve; neither establishing processes, nor controlling digital data is a guarantee of good design. Nevertheless, it opens up interesting possibilities for design, or the creation of new processes of creating architecture in a generative computational manner, informed equally by specific small-scale details as by large-scale design intentions.

Y. W.: When we look at the brick wall projects you have been working on in the past, and then when we look at Eladio Dieste's work, who was a structural engineer using bricks (he also created curved walls that are structurally informed and are highly resistant), I ask myself how your work might be informed by Dieste. You're not interested in the global form to begin with; you're more interested in the purpose. It's similar for engineers: they don't tend to be interested in the global form at first.

M. K.: Are you interested in the global form from the outset?

Y. W.: Yes, of course. Dieste looks at how the geometric positions of the bricks give structural efficiency to the wall, while you are looking at the process and how the process influences material and form. I would like to clarify this and to see how to synthesize these ideas. Dieste does not speak about the global form either.
I think his work is very impressive, because it includes content between global form, local geometry, and structure. This aspect, which has to do with synthesis, will also form the focus of our research within the NCCR.

M. K.: Our interest in the global form remains relatively strong. You see that a lot of our work has quite an expressive, formal capacity. However, our true interest lies in how we can explore and develop these expressions; it is not in the form per se, nor in a preconceived narrative or even a pictorial approach. In fact, I think that your origami structures are not far from that, as you're also not designing the global form first. But with the origami structures, the expression is inherently set. You are working with a far more constrained system and therefore the converse question would be: Why do you look for such tightly constrained systems, while we are always trying to extend the degree of design freedom?

Y. W.: Well, it's true that I already had two aspects in mind when we started the origami project. First, it grew from an ongoing fascination with folded structures. When I began my studies as a civil engineer, I was already drawn to folded structures. I was always interested in this and in the "origami aspect" of it. I like to inform the subjective design process with objective elements, which are more related to nature or scientific observation. The constraints of the developed tool are that you can flatten it.

M. K.: When you generate specific structural designs and when you develop the tools needed to generate them, do you have a preconceived idea of what they should look like? Or are you primarily interested in exploring the possibilities of an abstract, constrained space?

Y. W.: It's like a series of trial-and-error attempts in different directions. Geometrically speaking, there is no overall framework to it. There is still no answer to this. I am unsatisfied by the fact that the synthesis cannot be understood as a geometric synthesis in any way. It remains open—it's more like a body of work that is presented in a specific way.

M. K.: As an architect, you generally start to shift priorities during the design process. You might start with initial assumptions and then you choose some specific constraints as being important to your reading of a situation. This is what you're going to focus on, and then, after a while, perhaps after a few initial sketches, other aspects might capture your attention and become more important. Issues that you might have been interested in at the start become subsumed by new, constantly evolving perspectives. Therefore I ask myself: How do you deal with this when you use computational design processes? Because, generally speaking, in a computational design process you only define your system once. You have parameters that you can tweak, but basically your system is defined. You either re-program the entire system, or otherwise you develop the initial system further. Nowadays, I wonder if this is a limiting factor. Are we artificially constraining ourselves by not shifting perspectives within the design process? Or will we have computational design tools one day that are more open to shifting design perspectives?

Y. W.: I don't believe it, as the richness lies in the fact that you have something similar to case studies.

M. K.: What don't you believe in?

Y. W.: That you can do this, that you can have a computational framework, which you can change according to the progress of the design. It's more like you still have to choose, as a carpenter does, the right tool to do a specific job.

M. K.: But then, you're talking about having a sequence of tools?

Y. W.: Yes.

M. K.: If this is a given, it might also be contrary to the notion that you could have a more integral and consistent computational design process throughout an agency.

Y. W.: Mark Pauly spoke about this contradiction with regard to a case study. Although he found it very interesting, he said that if you want to analyze this in a global manner, it would be far too complex.

M. K.: As a computer scientist, he says it's an impossibility and, as an architect, you also believe that it's not clever to overload computational generative tools by having expectations that are too high.

Y. W.: My aim would have been to find some kind of geometric frame.

M. K.: Why geometric?

Y. W.: I believe it could put some things in order. But that order would not be based on a computational tool definition. In construction, it always comes back to geometry.

M. K.: Though, in order to provoke you a little, I would say this is a déformation professionnelle, a bias introduced by your profession. Because, rather than geometry, in the end it always comes back to physical construction and materials.

Y. W.: Well, you can say déformation professionnelle, but it's true that I have the impression that I can achieve the synthesis between structure and form within geometric constraints. If you look at the history of construction in architecture, you always have very important points. Sometimes, it's the geometry that gives us clues, sometimes it's on another level. It has remained on the geometric level several times.

M. K.: Clearly, geometric processing allows a high level of control—not only in architectural design, but also in structural analysis. Yet I wonder if the relationship between geometric calculation—taking geometry as a foundation for design decisions—and architecture is not outdated and superseded by the computer, which works according to numerical processing, and therefore establishes a different relationship between the technician and the real, material world. For example, if you have machines that can calculate how material behaves and at the same time work with these materials—is this idea of a geometric ontology still a powerful and central architectural model?

Y. W.: I wish that it had been possible. I spoke about our work with a few French mathematicians and they indicated that you could order these elements in some kind of geometric frame, but it has not been done yet.

M. K.: Is this all just an ideal?

Y. W.: Yes, it's an ideal. It's been done very intuitively.

M. K.: You care a lot about details in your work, while at the same time, you say that in the end it's just a geometric model. I would like to challenge this a little bit, having visited your lab, and seen all the models around and all the physical explorations. What role does physical and empirical exploration play in your view? Because, alternatively, you could also say that you concentrate on theoretical work and the exploration of geometric space, but I believe this wouldn't be of interest to you, right?

Y. W.: Yes, I think there is clearly an architectural motivation. But, this form needs to be structurally informed. Sometimes I have to indicate that there is structural information in our architecture that we have been able to demonstrate.

M. K.: What's the role of aesthetics for you? When you make decisions, for example, to what degree are you driven by performance and to what degree are you driven by aesthetics?

Y. W.: Clearly we are aware of the aesthetic quality. I think it's more a result of a synthesis of structure and form that we are able to achieve something stronger, for instance in its expression, but I'm unsatisfied with the word "tectonics," for example. It's related to mechanics, but it's unrelated to what you are saying about tectonics in architecture, which is more related to a kind of objective aesthetic criteria, where you can play with local and global geometry. We have that, but again, I can't explain it. If I had to explain it, I'd look at the geometry. The project "fractal geometry," for instance, was clearly a speculation about similarities at various different scales. I find this idea fascinating.

M. K.: My assumption would be that you deliberately focus on finding the limits of these architectural spaces, while at the same time you'd like to explore the respective aesthetic capacities. In contrast to our research, where we are less interested in constraining and more interested in maximizing degrees of freedom when designing. Honestly, we don't care so much about structural optimization. We look for project-specific drivers that are conceptually interesting for the design. When we work with our partners, we realize that a certain aspect that we are looking for is, in fact, unavailable. For example, when we were designing the robotically fabricated brick wall for the Venice Architecture Biennale (2007/08), we wanted to explore if it would be possible to extract specific rules regarding the structural behavior of this double-curved entity. However, we could not develop more than a rule of thumb with the engineer. It is precisely this detailed knowledge about how complex structures ultimately perform that would be interesting to analyze from a design perspective.

Y. W.: (Looking at the modular timber structure pavilion) We designed and built this wall structure as a prototype, then I used it as a basis for a civil engineering class, to explore design optimization and geometry structurally speaking. In the end, some students were able to change the global geometry in order to optimize the structure. I asked Mark Pauly, how we could do this in a different way; other than in an iterative way? Pauly said that we have to use a kind of iterative process, which is contrary to what you are requesting. You are requesting something beforehand, an earlier stage, some sort of information set.

M. K.: For me this is simply a matter of intellectual curiosity. Can computational design processes be radically distinct from classical iterative design procedures? If you define the scope of what you want to include in such a "synthetic process" and make it comprehensive and logical, is it actually feasible? Or do we run the risk of shifting architecture towards technocratic, determinist, and optimization-driven design procedures, which leave out key aspects of creativity? These are tricky questions.

Y. W.: Well, you have to clarify the vocabulary: What do you mean by "constraints," "degree," etc.? I could answer: "Well, you have a maximum of 6 degrees of freedom." It's like an engineer that would see this degree of freedom locally, but I'm sure that you have a larger interpretation of what you call "degrees of freedom," as you were saying earlier. Clarification of the vocabulary would be needed. "Constraints" are also different for a civil engineer than for a mathematician, for example.

M. K.: I agree. When I talk about "degrees of freedom," I'd like to refer to this in a very broad and open sense, or in other words, in a conceptual way as an architect. So, for example, for the Venice Biennale we left the form open right up to the day of production, because the form could be generated at any moment. This "degree of freedom" was built into our design concept. Nevertheless, architectural questions need to be answered: What would the installation look like? How would it appear to the public? What would its formal capacities in conjunction with its structural performance be? All these aspects had to be meticulously designed, even though the final geometric definition of the global form itself was deliberately left open. This is an example of what I mean by "conceptual degree of freedom." That's what I'm particularly interested in. But there are also physical degrees of freedom, which we need to figure out. That's the issue of new fabrication technologies. Where is it possible to liberate

ourselves from certain traditional constraints or models? Are these liberations just interesting as a limited gain, or do they have a disruptive potential for a factual impact on the overall building practice, the building culture, and its design?
But coming back to our previous discussion: What about the pavilion's structure you were talking about?

Y. W.: Yes, it was flexible.

M. K.: What do you mean by "flexible" here?

Y. W.: Well, according to wind load, we had to stiffen it, as we only adjust one, single contact point and then we analyze it. What I would like to do is to return to the global geometry observation and change the position of each piece and allow them to delve far deeper into each other, in order to increase a static height. Thus, it was actually the manipulation of the global form, in order to achieve structural optimization. It's an iterative project.

M. K.: I would like to go into more detail regarding timber. But first I would like to ask where you see your work being applied?

Y. W.: The most appropriate application would be for public and sports infrastructure. I have recently been collaborating on a project to build drone ports (airports for drones) throughout Africa. I thought we should develop a timber modular product in Rwanda using parametric design tools inspired by vernacular aspects of local patterns.

M. K.: Are you talking about aesthetic and decorative aspects, or of inherent construction aspects?

Y. W.: There isn't only one solution; even structurally, there are several. You can adapt the reading of the mechanically working connection—in aesthetic terms—to local patterns. So, they do both: they function and they integrate. The fact that you cover both arguments is fascinating. Low-tech parametric design could be used all over Africa. Then, people could adapt the file to specific local conditions. I think this is an architectural strategy.

M. K.: So, you are also conceptually working with degrees of freedom and adaptation. I think this is interesting, because there are indeed multiple levels where we could use these new technologies. Let's focus a bit more on the material research you are pursuing. You're using large-scale timber, mainly LVL and CLT panels.
Wood is basically a renewable resource, a natural material, but at the same time, you're using it in a highly processed, highly engineered form. As you mentioned, you're interested in mass-produced, large-scale-products, and how to work with them very efficiently. Where do you see important development paths, and what do you think the main advantage of these large-scale products would be?

Y. W.: I think they could provide very efficient constructions; for instance, speeding up construction time.

M. K.: Do they use these panels in Rwanda?

Y. W.: No, not yet.

M. K.: Could you tell me a bit more about the advantages of these engineered timber products in comparison to the use of local resources, which certainly have lower structural capacities? Are you interested in the topic of resources, and where do you see the advantages and potential developments?

Y. W.: I believe that timber has not been developed to its full potential because of its formal instability when you use massive wood.
In Switzerland some people have managed to use it for centuries, because they have a keen understanding of where to place pieces and how they could transition over time in order to prevent damage to the building due to movement.

But this is not possible in terms of thermal insulation and our construction methods today, in terms of ventilation. I believe these products, for example LVL panels, are the right solution, in order to have a formal stability. In addition, I appreciate the mechanical performance of these elements, because you remove the problem of natural anomalies as you have it in massive timber. This gives you triple the resistance values compared with massive timber.

M. K.: Do you think there will be an increase in the use of panel products in developed countries, and do you think they will be used as structurally integrated components?

Y. W.: Yes, and they would be understood by architects who would relate the panels to architectural concepts that are also suited to this material.

M. K.: Did you ever consider intervening in the material process chain? Are you interested in discussing how these products are made and maybe customizing the way in which these products are manufactured?

Y. W.: Of course. We tried to produce welded, cross-laminated timber panels. The welding process by friction proved to provide sufficient resistance. The shear resistance is less important than in a glued-laminated beam. We approach it from the construction side, rather than from the material science side.

M. K.: What about the role of customized timber joining techniques in your overall design research at IBOIS? How do you want to explore this further in the wake of Christopher Robeller's thesis?

Y. W.: When I look at Markus Hudert's work with wooden structures, we spent a lot of time trying to find and understand the geometry of the connection of those two panels. We would like to develop all this out of timber in a very simple way.

M. K.: Indeed, this simplicity is also very appealing to me. What's your strategy in pursuing this kind of research, and what new findings are you collecting there?

Y. W.: When you look at civil engineering structures, like in textbooks, you have massive steel beams, or concrete beams, or you have steel plates, etc. The idea is to take all of these out, because you pay for the beam and you pay for the connection, which is where so many problems occur. We deduct the price of the beam in timber constructions. When you construct a sports hall from glue-laminated beams, you halve the price in cubic meters of laminated timber and halve the price of assembly, as well as costs and hassle due to connections. This needs integration; there needs to be an integral attachment that's part of the system, rather than something added to it. The idea is to remove this when you have the structural drawings and the production drawings, so we are able to fuse these two drawings into one. This is something we have to convince industry about.

M. K.: In robotic fabrication, we attempt to collapse most processing steps into one, continuous robotic process. So by different means we are also looking for new efficiencies, new simplicities, integrating these processes in a complex product. I'm not so sure about large panels in this regard. How will you explore the efficiency of assembly?

Y. W.: I'm still analyzing the current situation, on how industry produces small attachments. I'm pragmatic about this; I don't really feel the necessity to conceptualize some kind of assembly procedure beforehand, and then to look at what we actually want to build.

M. K.: If I take the example of the curved brick wall that we built at the Architecture Biennale, we had to reinvent a new joining and support technique due to the fabrication process. This was not part of the original design proposition. Should the assembly not be a vital part of your development of integral attachments?

Y. W.: Yes, of course! We want to use thin panels that can't be screwed, as we don't have sufficient distance to the edge. That's the reason why everything breaks or pulls apart. The connections we are currently working on no longer have these problems.

M. K.: I was talking about flat joints, where you need to introduce extra pieces within the manufacturing process. That being the case, you are realistic and pragmatic about what's available.

Y. W.: Matthias, don't you think that the program of the pavilion has become obsolete?

M. K.: Yes, absolutely!

Y. W.: What do you propose?

M. K.: I propose to focus our energies on buildable, multi-story constructions. How else can we challenge ourselves to be more relevant to the actual building practice?
And furthermore your work—like ours—is primarily mono-material. Of course, we are all aware that buildings are not made of only one material. Thus, I would also put forward the question of how different materials and technical subsystems could become integrated. I think these questions are important because many people still only recognize the formal aspects in these kind of pavilion-explorations. The actual impact of these new technologies at a larger scale is what I think remains open and represents a larger, more fundamental question.
Let's return to robots and what is pragmatically possible. We were discussing very thin panels and—because we are working with very small-scale elements that could be handled by robots, such as bricks or small timber members—our focus is on the automated assembly of such small elements, while your focus is on the geometric conditioning of larger panels. Therefore, do you have ideas about how there could be synergies between these two research approaches?

Y. W.: We have often discussed discrete elements, like bricks for instance, and how to attach them. Also, regarding timber construction, we have a number of discrete elements that we need to assemble. We need to figure out how every piece can be mounted. This process can be helped by the use of robots. For example, if you take our woven structures, you could imagine two or three hands with robotic arms, because they are self-supported structures, they could be connected at one point with two arms, and then there would be another arm, which could be working a short distance away, etc.. It means that the form-finding procedure would be a single stage during the construction process. There, I see something very different that could be done by robots.

M. K.: In our research, the critical question is always whether robots add something to the construction process, or if they only replace human labor. Regarding bent elements, for example, robots could clearly bring an added value to this process. Robots could sense and validate how the deformation is physically performing, and then adapt the next panel to the actual deformation. There are substantial skills that robots can add in this particular case.
I have one more question about research culture: What aspects do you think are specific to the research culture of architects and engineers, where do they differ? What aspects do you think we should focus on in an architectural research culture?

Y. W.: I think it has to do with what you mentioned before; for example, that you allow yourself to have different interpretations on the degree of freedom. This opens up the discussion and also allows interdisciplinary collaboration. I've been writing a short article about this in the "Best of" publication, where I explained the advantages of working in an interdisciplinary manner on a scientific level to EPFL's scientific community. It is like removing certain parameters which are understood in a certain way in different disciplines and trying to connect them in a different, more organic, sometimes empirical way. There is a lot for the scientific community to learn about the way architects work if they work on a certain level.

M. K.: You also mentioned what can be learned from engineers and what kind of specific qualities we need as architects in particular, in research? When would you consider something relevant for research in architecture? In engineering, it seems that there are clear criteria, which we lack in architecture. What's your personal opinion on this?

Y. W.: It is true that architects are inspired by a broader spectrum of elements in design. There is a shift in the manner of how you judge, or find, or identify yourself with the production of architecture. I relate to architects and accept their openness to a broad range of influences. I think there is an attitude on the part of the architects, which is related to the fact that the individual position is more frequently questioned by existing techniques.

M. K.: It's a kind of contextualizing of the individual and also educating the individual in such a way that you are able to position yourself within a more complex world.

Y. W.: Yes, and I see the opportunity for that.

References

1 The National Centre of Competence in Research (NCCR) Digital Fabrication—Advanced Building Processes in Architecture is being hosted at ETH Zurich, directed by Prof. Matthias Kohler, in collaboration with 13 laboratories—8 laboratories from ETHZ, 2 laboratories from EPFL, 2 from EMPA, and 1 from BUAS. Included is the Laboratory for Timber Construction, IBOIS, directed by Prof. Yves Weinand.

2 Stotz, I. "Iterative geometric design for architecture." Thesis no. 4572, EPFL, Lausanne, 2009.

3 Hudert, M. "Timberfabric: Applying Textile Assembly Principles for Wood Construction in Architecture." Thesis no. 5553, EPFL, Lausanne, 2013.

4 Weinand, Y., "The Architecture Studio as a model?" In: Veillon C., and N. Maillard (eds.), *Best of 2012.* Editions Archizoom, Lausanne, 2011., pp. 15-16.

5.5

Curved-panel wood pavilion

Marielle Savoyat

Project design IBOIS—Laboratory for Timber Constructions/ EPFL—Swiss Federal Institute of Technology, Lausanne, Switzerland/Prof. Yves Weinand, Christopher Robeller, Sina Nabaei, and Hans Buri

Project execution Merk Timber GmbH, Germany

Completion 2013

Location Accademia di architettura (Academy of Architecture), Universita della Svizzera Italiana (University of Italian Switzerland), Mendrisio, Switzerland

Fig. 1 Detail of the connection

A recent research project undertaken by IBOIS, the Laboratory for Timber Constructions at the Swiss Federal Institute of Technology in Lausanne, was focused on complex geometries and their application to wood and, more specifically, on the curve potential of this material.

In order to explore these questions, a delicate prototype pavilion in cross-laminated timber (spruce) panels was created in autumn 2013 in the Academy of Architecture's gardens in Mendrisio, Switzerland. Such a slender rendering of panels, as adopted for this pavilion, had not yet been attained at this scale. The resulting level of resistance and lightness achieved by the curve factor was ground-breaking, displaying technical prowess and paving the way for new architectural forms. The lightweight construction spans over 13.5 meters, with a panel thickness of a mere 77 millimeters. It was presented in parallel with the Timber Project exhibition, held at the Accademia gallery from September 20 to October 27, 2013.

The initial concept behind the pavilion emerged as a result of Hans Buri's thesis on origami (completed in 2010), where he recognized that the Japanese art of folding could be transferred to curved folds, creating new possibilities for wooden structures. While wood does not lend itself to double curvature, new surfaces can be created based on curved lines. The researchers who succeeded him, Christopher Robeller and Sina Nabaei, demonstrated that by replacing two zigzag polylines with two wavy lines, new types of forms applicable to timber constructions could be generated. This realization led to further questioning and explorations, which were then rendered concrete in relation to wooden material through the study and construction of a prototype pavilion. The same software application developed for origami allowed the development of deformed or pre-deformed structures. As a result of this program, a greater comprehension of geometry was achieved through observing different parameters, such as the curve radius, overall dimensions, or number of connections. The various measurements that emerged through computer simulations during the modeling of the deformation processes, including those

Fig. 2 The pavillon by night

caused by natural elements such as wind or gravity, also offered a broad spectrum of variations for determining the dimensions of the form.

The other novelty of this pavilion, other than its curved, folded-form resistance, resides in the wooden dovetail joints, which allow a more precise connection between the curved elements without the use of metallic connectors. The geometry of the curves is rendered digitally, thus determining the form of the connections. A close collaboration between IBOIS and the German construction firm Merk pushed the limits of technology: the research laboratory proved theoretical results through digital calculations and simulations, while the construction firm confirmed the feasibility in the execution of the project. The elegance and slenderness achieved results in a controlled aesthetic, with a sense of coherence between form and structure mastered through their geometric parameters.

The Research Laboratory IBOIS at the EPFL Lausanne

The domination of steel, and later reinforced concrete during the past two centuries in the research, the practice of civil engineering, and materials science has led to a severe lack of research regarding wood as a construction material. The intuitive knowledge of carpenters and our skilled predecessors has been lost since the profession of "ingénieur des ponts et chausses" was evolved in the eighteenth century. Today, many engineers do not use timber as a building material as they assume a priori that timber is less strong than steel and concrete.

My duo profile as both architect and civil engineer allows me to focus on interdisciplinary aspects of architectural design and construction and thus to develop synergies between them. Since I have conducted pioneering research in both structural engineering as well as in construction, my perspective of various aspects differ from most theorists and practitioners who only specialize in one of these areas. Since I practice, teach, and research, I am able to provide students with skills claimed by architects, such as subjectivity and aesthetics, whilst at the same time I am also able to communicate in-depth structural and technical knowledge. My research focuses on technical, design, material-based, and structural issues that—with some exceptions since the Renaissance—have been neglected or delegated by architects in favor of achieving their aesthetic aims. I take into account the fundamental links between art and science, as well as the specific constraints of the observed phenomena and their concrete implementation. The effect of scale is often simply ignored in the field of structural analysis for building construction.

My approach regards the mechanical requirements of form and structure as attributes that can gain full meaning and sense only in the context of the geometrically scaled phenomenon on which they depend. I consider the use of the digital representation of architecture as an invaluable tool that ought to strengthen the integration of structure, form, and material within our design concept. However, digital modeling cannot replace the study of physical reality, which is crucial when designing form and space and, integral to them, structure.

Since 2004 I am professor and head of the IBOIS Laboratory for Timber Construction at the Ecole polytechnique fédérale de Lausanne (EPFL), where I direct an interdisciplinary group of architects, engineers, mathematicians, and computer scientists, who perform research work in the fields of timber rib shells, folded timber plate structures, woven timber structures, integral mechanical wood-wood connections, and robotically assembled timber structures.

I would like to thank my doctoral students Claudio Pirazzi, Hani Buri, Johannes Natter, Ivo Stotz, Markus Hudert, Sina Nabaei, Christopher Robeller, Andrea Stitic, and Stéphane Roche for their valuable work.

Yves Weinand, October 2016

Yves Weinand has served as professor and head of the department at the Laboratory for Timber Constructions IBOIS at the EPFL Lausanne. He also established the Bureau d'Etudes Weinand in Liege/Belgium. He is currently working on the Vidy theater project, where timber is used as a double-layered folded plate structure (pp. 29).

If not otherwise credited, the illustrations are taken from the IBOIS, the laboratory for timber construction of the EPFL, Lausanne.

Alain Herzog: Fig. 11, p. 55; Fig. 4, p. 73; Fig. 2, p. 123.
Architecture collective Valentiny-Weinand: Fig. 11, p. 196.
Barnsley M. F., Fractals Everywhere, Academic Press, 1988: Fig. 4, p. 60.
Barthel, Eladio Dieste: Form und Konstruktion, 2001: Fig. 10, p. 195.
Bureau d'Etudes Weinand : Fig. 8, p. 10; Fig. 4, 5, p. 15; Fig. 6, p. 16; Fig. 10 (a, b, c), p. 20; Fig. 2, 3, 4, p. 93.
Christopher Robeller / IBOIS: Fig. 16, p. 133.
C. Robeller and Y. Weinand, RobArch 2016: Fig. 33, p. 29; Fig. 34, 35, p. 30; Fig. 12, p. 197.
C. Robeller and Y. Weinand: Fig. 6b, p. 128; Fig. 14 (a, b, c, d, e), p. 132; Fig. 14, p. 198; Fig. 15, 16, 17, 18, 20, p. 199.
M. Dedjier and Y. Weinand: Fig. 19, p. 199.
Fred Hatt / IBOIS: Fig. 16, 17, 18, 19 (a, b), p. 22; Fig. 2, p. 45; Fig. 3, 4, 5, p. 46; Fig. 6, p. 47; Fig. 6, p. 187; Fig. 1, p. 230; Fig. 2, p. 234.
Gloss 1999: Fig. 2, p. 8.
Group of architects: Local Architecture / Danilo Mondada architecture firm and Shel – Architecture, Engineering and Production Design: Fig 1, p. 44; Fig. 7, 8, p. 47.
H. Buri, C. Robeller, and Y. Weinand: Fig. 16, p. 133.
H. Buri and Y. Weinand: Fig. 28, p. 26; Fig. 30 (a, b, c, d), p. 28; Fig. 31, p. 29; Fig. 6a, 7, p. 128; Fig. 13, p. 131.
Ivo Stotz / IBOIS: Fig. 8, p. 61; Fig. 21 (a, b), p. 67.
I. Stotz and Y. Weinand: Fig. 3, 4, p. 52; Fig. 5, 6, 7, p. 53; Fig. 5, p. 128.
I. Stotz, H. Buri, and Y. Weinand: Fig. 8, p. 130.
L. Humbert and Y. Weinand: Fig. 9, p. 130; Fig. 10, 12, p. 131.
Markus Hudert / IBOIS: Fig. 2, p. 89; Fig. 11, 12, 13, 16 (a, b), p. 98; Fig. 19–22, p. 99; Fig. 1–13, p. 110–116; Fig. 1, p. 122; Fig. 1, p. 135; Fig. 1, p. 184; Fig. 2, p. 185.
P. Niemz, 1993, and P. Gloss, 1981: Table 1, p. 8.
PhD Thesis C. Robeller, 2015: Fig. 32, p. 29; Fig. 7, 8, 9, 10, p. 38; Fig. 3, 4, 5, p. 191; Fig. 6 (a, b, c, d), p. 192; Fig. 7 (a, b, c), p. 194.
PhD Thesis H. Buri, 2010: Fig. 22 (a, b), p. 23; Fig. 26, 27, p. 25.
PhD Thesis I. Stotz: Fig. 2, p. 51.
PhD Thesis J. Natterer, 2010: Fig. 6, 7, p. 94; Fig. 8, 9, p. 96.
PhD Thesis M. Hudert, 2013: Fig. 20 (a, b, c), 21 (a, b, c), 22, p. 99.
Philibert de L'Orme, Traités d'architecture: Nouvelles inventions pour bien bastir et à petits fraiz (1561); Premier tome de l'architecture (1567): Fig. 6, p. 10.
P. Mayencourt, C. Robeller, Y. Weinand 2015: Fig. 8 (a, b, c), p. 194.
S. Nabaei and Y. Weinand: Fig. 10, p. 55; Fig. 23, p. 100; Fig. 3, 4 (a, b), p. 127; Fig. 14f, p. 132.
S. Nabaei, F. Del Zotto, C. Robeller, and Y. Weinand: Fig. 13 (a, b, c), p. 197.
Shel Architecture, Engineering, and Production Design: Fig. 7, 8, 9, p. 19; Fig. 10 (b, c), p. 20; Fig. 14, 15, p. 21.
Schickhofer 2004: Fig. 3, 4, p. 9.
W. Feyferlick, S. Fritzer, and Y. Weinand: Fig. 21, p. 23.
Yves Weinand / IBOIS: Fig. 8, p. 10; Fig. 2, 3 (a, b), p. 15; Fig. 36, 37 a, p. 30; Fig. 1 (a, b), p. 92.

Cover: PhD Thesis C. Robeller, 2015; Anna-Lena Walther

Concept Yves Weinand
Translation and Copyediting Anna Roos
Project management Alexander Felix and Lisa Schulze
Production Heike Strempel
Layout, cover design and typesetting Anna-Lena Walther
Paper 120 g/m² MultiOffset
Printing Grafisches Centrum Cuno GmbH & Co. KG, Calbe, printed in UltraHDprint®

Library of Congress Cataloging-in-Publication data
A CIP catalog record for this book has been applied for at the Library of Congress.

Bibliographic information published by the German National Library
The German National Library lists this publication in the Deutsche Nationalbibliografie; detailed bibliographic data are available on the Internet at http://dnb.dnb.de.

This publication is also available as an e-book
(ISBN PDF 978-3-0356-0490-0;
ISBN EPUB 978-3-0356-0513-6)
and in a German language edition (ISBN 978-3-0356-0560-0).

P.O. Box 44, 4009 Basel, Switzerland
Part of Walter de Gruyter GmbH, Berlin/Boston

Printed on acid-free paper produced from chlorine-free pulp. TCF ∞

Printed in Germany

ISBN 978-3-0356-0561-7

9 8 7 6 5 4 3 2 1 www.birkhauser.com